# CONTAMINANTS IN THE SUBSURFACE

## SOURCE ZONE ASSESSMENT
## AND REMEDIATION

Committee on Source Removal of Contaminants in the Subsurface

Water Science and Technology Board
Division on Earth and Life Studies

NATIONAL RESEARCH COUNCIL
*OF THE NATIONAL ACADEMIES*

THE NATIONAL ACADEMIES PRESS
Washington, D.C.
**www.nap.edu**

**THE NATIONAL ACADEMIES PRESS** 500 Fifth Street, N.W. Washington, D.C. 20001

NOTICE: The project that is the subject of this report was approved by the Governing Board of the National Research Council, whose members are drawn from the councils of the National Academy of Sciences, the National Academy of Engineering, and the Institute of Medicine. The members of the committee responsible for the report were chosen for their special competences and with regard for appropriate balance.

This study was supported by Contract Number DACA31-02-2-0001 between the National Academy of Sciences and the U.S. Department of the Army. Any opinions, findings, conclusions, or recommendations expressed in this publication are those of the author(s) and do not necessarily reflect the views of the organizations or agencies that provided support for the project.

International Standard Book Number 0-309-09447-X (Book)
International Standard Book Number 0-309-54664-8 (PDF)

Library of Congress Control Number 2004118026

*Contaminants in the Subsurface: Source Zone Assessment and Remediation* is available from the National Academies Press, 500 Fifth Street, N.W., Lockbox 285, Washington, DC 20055; (800) 624-6242 or (202) 334-3313 (in the Washington metropolitan area); Internet, http://www.nap.edu

# THE NATIONAL ACADEMIES
*Advisers to the Nation on Science, Engineering, and Medicine*

The **National Academy of Sciences** is a private, nonprofit, self-perpetuating society of distinguished scholars engaged in scientific and engineering research, dedicated to the furtherance of science and technology and to their use for the general welfare. Upon the authority of the charter granted to it by the Congress in 1863, the Academy has a mandate that requires it to advise the federal government on scientific and technical matters. Dr. Bruce M. Alberts is president of the National Academy of Sciences.

The **National Academy of Engineering** was established in 1964, under the charter of the National Academy of Sciences, as a parallel organization of outstanding engineers. It is autonomous in its administration and in the selection of its members, sharing with the National Academy of Sciences the responsibility for advising the federal government. The National Academy of Engineering also sponsors engineering programs aimed at meeting national needs, encourages education and research, and recognizes the superior achievements of engineers. Dr. Wm. A. Wulf is president of the National Academy of Engineering.

The **Institute of Medicine** was established in 1970 by the National Academy of Sciences to secure the services of eminent members of appropriate professions in the examination of policy matters pertaining to the health of the public. The Institute acts under the responsibility given to the National Academy of Sciences by its congressional charter to be an adviser to the federal government and, upon its own initiative, to identify issues of medical care, research, and education. Dr. Harvey V. Fineberg is president of the Institute of Medicine.

The **National Research Council** was organized by the National Academy of Sciences in 1916 to associate the broad community of science and technology with the Academy's purposes of furthering knowledge and advising the federal government. Functioning in accordance with general policies determined by the Academy, the Council has become the principal operating agency of both the National Academy of Sciences and the National Academy of Engineering in providing services to the government, the public, and the scientific and engineering communities. The Council is administered jointly by both Academies and the Institute of Medicine. Dr. Bruce M. Alberts and Dr. Wm. A. Wulf are chair and vice chair, respectively, of the National Research Council.

**www.national-academies.org**

*v*

# Preface

Remediation of contaminated groundwater sites has been the subject of thousands of research studies (bench experiments) and both pilot and full-scale field projects over the past two decades, consuming billions of dollars; however, the effectiveness of such efforts is largely unknown. A landmark 1994 National Research Council (NRC) study, *Alternatives for Ground Water Cleanup*, reviewed data on the performance of remediation projects available at that time and stated, "As a result of these studies, there is almost universal concern among groups with diverse interests in groundwater contamination . . . that the nation may be wasting large amounts of money on ineffective remediation efforts."

A number of more recent studies by the NRC and government agencies have concluded that while various technologies have been demonstrated to be effective at removing contaminant mass from the subsurface under certain conditions, their performance is so site specific that it is difficult to make meaningful generalizations. It has also been concluded that restoration to drinking water standards is unlikely to be achieved at complex sites in a reasonable period of time (e.g., 100 years), particularly when there is a source zone (a highly contaminated area that is defined in Chapter 1) present. Hence, it is currently difficult to determine when and if remediation of source zones is appropriate.

The Army, like other branches of the military and many private industrial operations, has a large number of complex sites at which there is reason to expect that source zones are present. In view of the high cost of remediation of such sites (the Army's remaining liability alone is estimated at almost $4 billion), the question of what source zone remediation can accomplish and whether it is appropriate for individual sites is critical.

This report, the result of a study undertaken at the request of the Army, develops a logical basis on which to evaluate source zone remediation on a site-specific basis. It puts the technical questions of technology selection and probable performance in the context of site characteristics, remediation objectives, and metrics. This structure reflects the fact that whether a remediation project "works" or not is a function of the objectives of the project, the technology selected, and the site characteristics.

The report discusses how the diverse aspects of stakeholder and regulatory concerns, site hydrogeology, technology selection, and performance monitoring can be incorporated in the decision-making process, and thus is intended to inform decision makers within the Army, the rest of the military, and many other government agencies and the private sector about potential options for their sites contaminated with dense nonaqueous phase liquids (DNAPLs) and chemical explosives. The necessity of using a formal decision-making process derives from the influence of site-specific parameters on remediation performance, the public's desire for aggressive remediation, the high cost of remediation, and the implausibility of complete restoration in most cases, as emphasized in earlier studies.

In developing this report, the committee benefited greatly from the input of Army liaisons and remedial project managers (RPMs) who provided valuable information on Army cleanup efforts and assisted the committee in collecting relevant data and information. In particular, we would like to thank Laurie Haines of the Army Environmental Center, who gave two presentations to the committee, helped distribute and collate a survey for Army RPMs, and collected a significant amount of information for the committee's perusal over the last two years. The committee was fortunate to have received presentations from Susan Abston, Joe Petrasek, and Terry Delapaz, U.S. Army; Corinne Shia, SAIC; Greg Daloisio, Weston; Ken Goldstein, Malcolm Pirnie; Doug Rubingh and Tom Zondlo, Shaw E&I; John Blandamer, RSA; Wes Smith and Kira Lynch, Army Corps of Engineers; Ira May, Army Environmental Center; Hans Stroo, The Retec Group, Inc.; Erica Becvar, Air Force Center for Environmental Excellence; Robert Siegrist, Colorado School of Mines; James Spain, U.S. Air Force; Hans Meinardus, INTERA; Charles Newell, Groundwater Services, Inc.; Suresh Rao, Purdue University; Lawrence Lemke, University of Michigan; and Tissa Illangasekare, Colorado School of Mines. Doug Karas of the Air Force Real Property Agency organized and ran a field trip of Kelly Air Force Base during the committee's second meeting. The committee was ably served by the staff of the Water Science and Technology Board, including study directors Laura Ehlers and Stephanie Johnson and project assistants Jon Sanders and Anita Hall.

This report has been reviewed by individuals chosen for their diverse perspectives and technical expertise, in accordance with the procedures approved by the NRC's Report Review Committee. The purpose of this independent review is to provide candid and critical comments that will assist the authors and the NRC in making the published report as sound as possible and to ensure that the report meets institutional standards of objectivity, evidence, and responsiveness to the study charge. The reviews and draft manuscripts remain confidential to protect the integrity of the deliberative process. We thank the following individuals for their participation in the review of this report: Elizabeth Anderson, Sciences International, Inc.; John Hopkins, Los Alamos National Laboratory; Michael Kavanaugh, Malcolm Pirnie, Inc.; Douglas Mackay, UC Davis; Jeffrey Marquesee, SERDP/ESTCP Program Office; Richard Martel, Université du Québec; Suresh Rao, Purdue University; William Walsh, Pepper Hamilton LLP; and Charles Werth, University of Illinois. Although the reviewers listed above have provided many constructive comments and suggestions, they were not asked to endorse the conclusions or recommendations, nor did they see the final draft of the report before its release. The review of this report was overseen by Randall Charbeneau, University of Texas. Appointed by the NRC, he was responsible for making certain that an independent examination of this report was carried out in accordance with institutional procedures and that all review comments were carefully considered. Responsibility for the final content of this report rests entirely with the authoring committee.

John Fountain, *Chair*

# Contents

EXECUTIVE SUMMARY                                                    1

1   INTRODUCTION                                                    16
    The Status of Cleanup in the United States, 18
    Army Cleanup Challenges and the Army's Request for the Study, 20
    Characteristics and Distribution of DNAPLs and Chemical
        Explosives, 24
    Defining the Source Zone, 26
    Report Roadmap, 29
    References, 32

2   SOURCE ZONES                                                    34
    Hydrogeologic Settings, 34
    DNAPLs, 46
    Chemical Explosives, 65
    Summary, 74
    References, 75

3   SOURCE ZONE CHARACTERIZATION                                    79
    Key Parameters of Source Zone Characterization and the Tools to
        Measure Them, 81
    Approach to Source Zone Characterization, 102
    Repercussions of Inadequate Source Characterization, 113
    Conclusions and Recommendations, 118
    References, 119

4   OBJECTIVES FOR SOURCE REMEDIATION                                    125
    Formulating Objectives, 127
    Commonly Used Objectives, 130
    Existing Frameworks, Their Objectives, and Associated
        Metrics, 159
    Conclusions and Recommendations, 173
    References, 174

5   SOURCE REMEDIATION TECHNOLOGY OPTIONS                                178
    Conventional Technologies, 180
    Extraction Technologies, 187
    Chemical Transformation Technologies, 206
    Soil Heating Technologies, 223
    Biological Technologies, 250
    Integration of Technologies, 269
    Comparison of Technologies, 273
    Explosives Removal Technologies, 288
    Technology Cost Considerations, 291
    Conclusions and Recommendations, 292
    References, 295

6   ELEMENTS OF A DECISION PROTOCOL FOR
    SOURCE REMEDIATION                                                   306
    Review Existing Site Data and Preliminary Site Conceptual
        Model, 311
    Identify Absolute Objectives, 311
    Identify Functional Objectives and Performance Metrics, 312
    Identify Potential Technologies, 319
    Select Among Technologies and Refine Metrics, 324
    Design and Implement Chosen Technology, 329
    Conclusions and Recommendations, 330
    References, 332

APPENDIXES
A   Tables on Contaminants at Army and Other Facilities                 335
B   Abbreviations and Acronyms                                          349
C   Biographical Sketches of Committee Members and NRC Staff            352

# Executive Summary

At hundreds of thousands of commercial, industrial, and military sites across the country, subsurface materials including groundwater are contaminated with chemical wastes. Although many hazardous waste sites have been cleaned up since enactment of the Superfund regulations in the 1980s, many sites with recalcitrant organic contaminants remain in exceedance of water quality and soil standards. For a number of reasons, these sites have proven resistant to early remediation efforts such as pump-and-treat technology. Subsurface heterogeneities, contaminant sorption onto aquifer solids, and contaminant diffusion into low-permeability zones combine to make pump-and-treat much less efficient than originally envisioned. Furthermore, many organic pollutants have low water solubility and tend to remain in the subsurface as either a separate organic phase liquid (nonaqueous phase liquid or NAPL) or separate solid phase. Where they are present, separate phase or sorbed contaminants serve as a long-lived contamination source to groundwater.

The technical difficulties involved in characterizing and remediating source zones and the potential costs are so significant that there have been no reported cases of large DNAPL (dense nonaqueous phase liquid) sites where remediation has restored the site to drinking water standards. Nonetheless, pressure from the affected public to clean up these sites and a desire on the part of responsible parties to reach site closure remain, such that in the last few years, certain technologies capable of significant source remediation are being increasingly utilized by large responsible parties, like the U.S. military. In particular, the Army Environmental Center, which coordinates the Army's efforts to restore thousands of contaminated sites at installations across the country, requested the National

1

Research Council's (NRC) input to help determine the usefulness and applicability of source remediation as a cleanup strategy, including what can be accomplished by more aggressive technologies in terms of the total contaminant mass removed, risk reduction, and other metrics. Although chlorinated solvent DNAPLs are the primary focus of the report, chemical explosives are also considered in depth. The statement of task is provided below:

1. What is a meaningful definition of a "source" for the purpose of this study? How important is the source delineation step to the effectiveness of mass removal as a cleanup strategy? What tools or methods are available to delineate sources of organics contamination in complex sites? How should the uncertainty of these characterizations be quantified, in terms of both total mass and mass distribution?

2. What are the data and analytical requirements for determining the effectiveness of various source removal strategies, and how do these requirements change for different organic contaminant types or hydrogeologic environments? Effectiveness would consider the metrics of groundwater restoration, plume shrinkage and containment, mass removed, risk reduction, and life cycle site management costs.

3. What tools or techniques exist today, and what tools would need to be developed in the future, to help predict the likely benefits of source removal?

4. What would be the most important elements of a well-designed protocol to assist project managers in the field to assess the effects of source removal?

5. What can be concluded about the ability of source removal efforts to bring about substantial water quality benefits and to meet various cleanup goals? (For example, when can these efforts remove enough of the source to then rely on monitored natural attenuation?)

6. What have been the results of source removal activities at Army and other facilities to date? More generally, what can be said about the future use of source removal as a cleanup strategy and the specific technologies investigated during the study?

## SOURCE ZONES

As a preliminary step, the NRC committee formed to conduct the study created a definition of "source" that would capture the essence of a source as a reservoir of contamination while making a distinction between the source zone and the plume of contaminated groundwater. In addition, to better capture the properties of chemical explosives, the definition encompasses pure solid sources, as shown below:

*A source zone is a saturated or unsaturated subsurface zone containing hazardous substances, pollutants, or contaminants that acts as a reservoir that sustains a contaminant plume in groundwater, surface water, or air, or acts as a source for direct exposure. This volume is or has been in contact with separate phase contaminant (NAPL or solid). Source zone mass can include sorbed and aqueous phase contaminants as well as contamination that exists as a solid or NAPL.*

Understanding the characteristics of subsurface source zones is critical to effectively conducting both site characterization and remediation. The nature of the hydrogeologic environment, the composition and release of the chemical contaminants, and subsequent transport and transformation processes in the subsurface combine to determine how contaminants are distributed within source zones. Five hydrogeologic settings, described in Chapter 2, are broadly representative of the common conditions of concern:

- Type I granular media with low heterogeneity and moderate to high permeability
  - Type II granular media with low heterogeneity and low permeability
  - Type III granular media with moderate to high heterogeneity
  - Type IV fractured media with low matrix porosity
  - Type V fractured media with high matrix porosity

These settings differ in their permeability, heterogeneity, and porosity—parameters that control how contaminants are stored and released from source zones under natural and engineered conditions. For example, fractured media sites characterized by high matrix porosity (Type V) tend to store contaminants in stagnant aqueous zones and sorbed to aquifer solids. Reverse diffusion of contaminants from these areas can sustain elevated contaminant concentrations in groundwater for long periods of time. The scale of the representative hydrogeologic settings is in the range of a few meters, whereas the size of an entire source zone can be on the order of tens of meters. Source zones can occur within a single hydrogeologic setting (e.g., a sand dune deposit) or can include multiple hydrogeologic settings (e.g., alluvium overlying fractured crystalline rock). In addition to determining the overall subsurface distribution of contamination, the existing hydrogeologic setting limits both the types of tools that can be used to characterize the source zone and the technologies that might achieve reductions in source mass.

Organic contaminants are typically released to the subsurface as constituents of a liquid phase, such as a dilute aqueous solution, a concentrated aqueous solution (leachate), or an organic liquid (NAPL) that is immiscible with water. In the case of chlorinated solvents that form DNAPLs, migration in the subsurface is controlled by the liquid's density and viscosity and its interfacial tension with the pore water, as well as by properties of the formation solids, including texture and wettability. DNAPLs can form pools in the subsurface, or they can exist as small globules or ganglia retained within the aquifer pores. Among the many distinguishing features of DNAPL sites is the fact that the distribution of DNAPL in the subsurface is typically sparse and highly heterogeneous. Once contaminants are in the subsurface, processes such as dissolution, sorption, and biodegradation work to further affect contaminant distribution by redistributing mass locally among phases as well as carrying the contaminant away from the site of initial release. Depending on the hydrogeologic setting, a portion of the contaminant mass released to the subsurface as a DNAPL may diffuse into stagnant zones as either sorbed or dissolved phase contamination.

Compared to DNAPLs, the characteristics of chemical explosive source zones are less well understood, partly because of the safety issues involved in characterizing these sites. Nonetheless, it is thought that most chemical explosives from production and manufacturing process discharges are released to the environment as aqueous mixtures, from which the compounds precipitate out, usually within 6 m (20 ft) of the soil surface. Some highly concentrated wastes in production process discharges might act like DNAPLs or dense miscible phase liquids. However, even these explosive materials are likely to undergo significant change once they are introduced into soil, where environmental conditions would tend to decrease both temperature and acidity, promoting the creation of a separate solid phase material. Recharge of the subsurface during rain events can lead to dissolution of solid phase explosives and subsequent transfer of explosives mass to soil pore water and perhaps groundwater.

For both chlorinated solvent and explosives source zones, contaminant plumes develop downgradient of the source material in cases where the contaminants are soluble in water and are resistant to natural biodegradation. In general, groundwater plumes tend to have larger spatial extents and to be more continuous in nature in comparison to contaminant mass distributions within source zones. Over time, biogeochemical processes in the plume can result in a contaminant mixture with a very different composition than the original release materials. It should be noted that sorption or diffusion of contaminants *from the plume* onto aquifer solids (and subsequent reverse diffusion) is common in many hydrogeologic settings. Although this sorbed or stagnant-zone mass can be a chronic supply of aqueous phase contamination, it does not constitute a source zone (as herein defined) because it does not exist where the DNAPL or solid phase was present. Therefore, not all groundwater contaminant plumes imply the presence of a source zone.

## SOURCE ZONE CHARACTERIZATION

The hydrogeologic environment and contaminant distribution of a hazardous waste site are revealed through site characterization—a continuous, dynamic process of building and revising a site conceptual model that captures all aspects of the site, including the source zone. Chapter 3 addresses several aspects of source zone characterization, including the potential ramifications of inadequate characterization, characterization methods and tools, the importance of source zone characterization to determining cleanup objectives, scale issues, and coping with uncertainties during source characterization. A recurring theme is that source zone characterization should be carried out in a manner that best informs the entire source remediation process. Decisions regarding the objectives of remediation and the remediation technologies selected will have a strong impact on the source zone characterization strategy and vice versa.

Although it is impossible to prescribe a specific step-by-step source zone characterization process because of differing conditions from site to site, there are four broad categories of information that are critical to characterizing all source zones:

1. **Understanding source presence and nature**. What are the components of the source, whether a DNAPL or explosive material, and what is the expected behavior of the individual components based on known information?

2. **Characterizing hydrogeology.** What are the lithology of the subsurface and groundwater flow characteristics as they pertain to the source zone? Are there multiple aquifers at the site, and how are they connected? What are the properties and connectedness of the low-permeability layers or zones? Can the flow system be described at the specific site and at a larger scale? Can the groundwater velocity and direction (and the spatial and temporal variation in both) be measured?

3. **Determining source zone geometry, distribution, migration, and dissolution rate.** Where is the source with respect to lithology? Is it present as pooled DNAPL, distributed as residual saturation, or both? Is it crystalline explosive material or is it sorbed? What is the current vertical and lateral extent of the source material, and what is the potential for future migration based on the hydrogeologic characteristics of the site? How fast is the source dissolving?

4. **Understanding the biogeochemistry.** What roles do transport and transformation processes play in attenuating the source zone and the downgradient plume? How will possible remediation strategies affect the geochemical environment (e.g., by releasing other toxic substances, or by adding or removing substances upon which microbial activity and contaminant degradation depend)?

Although there may be an overall work plan directing that source characterization activities be conducted in a particular order, each of the activities is related to the others, and a good deal of iteration between the general categories is not only desirable but critical to the process.

For each of these four primary categories of information, Chapter 3 outlines characterization methods and tools that can be used, including noninvasive characterization approaches (ranging from collecting historical information to certain geophysical techniques), invasive sampling tools, methods for laboratory analysis, and tools that represent a combination of the above. Some of the tools are approaches to removing contaminant samples from the subsurface, some measure specific chemicals either in situ or following sample extraction, some perform both functions, and some do neither. Most of the tools have been developed and utilized at sites with unconsolidated geologic media and thus do not apply to fractured media or karst. Indeed, the tools that are available for use in fractured rock systems often provide limited (i.e., point-specific) information because of the high degree of spatial variability at these sites.

At many DNAPL or explosives-contaminated sites, there is inadequate site characterization to support the remediation strategies and success metrics chosen. This is most likely due to unclear objectives, financial constraints, or pressure to show progress and meet deadlines. Despite its technical challenges, some level of source zone characterization is indispensable for the effective management of an environmental remediation effort. Severe overestimation of the source size may inflate the cost of remediation efforts to exorbitant levels. Conversely, missing the source material will jeopardize the success of the cleanup and will require additional characterization and remediation work. The following conclusions and recommendations regarding source characterization are made.

**Source characterization should be performed iteratively throughout the cleanup process to identify remedial objectives, metrics for success, and remediation techniques.** All sites require some amount of source characterization to support the development and refinement of a site conceptual model. In general, successful source remediation requires information on the nature of the source material, on the site hydrogeology, on the source zone distribution, and on the site biogeochemistry. However, the level of characterization effort required and the tools used at any given site are dependent on site conditions, on the cleanup objectives chosen, and on the technology chosen to achieve those objectives.

**An evaluation of the uncertainties associated with the conceptualization of the source strength and location, with the hydrogeologic characteristics of the subsurface, and with the analytical data from sampling is essential for determining the likelihood of achieving success.** This is often accomplished through the use of statistical, inverse, and stochastic inverse methods. Unfortunately, quantitative uncertainty analysis is rarely practiced at hazardous waste sites. Obtaining a better handle on uncertainty via increased source characterization would allow eventual remediation to be more precise. It is likely that at most sites, there is not an optimum combination of resources and effort expended on source characterization and thus uncertainty reduction vs. remedial action.

## OBJECTIVES FOR SOURCE REMEDIATION

The success of source remediation requires the specification of remedial objectives with clarity and precision. This includes knowing the full range of site remedial objectives, their relative priorities, and how they are defined operationally as specific metrics. Unfortunately, failure to unambiguously state remedial objectives appears to be a significant barrier to the use of source remediation. Too often, either data presented on the effects of source remediation are irrelevant to the stated objectives of the remedial project, or the objectives are stated so imprecisely that it is impossible to assess whether source remediation contributes to achieving them. For example, the committee is aware of situations in which an explicit operational statement of site objectives, if made prior to beginning source remediation, might have led to a decision not to attempt source remediation.

This widespread problem of vaguely formulated remedial objectives is compounded by the fact that multiple stakeholders at a site not only may have very different objectives, but may also use similar language to describe those very different objectives. Moreover, a particular performance metric may potentially correspond to a variety of different objectives and accordingly be viewed quite differently by different stakeholders. Finally, both the DNAPL problem and the effects of source remediation efforts raise temporal issues that are very poorly addressed by conventional analytical frameworks for assessing risks to human health and the environment.

Chapter 4 describes a variety of objectives possible at sites for which source remediation is a viable option. A distinction is made between absolute objectives, which are important in and of themselves, and functional objectives, which are a means to an end. For example, the objective of reducing contaminant concentrations in groundwater to a specified level at a particular point in time and space may be mandated under a particular regulatory framework as a necessary feature of a successful remediation, in which case it represents an absolute objective. The identical criterion, however, could be selected as a means of ensuring that human health risks have been reduced to an acceptable level. In this case, the objective is functional, because other objectives may achieve a comparable degree of health protection, such as precluding the use of contaminated groundwater.

Physical objectives are discussed first, including mass removal, concentration reduction, mass flux reduction, reduction of source migration potential, plume size reduction, and changes in toxicity or mobility of residuals. Objectives relating to risk reduction, cost minimization, and scheduling are also discussed, many of which have been institutionalized within regulatory, risk assessment, and economic frameworks for site cleanup. The following conclusions and recommendations regarding objectives for source remediation are made.

**Remedial objectives should be laid out *before* deciding to attempt source remediation and selecting a particular technology.** The committee observed that remedies are often implemented in the absence of clearly stated objectives, which are necessary to ensure that all stakeholders understand the basis of subsequent remediation decisions. Failure to state objectives in advance virtually guarantees stakeholder dissatisfaction and can lead to expensive and fruitless "mission creep" as alternative technologies are applied. This step is as important as accurately characterizing source zones at the site.

**A clear distinction between functional and absolute objectives is needed to evaluate options**. If a given objective is merely a means by which an absolute objective is to be obtained (i.e., it is a functional objective), this should be made clear to all stakeholders. This is particularly important when there are alternative methods under consideration to achieve the absolute objectives, and when it is known or is likely that different stakeholders have a different willingness to substitute objectives for one another.

**Each objective should result in a metric; that is, a quantity that can be measured at the particular site in order to evaluate achievement of the objective.** Objectives that lack metrics should be further specified in terms of subsidiary functional objectives that do have metrics. Furthermore, although decisions depend upon both technical and nontechnical factors, once a decision has been made, the focus should be on the technical metric to determine if remediation is successful.

**Objectives should strive to encompass the long time frames characteristic of many site cleanups that involve DNAPLs.** In some existing frameworks, timeframes are very short (rarely longer than 30 years) relative to the persistence of DNAPL (up to centuries), such that alternative actions with significant differences in terms of the speed with which a site can be remedied cannot be distinguished. Within life cycle cost analysis, the chosen timeframes and discount rate can significantly affect cost estimations for different remedies. Decision tools with a more realistic temporal outlook have been developed in other areas of environmental science (e.g., storage and disposal of radioactive materials). Their application to DNAPL problems needs to be considered by the Army and by the site restoration community as a whole.

## SOURCE REMEDIATION TECHNOLOGY OPTIONS

Chapter 5 presents those technologies that have surfaced as leading candidates for source zone remediation, including a description of each technology, a discussion of the technology's strengths and weaknesses, and special considerations for each technology. The discussion of chlorinated solvents focuses on contamination of the saturated zone, as this medium presents the greatest difficulties in terms of site cleanup.

Two technologies commonly used for source remediation function primarily by physically extracting the contaminants from the subsurface. *Multiphase extraction* employs a vacuum or pump to extract NAPL, vapor, and aqueous phase contaminants, which may then be disposed of or treated. *Surfactant* and *cosolvent flushing* involve introducing a liquid into the subsurface into which the contaminant partitions, and then the mixture is extracted out of the subsurface and is subsequently treated. Two technologies that attempt to transform subsurface contaminants in situ include *chemical oxidation* and *chemical reduction*. In both cases, chemicals introduced into the subsurface react with the compounds of concern, leading to their transformation or degradation into less toxic breakdown products. The three most widely applied soil heating methods used for source remediation are *steam flooding*, *thermal conduction heating*, and *electrical resistance heating*—all of which are intended to increase the partitioning of organic chemicals into the vapor or gas phase where they can be extracted under vacuum. In addition, these remedies can achieve destruction of many organic contaminants in situ at sufficiently high temperatures. Two DNAPL remediation technologies either directly or indirectly invoke biological processes to degrade contaminants in situ. *Air sparging* accomplishes contaminant removal primarily by stripping volatile compounds from the subsurface while simultaneously supporting in situ biodegradation of contaminants. *Enhanced bioremediation* refers to any in situ treatment in which chemicals are introduced into the subsurface with the goal of stimulating microorganisms that can degrade or transform the contaminants of concern. Although excavation, containment, and pump-and-treat are considered conventional approaches for addressing DNAPL contamination, they are briefly discussed for comparison purposes.

A comparison of the technologies is given in Table 5-7, the goal of which is to help identify a list of the most viable technologies that should be thoroughly evaluated for use under site-specific conditions. The table assesses the types of contaminants for which each technology is suitable and then qualitatively evaluates each technology's relative potential for mass removal, local aqueous concentration reduction, mass flux reduction, source migration, and changes in toxicity. This evaluation is presented for each of the five hydrogeologic settings described in Chapter 2. The table provides a rank of "high," "medium," "low," or "not applicable" to describe the likelihood that a given technology would be effective at achieving the listed objective. It should be kept in mind that the performance of a given technology is extremely site-specific, as are the objectives associated with any remediation strategy. Thus, the scores are somewhat subjective and should be considered more relative (one technology compared to another) than absolute. Furthermore, a single site may encompass several media settings, or it may not clearly fall into one of the five settings.

The table entries are based, when possible, on reported case studies; more frequently, the entries are based on the committee's best professional judgment (due to a lack of comprehensive full-scale demonstrations). Thus, few of the metrics in Table 5-7 have been measured in the field for the technologies. Nonetheless, some generalizations can be made. Some source remediation technologies have been demonstrated to achieve substantial mass removal across a range of sites and contaminants. A number of these studies have also demonstrated concentration reductions (at only one or a few wells), but the meaning of these measurements is highly debatable. Mass flux reduction, reduced migration of the source, and changes in toxicity have not yet been demonstrated at any of the source remediation case studies reviewed. This is partly because of the difficulty in making such measurements. Furthermore, there are few field data to support the hypothesis and existing laboratory data that suggest that partial mass removal can affect local concentration and down gradient mass flux. **Thus, available data from field studies do not demonstrate what effect source remediation is likely to have on water quality.** The following additional conclusions and recommendations are made regarding technologies for source remediation.

**Performance of *most* technologies is highly dependent on site heterogeneities.** In general, the efficiency of flushing methods decreases as the heterogeneity increases, although the degree of impact depends on the specific site characteristics and on the operative processes. In the case of surfactant flushing, foam generated by air injection has emerged as a viable way to mitigate heterogeneities. Steam flushing is affected by preferential flow of the steam, but conduction mitigates this impact to some degree. Soil heating by conduction is least sensitive to heterogeneities because thermal conductivity varies very little with media properties. Chemical oxidation and enhanced bioremediation are more sensitive to heterogeneities than are thermal methods, and air sparging is the most sensitive to heterogeneity because there are no mitigating factors preventing the preferential flow of low-viscosity air and the bypassing of the target DNAPL. Heterogeneities are more likely to affect a technology's ability to achieve mass removal and local aqueous concentration reductions compared to mass flux from the source zone.

**Most of the technologies are not applicable in, are negatively impacted by, or have not been adequately demonstrated in low-permeability or fractured materials.** The effectiveness of flushing technologies in low-permeability settings (Type II) is limited due to the difficulty in moving flushing solutions (surfactants, oxidants, reductants, steam) through low-permeability formations. Technologies which do not use fluid flow as a delivery mechanism, such as conductive heating and electrical resistance heating, have greater potential in Type II settings. Applications of source remediation technologies in fractured media (Types IV and V) have been limited due to difficulties in and cost of characterizing the fracture networks and delineating the source zone. In addition, channeling along high-permeability fractures results in poor removal of mass from lower-permeability matrix zones for most technologies, with the possible exception of conductive heating since heat can be conducted efficiently through the rock matrix.

**Each technology has the potential to produce negative side effects that need to be accounted for in the design and implementation of that tech-**

**nology.** Examples of potential side effects include surfactant/cosolvent/steam-induced vertical migration of DNAPL, alteration of the redox potential by chemical oxidants or reductants (potentially serving to release previously bound nontarget compounds into the groundwater), and changes in the indigenous microbial population due to chemical or thermal treatment. These side effects can at times be avoided by an experienced design/implementation team. In other cases, the negative side effects should be factored into the design/implementation process.

**Almost all of the source remediation technologies evaluated require more systematic field-scale testing to better understand their technical and economic performance.** Of the innovative technologies reviewed, only surfactant flooding has amassed a substantial number of field-scale studies in the peer-reviewed literature. Because full-scale applications of source remediation technologies are scarce, there is insufficient information to thoroughly evaluate most technologies, especially with regard to the long-term impact of mass reduction in the source zone. Furthermore, due to insufficient economic data and the site-specific nature of both performance and cost, it is not possible to generically predict the impact of source remediation technologies on life cycle costs.

**The level and type of source zone characterization required to design, implement, and monitor the performance of remedies is dependent on the chosen objectives and the remediation technology.** For example, in situ chemical oxidation requires accurate estimates of source zone mass and composition and matrix oxygen demand, or else the remedy could be plagued by stoichiometric limitations or by the consumption of oxidant by unidentified co-contaminants. The properties of the source material (e.g., composition, viscosity, density, interfacial tension) should be determined for field-weathered samples in order to assess such remedies as surfactant-enhanced flushing. The location and geometry of source zone materials should be known to some level of certainty in order to design containment systems. For example, the most effectively designed slurry wall will have less effect on downstream mass flux if it is placed across the source zone rather than around it. With respect to performance monitoring, judging the effectiveness of in situ chemical oxidation by monitoring mineralization products or by monitoring the consumption of oxidant could overestimate treatment effectiveness in cases where alternate contaminants are present.

**Development of treatment technologies for explosives source zones is in its infancy because the characterization of explosive source materials and of their interactions with geologic media lags far behind the knowledge base that exists for DNAPLs.** Before one can understand the utility or performance characteristics of treatment technologies for explosives contamination, one should understand the chemical and physical nature of the explosives source zones. Furthermore, source areas containing high concentrations of explosives have the potential for dangerous explosions during remediation, which will necessitate laboratory and field assessment of explosives source zones within specialized facilities.

## ELEMENTS OF A DECISION PROTOCOL FOR SOURCE REMEDIATION

Investments in source remediation technologies have often failed to achieve the desired reductions in risk and/or site care requirements, partly because historical releases of DNAPLs and explosives are technically difficult to clean up, but also because of how source zones are managed. The design and implementation of a successful source remediation project involve the iterative characterization of the source zone, development of remediation objectives, and evaluation of technologies—a process that is sufficiently complex to warrant a formal protocol. Chapter 6 describes the elements of a protocol to assist project managers in designing, implementing, and assessing the effects of source remediation. These elements are laid out in Figure ES-1, which depicts a six-step process that includes activities (white boxes), data and information collection (gray boxes), and decision points (gray diamonds). The six steps are taken sequentially as source remediation moves forward. As can be seen from the figure, however, there may be multiple iterations of each step until a decision can be made to proceed to the next step.

Each of these six steps is described in detail, using a hypothetical example of a typical hazardous waste site and drawing on information from previous chapters. Steps 2 and 3 in Figure ES-1 focus on identifying absolute and functional objectives that are clearly articulated and verifiable (one of the distinguishing features of the framework). The figure emphasizes the role of managing data gaps and uncertainty via site-specific data collection (as exemplified by "collect data and refine site conceptual model" shown in gray boxes). As discussed in Chapters 2 and 3, considerable uncertainty exists at almost every hazardous waste site with respect to the location and extent of contamination. The protocol focuses users on recognizing the limitations of their current understanding, on the importance of collecting the necessary information to effectively make decisions, and on managing plausible variations from perceived conditions. Step 4 in Figure ES-1 involves referring back to the comparison table of technologies in Chapter 5 (Table 5-7) to determine which technologies might be viable given the contaminant type present, the hydrogeologic setting, and the chosen functional objectives. Collection of adequate data to characterize the source, development of clear absolute and functional objectives and their metrics, and remedy evaluation have seldom been done at the level described in this report. If all of these steps are not included, source remediation at an individual site will have a low probability of success.

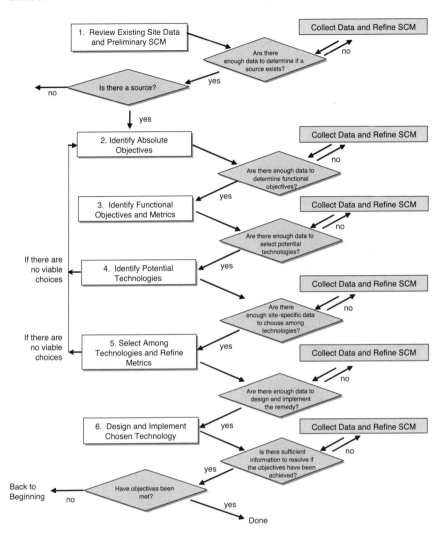

FIGURE ES-1  Six-step process for source remediation. SCM = site conceptual model.

**The Army should develop and use a detailed protocol consistent with the elements prescribed in Figure ES-1.** A protocol specific to source zones is needed to aid stakeholders in optimizing the benefits derived from investments in remediating source zones. The key attributes that need to be addressed are pursuing actions that effect intended changes, understanding the extent to which objectives are attainable, and being able to measure progress toward desired objectives. The protocol will need to be integrated into the existing remedy selection frameworks used by the Army at individual sites, including Superfund, RCRA, relevant state laws, or the Base Realignment and Closure program.

**Involvement of potentially affected parties is essential to the success of source remediation.** Stakeholder participation is needed to better understand the range of absolute objectives at a given site, to develop functional objectives, and to gain consensus on appropriate actions. Without adequate public participation, critical elements of solutions may be missed, a subset of the involved parties may feel that their needs have been ignored, and/or false expectations may develop as to what can be achieved. As for all relevant stakeholders, knowledge acquisition by the public is essential to making decisions about source remediation.

With respect to the future use of source removal as a cleanup strategy, an important conclusion that can be made from reviewing source zone remediation attempts to date is that the data are inadequate to determine how effective most technologies will be in anything except the simpler hydrogeologic settings. Furthermore, it is unlikely that available source remediation technologies will work in the most hydrogeologically complex settings such as karst. Beyond defining these extremes, it is difficult to make generic statements about source remediation, as most field studies have not provided quantitative information on the ability of the technologies to meet most remediation objectives. In a few carefully documented cases, removal or destruction of a large fraction of the contaminant mass present in the source zone was achieved. There have also been several well documented remediation projects in which concentrations in monitoring wells were reduced to a small fraction of pre-project concentrations, although few of these cases provided long-term data that might reveal rebound. Almost none of the dozens of projects reviewed by the committee contained quantitative mass flux measurements.

Despite these drawbacks, **by following the elements of a source remediation protocol illustrated in Figure ES-1, project managers will be able to make critical decisions regarding whether and how to attempt source remediation and thereby accomplish a more beneficial distribution of resources.** The steps presented in Figure ES-1—determining whether a source exists; developing absolute and functional objectives and their metrics; selecting, designing, and implementing a technology; and collecting data to support all these decisions— have seldom been conducted in the manner described in this report. Not following these steps has led to source remediation technologies being prematurely scaled up at poorly characterized sites, at sites where there is known complex hydrogeology, and where there is no clear reason for proceeding with the project.

Finally, several technologies show enough promise, in terms of demonstrated mass removal and concentration reduction in simple hydrogeologic settings, to warrant further investigation to determine their long-term effects on water quality, especially if objectives such as mass flux reduction become more prevalent. Thus, future work should attempt to determine the full range of conditions under which these technologies can be successfully applied, and to better understand how mass removal via these technologies affects water quality.

# 1

# Introduction

Concern about polluting water supplies began over 100 years ago when the industrial revolution resulted in noticeable changes in the quality of surface waters. Because groundwater was generally believed to be protected from such pollution, however, groundwater contamination did not become a major issue until the 1970s (NRC, 1994; Pankow and Cherry, 1995). Strict regulation of point-source discharges, including the development of more effective sewage treatment plants, has led to substantial improvements in water quality for some surface waters in the United States (EPA, 2000). The same cannot be said of groundwater, for which cleanup efforts represent a significantly more difficult challenge.

Attempts at large-scale groundwater cleanup began in earnest in the 1980s after passage of the Comprehensive Environmental Response, Compensation, and Liability Act (CERCLA) and the Resource Conservation and Recovery Act (RCRA). Results of early remediation efforts seldom produced the expected reduction in contamination levels. Studies by the U. S. Environmental Protection Agency (EPA) (EPA, 1989, 1992) found that the commonly used pump-and-treat technologies rarely restored sites that had contaminated groundwater to background conditions. This was confirmed in a much more extensive 1994 National Research Council (NRC) study that explicitly reviewed 77 sites across the country where full-scale pump-and-treat was being used.

The inherent difficulty of groundwater cleanup results directly from fundamental aspects of hydrogeology and chemistry. First, heterogeneities in the subsurface, sorption of contaminants onto solid organic matter, and contaminant diffusion into low-permeability zones combine to make pump-and-treat much

16

less efficient than originally envisioned (Mackay and Cherry, 1989). Second, most common organic pollutants in the subsurface have low solubilities in water and tend to remain as either a separate organic phase liquid in the subsurface (nonaqueous phase liquid or NAPL) as in the case of chlorinated solvents or a separate solid phase as where chemical explosives have precipitated in the subsurface. Organic liquids that are denser than water are referred to as dense nonaqueous phase liquids (DNAPLs). During the late 1980s, it was recognized that the presence of DNAPLs made a site particularly difficult to remediate (Feenstra and Cherry, 1988; Mackay and Cherry, 1989; Mercer and Cohen, 1990; NRC, 1994). Before it was understood that DNAPLs commonly exist in source areas, it was assumed that by removing a few pore volumes of contaminated groundwater, the majority of the total contamination could be extracted.

At sites where they are present, separate phase or sorbed contaminants serve as a long-lived contamination source. That is, groundwater that flows through the volume of subsurface containing the contaminant—termed the source zone—will be contaminated by the small amount of contaminant that dissolves. This suggests that groundwater remediation to background levels will not be achieved unless the contaminant source is removed or physically isolated from flowing groundwater (NRC, 1994). Unfortunately, due to a lack of effective characterization tools and the tendency of DNAPLs to have a spatially limited but extremely heterogeneous distribution, it is very difficult to find contaminant sources within the subsurface. In addition, although numerous new technologies have been developed to remediate source zones, the difficulty in evaluating these technologies (due to the lack of data from pilot studies) makes prediction of their effectiveness for full-scale applications problematic.

Several NRC reports extend the findings of the 1994 report on pump-and-treat systems to include more comprehensive analysis and encompass new remediation technologies (NRC, 1997, 1999a, 2003). These reports have noted the general paucity of data available for evaluating remediation technology performance (including technologies for DNAPL sites). These and many other recent studies (e.g., ITRC, 2000, 2002; SERDP, 2002; EPA, 2003) have demonstrated that restoration of sites with DNAPL contamination to pre-contamination levels is rare and may not be practically achievable. Indeed, there are no reported cases of large DNAPL sites where remediation has restored the site to drinking water standards. At this time, most DNAPL sites have pump-and-treat systems in place to contain the dissolved phase plumes and thus minimize risk to the public. At only a small fraction of these sites has remediation of the DNAPL source actually been attempted.

Layered onto this issue of technical impracticability are the opinions of stakeholders, including those who live or work near contaminated sites, as well as the high cost associated with remediation efforts. There is often pressure from the public to remediate when pollution is found. This pressure to clean up sites is contrasted by the fact that remediation technologies for DNAPL sites are understandably expensive, such that there have been relatively few large-scale remediation attempts. Whether it is worth the expense to undertake remediation at DNAPL sites depends on the objectives of the remediation project, on what can be achieved (which is often unknown), and on the competing needs of other critical sites. Because the cost of remediating the nation's contaminated groundwater has been estimated to range from a few to several hundred billion dollars (NRC, 1999b), giving priority to sites where remediation efforts can make the most impact is essential.

Unlike previous NRC reports, this report focuses on active remediation of source zones and the effect of that remediation on a number of factors including groundwater quality. It addresses what can be achieved given the fact that DNAPL is present at many sites (Villaume, 1985; Feenstra and Cherry, 1988; Mercer and Cohen, 1990; Pankow and Cherry, 1995) and given the findings of prior studies that remediation of DNAPL sites may not provide complete restoration (see NRC, 1994, 1999a; numerous case studies in Chapter 5). Certain technologies capable of significant source remediation are being increasingly utilized by large responsible parties, like the U.S. military. Just what can be accomplished by these more aggressive technologies, in terms of the percentage of total contaminant mass removed, risk reduction, and other metrics, is uncertain. The study was initiated at the request of the U.S. Army Environmental Center, which coordinates the Army's efforts to restore thousands of contaminated sites at installations across the country. Although chlorinated solvent DNAPLs are the primary focus of the report, chemical explosives are also considered in depth because of the Army's large potential liability in subsurface sites contaminated by explosives.

## THE STATUS OF CLEANUP IN THE UNITED STATES

During the past two decades of cleaning up hazardous waste sites in the United States, there has been an evolution of activities, from the initial stages of the CERCLA or Superfund process to later remediation stages. Thus, a large percentage of sites have moved from initial characterization and investigation—activities embodied in the remedial investigation and feasibility study (RI/FS)—to remedy selection, remedy implementation, and, in some cases, site closure.

The remedies chosen at hazardous waste sites across the country have also evolved from an initial emphasis on source treatment (reflecting the preference of the National Contingency Plan to treat so-called principal threats) to containment measures. In large part, this change in emphasis reflects the technical difficulty of cleaning up many of the more complex and recalcitrant hazardous waste sites as well as the limited resources available for cleanup. In the early 1980s, the limitations of remediation technology were unclear to Congress. In 1986, CERCLA was amended to provide a preference for attaining drinking water standards in groundwater, such that the number of remedies relying on treatment dramatically increased. Since that time, however, it has become widely known that at many contaminated sites it is not feasible to reduce groundwater concentrations to drinking water standards with pump-and-treat technology in a reasonable time-frame (e.g., decades) (NRC, 1994). Several government agencies have estimated the long-term costs of continuing to operate pump-and-treat systems, despite their ineffectiveness, with projected annual costs in the hundreds of millions of dollars and life cycle costs in the billions of dollars[1] (e.g., DoD, 1998). In response to the rising costs of contaminated site cleanups and the growing recognition of the limitations of technology, federal and state regulatory agencies issued a number of explicit policies that led to the acceptance of more containment. For example, EPA released guidance in 1996 to select pump and treat as a presumptive remedy for DNAPL sites, reflecting the continuing debate at the time on whether it would be technically feasible to clean up these sites (EPA, 1996a). Although treatment as a source area remedy at Superfund sites increased from 14 percent to 30 percent during the 1982–1986 period to a peak of 73 percent in 1992, it has decreased ever since. Monitored natural attenuation (MNA) alone or in conjunction with other remedial actions increased from 0 percent in 1982 to between 28 percent and 48 percent in the 1998– 2001 period (EPA, 2004). EPA's 1990 Superfund remedy rules state that even though permanent remedies are preferred, EPA expects to use treatment to address the principal threats posed by a site wherever "practicable," and engineering controls, such as containment, for sites that pose a relatively low long-term threat (EPA, 1991).

Despite these trends toward containment and MNA, remedial actions and monitoring activities at many sites regulated under CERCLA and RCRA (which encompass almost all military sites) cannot legally be terminated unless the chemicals remaining at the site are reduced to levels that allow unrestricted use of the property. At the vast majority of sites, this goal corresponds to groundwater contaminant concentrations that are equal to or less than drinking water maximum contaminant levels (MCLs) within the source zone or at some specified location in the plume. NRC (1994) estimated that given such criteria, cleanup times will extend from a few years to thousands of years, with the actual treatment time being highly uncertain. Because it is a primary goal for the military to achieve site closeout at as many sites as possible within the next 10–15 years, there have been renewed efforts to reduce the time required for remedy operation and monitoring by attempting to remove a significant portion of contaminant mass at many hazardous waste sites with more aggressive source remediation technologies.

Source remediation can involve ex situ and in situ technologies, both conventional and innovative. As of FY2002, 58 percent of all Superfund source remediation actions used ex situ technologies (EPA, 2004), and trends at military

---

[1]Life cycle cost estimated by assuming that the average life cycle cost for a pump-and-treat site is $9.8 million and that 10 percent of the 3,000 DoD sites have or will have full-scale pump-and-treat systems (Quinton et al., 1997).

facilities are expected to be the same. Of the 42 percent of Superfund sites where in situ source remediation technologies were used, over half utilized soil vapor extraction, with the remainder being composed primarily of solidification/ stabilization, bioremediation, and soil flushing. However, several additional innovative in situ technologies, which are the focus of this report, have recently demonstrated potential for effecting at least partial depletion of the source. Although comprehensive data on most of these innovative technologies are not available, the EPA has compiled information on in situ chemical oxidation and in situ thermal treatment (which includes steam injection, electrical resistance heating, conductive heating, radio-frequency heating, and hot air injection). Of the 69 thermal projects in the EPA database, 49 were completed in the last five years or are ongoing (www.cluin.org/products/thermal); similar upward trends in usage were observed for in situ chemical oxidation (www.cluin.org/products/chemox). Use of these more aggressive source remediation technologies at Superfund sites has increased substantially in the past six years despite an overall trend toward less private investment in innovative technologies during the 1990s (NRC, 1997).

## ARMY CLEANUP CHALLENGES AND
## THE ARMY'S REQUEST FOR THE STUDY

The goals of the Army's environmental restoration program are to "protect human health and the environment, to clean up contaminated sites as quickly as resources permit, and to expedite cleanup to facilitate disposal of excess Army properties for local reuse" (Department of the Army, 1997). In addition, the program aims to optimize risk reduction per dollar spent (Haines, 2002).

Activities within the Army's Installation Restoration Program mirror the trends discussed above for the nation in general, in that the majority of sites are now in the latter stages of cleanup. As of September 30, 2003, the Army had identified 10,367 sites at active bases and 1,899 sites at closing bases (DoD, 2003). For both type of bases, about 88 percent of the identified sites have reached "remedy-in-place/response complete," which is a military milestone in the cleanup process that indicates the end of remedy construction or completion of cleanup activities. These numbers do not include sites contaminated with unexploded ordnance (UXO), discarded military munitions, or munitions constituents, of which there are 177 sites located at 26 closing bases and 819 sites located at 166 active bases. The Army estimates that the remaining cumulative cost to reach remedy-in-place/response complete in today's dollars is $3.1 billion at active bases (DoD, 2003) and is $439 million at closing bases (not including UXO cleanup). It is expected that funding for the Active Installation Restoration Program will hold steady at around $400 million for FY2005 and FY2006.

Like other branches of the military and large private responsible parties, the Army is responsible for hazardous waste sites that reflect a broad range of activities over the last century. Perhaps the most distinct characteristic of these facilities is the wide range of contaminant types—often present as mixtures of unknown composition and with no clear indication of how they were disposed. As summarized in Appendix A and other documents (NRC, 1999b), petroleum hydrocarbons are the most frequently reported organic compounds at Army and other military facilities, due to the high prevalence of large-scale transportation and industrial activities that utilize fuel. Petroleum hydrocarbons include components of gasoline [benzene, toluene, ethylbenzene, and xylene (BTEX) and oxygenates such as methyltertbutylether (MTBE)] as well as other fuels. Because many petroleum hydrocarbons are amenable to natural degradation processes, they are less likely to present long-term contamination problems that might eventually necessitate aggressive source remediation.

The greater concern at military facilities is with recalcitrant organic compounds such as the chlorinated solvents perchloroethene (PCE), trichloroethene (TCE), and trichloroethane (TCA) and their degradation products vinyl chloride, dichloroethene (DCE), and dichloroethane (DCA)—all of which can be present in DNAPLs in the subsurface. Chlorinated organic solvents were widely used for cleaning and degreasing military equipment and were typically disposed of at the land surface or in drums. Within the Department of Defense (DoD), there are approximately 3,000 individual sites that require cleanup of chlorinated solvents (Stroo, 2003). The EPA has estimated that approximately 5,000 DoD, Department of Energy (DOE), and Superfund sites are contaminated with chlorinated solvents (EPA, 1996b), although DNAPL may not exist at all of these sites. Additionally, there are an estimated 20,000 solvent-contaminated drycleaner sites in the United States (Jurgens and Linn, 2004).

Other frequently reported hard-to-treat organic compounds at military sites are polychlorinated biphenyls (PCBs), polycyclic aromatic hydrocarbons (PAHs), creosote, and coal tar. Mixtures of PCBs (the most common were Aroclor 1254 and 1260) were used as dielectric fluids in electrical transformers and capacitors before their use was restricted. PAHs are components of petroleum products, whereas creosote and coal tar, which were commonly used to treat wood, are mixtures of hundreds of compounds that include phenols, naphthalene and other PAHs, and nitrogen-heterocyclic compounds. Pesticides and herbicides are also frequently reported at military sites, as are heavy metals (particularly lead), paints, perchlorate, bis (2-ethylhexyl) phthalate, and nitrates.

Of the military services, the Army has the largest number of sites affected by chemical explosives. The chemical explosives 2,4,6-trinitrotoluene (TNT), 2,4-dinitrotoluene (DNT), hexahydro-1,3,5-trinitro-1,3,5-triazine (RDX), and octahydro-1,3,5,7-tetranitro-1,3,5,7-tetrazocine (HMX) are reported at military sites where the contaminants were manufactured or at depots where they were disposed of. The Army has 42 installations that contain 230 sites with chemical explosives as contaminants (Haines, 2003), although the number of explosives sites that require source remediation may increase significantly when source zones become more fully characterized.

This diversity of compounds reflects the wide array of activities typical at Army and other military installations. Activities include providing services, materials, and equipment to support military operations, designing and manufacturing weapons systems, and painting (which tends to release heavy metals and solvents). Military installations are characterized by industrial landfills, waste disposal pits, aboveground and underground storage tanks, and spill sites. In addition, they are also burdened with typical domestic waste streams, such as from municipal solid waste landfills, wastewater treatment plants, hospitals, laundries, golf courses, and underground storage tanks for automobile and truck fuels.

Despite the breadth of contamination problems discussed above, the Army Environmental Center's request to the NRC was specifically focused on contamination by recalcitrant organic compounds. This was further defined to encompass those organic compounds that can potentially exist in the subsurface as DNAPLs (primarily solvents) and those that can form pure solid phases (chemical explosives). Table 1-1 summarizes the number of Army installations at which these key recalcitrant organic chemicals are found (with details provided in Appendix A, Table A-3). It should be noted that in the remainder of this report these compounds—as well as NAPL and DNAPL—are referred to exclusively by their abbreviations.

TABLE 1-1 Prevalence of Organic Contaminants of Concern at Army Installations[a]

|  | Chlorinated Solvents | | | | | Explosives | | | |
|---|---|---|---|---|---|---|---|---|---|
| Contaminant Prevalence | PCE | TCE | cis-1,2-DCE[b] | 1,2-DCA | TCA[c] | DNT | TNT | HMX | RDX |
| Total number of installations with contaminant | 51 | 74 | 32 | 24 | 35 | 26 | 30 | 19 | 14 |
| Percentage of all installations | 37% | 54% | 23% | 17% | 25% | 19% | 22% | 14% | 10% |

NOTE: An installation many contain many individual hazardous waste sites.

[a]Number of Base Realignment and Closure Act (BRAC) installations – 23; Number of active installations – 115
[b]Does not include other DCE isomers.
[c]Includes 1,1,1-TCA and 1,1,2-TCA

SOURCE: Compiled by Laurie Haines, Army Environmental Center.

Given the technical difficulties inherent in source area cleanup and the potentially high costs associated with investigating and remediating such sites, the Army (like other branches of the military—see NRC, 2003) is concerned about its long-term management and cost responsibilities and its ability to reach site closure throughout the Installation Restoration Program. During 2001, the Army Environmental Center oversaw independent technical reviews at seven of its facilities where DNAPLs are present in hydrogeologically complex locations. It was observed that certain aggressive source remediation technologies were being pursued at these sites with little understanding of (1) the ability of the technology to achieve substantial mass removal and (2) the relationship between removal of contaminant mass from these sites and its long-term impact on ground-water contamination. There was also considerable uncertainty among Army managers about whether the costs of these efforts were commensurate with the risk reduction achieved. In addition, it was found that remedial project managers (RPMs) at the sites sometimes failed to form contingency plans and exit strategies in the event of remedy failure, which led to numerous iterations of aggressive source remediation efforts.

To counteract these trends, a new approach to these complex, high-cost sites was recommended by the Army Environmental Center, which included (1) protecting receptors directly (with alternate water supplies, well-head treatment, etc.), (2) considering the need for a technical impracticability waiver early in the project, and (3) documenting the cost of the remedy as well as its risk reduction benefits. According to the Army Environmental Center, this approach has been met with considerable resistance, primarily from those stakeholders (e.g., regulators and the public) who desire mass removal and thus prefer more aggressive source remediation strategies and view technical impracticability as an excuse for no action. On a more practical level, however, there is a lack of scientific expertise and tools to do the recommended analyses. Thus, the Army Environmental Center requested the NRC's input on several technical issues to help determine the usefulness and applicability of source remediation as a cleanup strategy. Several key questions served to guide the work of the committee. It should be noted that the term "source removal" (which appears below in the committee's charge) is replaced by the term "source remediation" for the remainder of this report (as discussed at the end of this chapter).

1. What is a meaningful definition of a "source" for the purpose of this study? How important is the source delineation step to the effectiveness of mass removal as a cleanup strategy? What tools or methods are available to delineate sources of organics contamination in complex sites? How should the uncertainty of these characterizations be quantified, in terms of both total mass and mass distribution?

2. What are the data and analytical requirements for determining the effectiveness of various source removal strategies, and how do these requirements

change for different organic contaminant types or hydrogeologic environments? Effectiveness would consider the metrics of groundwater restoration, plume shrinkage and containment, mass removed, risk reduction, and life cycle site management costs.

3. What tools or techniques exist today, and what tools would need to be developed in the future, to help predict the likely benefits of source removal?

4. What would be the most important elements of a well-designed protocol to assist project managers in the field to assess the effects of source removal?

5. What can be concluded about the ability of source removal efforts to bring about substantial water quality benefits and to meet various cleanup goals? (For example, when can these efforts remove enough of the source to then rely on monitored natural attenuation?)

6. What have been the results of source removal activities at Army and other facilities to date? More generally, what can be said about the future use of source removal as a cleanup strategy and the specific technologies investigated during the study?

## CHARACTERISTICS AND DISTRIBUTION OF DNAPLS AND CHEMICAL EXPLOSIVES

Chapter 2 of this report describes in detail the physical properties of DNAPLs and chemical explosives that affect their distribution and persistence in the subsurface. However, some brief comments are warranted here. Most NAPLs include several different chemical compounds due to a combination of mixing before release, reaction with aquifer and soil solids, and partial biodegradation of specific components. A DNAPL consisting of more than one compound (the general case) is a multicomponent DNAPL. A distinction should thus be made between a component of the DNAPL and the DNAPL itself.

Chlorinated solvents are the most common DNAPL components, particularly PCE, TCE, 1,1,1-TCA, 1,2-DCA, cis-1,2-DCE, methylene chloride, and chloroform. These compounds vary from slightly soluble to moderately soluble in water, causing plumes of contaminated groundwater that migrate away from the source material.

When released to the subsurface, DNAPLs flow downward through the vadose zone, typically traveling vertically with little spreading. When the water table is reached, capillary forces tend to produce horizontal spreading. In both vertical and horizontal flow, DNAPLs tend to be restricted to pathways of maximum permeability. As a result of this process, DNAPLs generally follow a very narrow, highly irregular path resulting in a source zone that contains narrow vertical pathways connected to thin, laterally extensive horizontal layers (see Figure 1-1). The limited and extremely heterogeneous distribution of DNAPLs makes both detection and remediation difficult. Given the many forces affecting

Entry Point

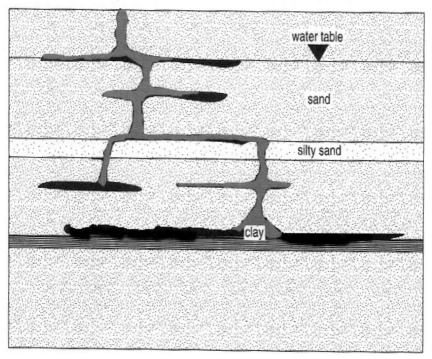

FIGURE 1-1 Typical distribution of DNAPL. Gray areas show residual saturation of DNAPL, while black areas are pools of DNAPL (see Chapter 2 for further explanation). SOURCE: Adapted from NRC (1999a).

DNAPL distribution discussed above, it is not surprising that only a small fraction of the subsurface volume at a contaminated site actually contains DNAPLs.

Explosives are a class of chemicals that can undergo rapid oxidation—a process called detonation—which releases a tremendous amount of energy. Explosives are divided into organic and inorganic chemical classes, though the organic explosive class has caused the greatest environmental risk. Most chemical explosives have melting points well above near-surface soil temperatures, and thus exist as solids at environmentally relevant temperatures. Unlike DNAPLs, when explosives are deposited on the ground surface as solids, they do not migrate into the subsurface.

Explosive compounds exhibit low solubility in water. Nonetheless, surface deposits of solid phase explosives can dissolve into percolating rainwater and can present a long-term contaminant source that threatens groundwater quality. A more common scenario for groundwater contamination by explosives is discharge of wastewater containing these compounds to the environment. At sites impacted by such discharges, the bulk of the explosives contaminant mass is usually present as sorbed and aqueous phase contamination. However, when effluents with high aqueous concentrations are discharged into a significantly colder environment, the explosive compounds may precipitate out of solution and form a separate solid phase in the soil system. These surface deposits are often removed by excavation to practical depths, leaving deeper source areas to be treated in situ.

## DEFINING THE SOURCE ZONE

In order to evaluate aggressive source remediation as a cleanup strategy and differentiate it from other activities, the term "source" must be defined. While seemingly simple, the term "source" can be defined from several perspectives. Implicit in all of the perspectives below, including the one endorsed by the committee, is that sites at which DNAPLs or explosives were released typically have a zone in which the mass of contamination is originally concentrated. This volume, termed the source zone, serves as the source for the development of a dissolved phase plume. As long as the source remains, a dissolved phase plume will continue to develop; hence, removal (or isolation) of the source zone is required to halt creation of the dissolved phase plume.

One approach to defining "source" is to consider the phases in which the contamination may exist. An uncontaminated soil system contains a solid phase, an aqueous phase, and a vapor phase. Any contamination that is present can exist as a separate liquid or pure solid phase, it can be dissolved in the aqueous phase, it can be associated in some fashion with the soil solid phase, or it can be volatilized. For the purposes of this report, the committee agreed that separate solid or liquid phase contaminants were indeed sources. Volatilized contaminants were considered *not* to be sources because the percentage of total contaminant mass in the gas phase per unit volume of subsurface is generally insignificant compared to other phases. The challenge, then, was to determine to what extent sorbed or dissolved contaminants are considered to be sources. Two definitions, in particular, shed light on this argument.

The EPA's regulatory definition of "source material" (EPA, 1991) is:

*Material that includes or contains hazardous substances, pollutants or contaminants that act as a reservoir for migration of contamination to ground water, to surface water, to air, or acts as a source for direct exposure.*

The term "reservoir" suggests a large supply of contamination, although this is not further clarified. The phrase "for migration of contamination to ground water, to surface water, to air" implies that contamination can move from its point of origin. EPA (1991) goes on to state that "contaminated groundwater generally is not considered to be a source material." EPA examples of source materials and nonsource materials are shown in Table 1-2.

Although contaminated groundwater is explicitly excluded from the EPA definition, sorption always results in contamination of solid phases that are contacted by contaminated water. In fact, the amount of sorbed contaminant may greatly exceed the amount of contaminant in solution. The extent of this association depends upon both the contaminant and the soil characteristics. Is this newly contaminated soil now considered source material? This seems unlikely, and it is probable that EPA intended for "contaminated soil" to mean only in-place contaminated soil and debris in the same context as drummed waste. However, nothing in the EPA definition allows one to make a clear distinction between source zones and the solids impacted by dissolved phase plumes. For this reason, the committee explored alternate definitions. One definition that includes such a distinction was made by a recent EPA expert panel (EPA, 2003) concentrating on DNAPL sites, which chose to define the term "source zone" as:

*the groundwater region in which DNAPL is present as a separate phase, either as randomly distributed sub-zones at residual saturations or "pools" of accumulation above confining units. . . . This includes the volume of the aquifer that has had contact with free-phase DNAPL at one time.*

The EPA panel excluded vadose zone issues from its definition and emphasized DNAPLs (chlorinated solvents, solvent/hydrocarbon mixtures, and coal tars/creosotes) rather than light nonaqueous phase liquids (LNAPLs). The EPA panel

TABLE 1-2 EPA Examples of Source and Nonsource Materials

| Source Materials | Nonsource Materials |
| --- | --- |
| • Drummed wastes | • Groundwater |
| • Contaminated soil and debris | • Surface water |
| • Pools" of dense nonaqueous phase liquids (NAPLs) submerged beneath groundwater or in fractured bedrock | • Residuals resulting from treatment of site materials |
| • NAPLs floating on groundwater | |
| • Contaminated sludges and sediments | |

SOURCE: EPA (1991).

definition provides a clear distinction between the region containing the dissolved phase plume (has not had contact with DNAPL) and the source zone (has had contact with DNAPL). The committee combined and modified the prior definitions in order to develop a definition that captures the essence of a source as a reservoir of contamination, while making a distinction between the source and the plume, and that encompasses pure solid sources:

> *A source zone is a saturated or unsaturated subsurface zone containing hazardous substances, pollutants or contaminants that acts as a reservoir that sustains a contaminant plume in groundwater, surface water, or air, or acts as a source for direct exposure. This volume is or has been in contact with separate phase contaminant (NAPL or solid). Source zone mass can include sorbed and aqueous-phase contaminants as well as contamination that exists as a solid or NAPL.*

Recognition of a source zone may be accomplished either from direct observation of the separate phase contaminant (NAPL or solid) or from inference. Because of equilibrium partitioning theory, certain soil phase and aqueous phase concentrations of contaminants imply that a separate pure solid or liquid phase exists in the subsurface, even if one cannot be discovered. Thus, for example, in subsurface areas where the groundwater contaminant concentrations are at or near the temperature-dependent aqueous solubility limit, the presence of separate phase material can be inferred.

A similar approach may be used to recognize a source zone that once contained DNAPL, but no longer does. For example, the most commonly cited example of such a source zone is the case in which DNAPL has been depleted due to matrix diffusion. That is, if DNAPL is trapped on top of a clay layer, or within a fracture in a clay unit, there will be a large concentration gradient established between the saturated water immediately in contact with the DNAPL and the water in the matrix. This will lead to diffusion of the contaminant into the matrix and subsequent sorption to the matrix solids. This relatively immobile contaminant mass is significant because if the source zone is treated by a flushing or chemical treatment that removes the NAPL (e.g., surfactants or in situ oxidation), the contamination in the matrix will remain largely unaffected, and at a later date it will diffuse back into the more permeable zones and recontaminate the groundwater. The presence of this source zone can be inferred from rebounding aqueous concentrations after treatment or recognized from the high concentrations within the matrix.

The charge to this committee included the term "source removal," but this term was abandoned in the course of writing this report because while many technologies involve contaminant mass removal (e.g., excavation and surfactant/cosolvent flooding), others do not. Several approaches involve contaminant mass *destruction* (chemical oxidation, reduction, or biodegradation), or they combine removal and destruction (e.g., steam treatment). Others such as containment or immobilization do not involve any contaminant mass removal. In almost all cases, removal of the source, whether through physical removal or reaction, will not be complete. Thus, in this report the term "source remediation" is used, defined as any approach to reduce the problem associated with source zones.

## REPORT ROADMAP

Among potentially responsible parties, scientists and engineers, regulatory agencies, and other stakeholders, rapidly growing interest in using more aggressive source remediation technologies has generated numerous questions and uncertainties that this report attempts to address. The strengths and weaknesses of source zone remediation as a strategy for hazardous waste remediation are discussed, focusing on recalcitrant organic contaminants (e.g., solvents, other DNAPLs, explosives). In order to make informed statements about site characterization and the various technologies of interest to the Army, the committee analyzed the results of source remediation activities completed to date at dozens of Army and other facilities. The 11 Army sites for which the committee heard presentations and reviewed extensive reports are listed in Table 1-3. These sites are contaminated with either chemical explosives or chlorinated solvents, and they span a range of hydrogeologic conditions. Numerous source remediation activities at non-Army sites were also reviewed, including almost 100 technology-specific cases cited in Chapter 5 and five additional sites discussed in the EPA expert panel report (EPA, 2003). Throughout the report, case studies of Army and other sites where source characterization and source remediation have occurred are presented.

A central theme to this report is the importance of understanding the relationship between the hydrogeologic setting of the site, the different objectives for remediation, and the effectiveness of source treatment technologies. These three independent variables were judged by the committee to be central to decision making at hazardous waste sites, and their multidimensional relationship is illustrated by the cube shown in Figure 1-2. (Other factors that are not included in Figure 1-2, but which play a role in source remediation, include the type of DNAPL and the magnitude of its release.) Envisioning source remediation as a multidimensional problem that links the appropriate source technology with the hydrogeologic setting at a given site and the cleanup objective is critical to maximizing the potential for success. The concepts embodied in this diagram are used throughout this report to illustrate the importance of the relationship of these independent variables to cleanup success.

To elaborate on the physical setting, Chapter 2 presents a more comprehensive picture of DNAPL and explosives sites than is presented in this chapter, including information about DNAPL architecture and contaminant characteristics.

TABLE 1-3 Army Sites Where Source Characterization and Source Remediation Were Reviewed by the Committee

| Installation and Site | Technology | Scale | Contaminants |
| --- | --- | --- | --- |
| Anniston | In situ chemical oxidation—Fenton's | Pilot, then Full | TCE + others |
| Letterkenny OU3 | In situ chemical oxidation—Fenton's | Pilot | DCE, TCE, VC, PCE |
| Watervliet | In situ chemical oxidation—$KMnO_4$ | Pilot | TCE, PCE |
| Letterkenny OU11 | In situ chemical oxidation—peroxone | Pilot | TCE + others |
| Pueblo | In situ chemical oxidation—Fenton's | Pilot | TNT, TNB, RDX |
| Milan | In situ chemical oxidation—Fenton's | Pilot | TNT, RDX, HMX |
| Letterkenny OU10 | Enhanced bioremediation | Pilot | TCE, TCA, DCE, DCA, VC |
| Badger | Enhanced bioremediation | Pilot, then Full | DNTs |
| Redstone | No remediation planned yet | NA | TCE, TCA, perchlorate |
| Ft. Lewis | Thermal treatment planned | NA | TCE |
| Volunteer | Monitored natural attenuation planned | NA | DNT, TNT |

Chapter 3 then discusses the role of source characterization, and it stresses the importance of understanding the nature of the source zone prior to making decisions about cleanup strategies. Without adequate characterization of the size, nature, and distribution of contamination, source cleanup is unlikely to succeed. The various objectives of source zone remediation are outlined in Chapter 4, with a focus on making sure project managers choose metrics that are appropriate for measuring whether their stated objectives are met. In many instances of source remediation observed by the committee, objectives for cleanup are not adequately defined in advance, leading to misunderstandings regarding the expected outcomes.

Chapter 5 discusses the specific technologies that are commonly associated with aggressive source remediation, including the conditions under which they are optimal, their limitations, and their effectiveness for meeting the objectives outlined in Chapter 4. It is here that conclusions are made about the ability of source remediation to bring about changes in water quality. Finally Chapter 6 presents a decision-making framework for source remediation based on Figure 1-2, as well as the committee's conclusions about the future of source remediation as

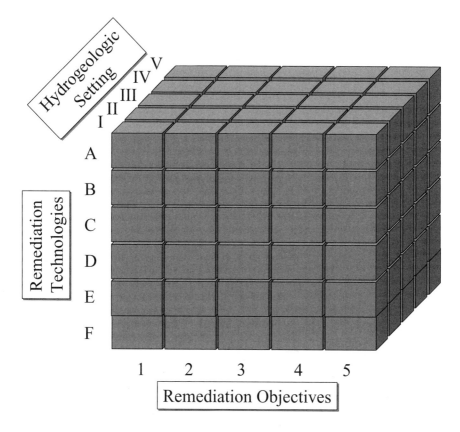

FIGURE 1-2 The success of source remediation is envisioned to depend on the physical setting, the chosen cleanup goal, and the selected remedy.

a cleanup strategy at hazardous waste sites containing DNAPLs and chemical explosives.

It is important to note, as discussed in detail in Chapter 3, that the process of recognizing, characterizing, and remediating a site is not linear, in that it does not follow the successive stages listed in these chapters, flowing directly from initial studies to final remediation. Rather, it is inherently iterative, requiring continual feedback from each stage of study. Each aspect, from the conceptual model to the objectives selected, the degree and methods of characterization, the remediation project design, and the basis for performance assessment, must be continually refined and reevaluated as understanding of the site develops.

The report is intended to inform decision makers within the Army, the rest of the military, and many other government agencies and the private sector about potential options for their sites contaminated with DNAPLs and chemical explosives. The scientific information contained herein should help in the prioritization of cleanup efforts, which is clearly essential given the conflicting forces of reconciling the public's desire for aggressive remediation, the apparent inability of current technologies to achieve aquifer restoration, and the high cost of remediation.

## REFERENCES

Department of the Army. 1997. Environmental Protection and Enhancement. Army Regulation 200-1. http://www.usapa.army.mil/pdffiles/r200-1.pdf.
Department of Defense (DoD). 1998. Evaluation of DoD waste site groundwater pump-and-treat operations. Report No. 98-090, Project No. 6CB-0057. Washington, DC: Office of the Inspector General.
DoD. 2003. Defense Environmental Restoration Program Annual Report to Congress—Fiscal Year 2003. Washington, DC: DoD. http://63.88.245.60/DERPARC_FY03/do/report.
Environmental Protection Agency (EPA). 1989. Evaluation of Ground-Water Extraction Remedies: Volumes 1 and 2. Washington, D.C.: EPA Office of Emergency and Remedial Response.
EPA. 1991. A Guide to Principal Threat and Low Level Threat Wastes. Publication 9380.3-06FS. Washington, DC: EPA Office of Solid Waste and Emergency Response.
EPA. 1992. Evaluation of Ground-Water Extraction Remedies: Phase II, Volume I—Summary Report. Publication 9355.4-05. Washington, DC: EPA Office of Emergency and Remedial Response.
EPA. 1996a. Presumptive Response Strategy and Ex Situ Treatment Technologies for Contaminated Ground Water at CERCLA Sites. EPA 540-R-96-023. Washington, DC: EPA Office of Solid Waste and Emergency Response.
EPA. 1996b. A Citizen's Guide to Treatment Walls. EPA 542-F-96-016. Washington, DC: EPA.
EPA. 2000. U.S. EPA National Water Quality Inventory 2000 Report. Washington, DC: EPA Office of Water.
EPA. 2003. The DNAPL Remediation Challenge: Is There a Case for Source Depletion? EPA 600/R-03/143. Washington, DC: EPA Office of Research and Development.
EPA. 2004. Treatment technologies for site cleanup: annual status report (11th edition). EPA 542-R-03-009. Washington, DC: EPA Office of Solid Waste and Emergency Response.
Feenstra, S., and J. A. Cherry. 1988. Subsurface contamination by dense non-aqueous phase liquid (DNAPL) chemicals. In: Proceedings of the International Groundwater Symposium, International Association of Hydrogeologists, May 1–4, Halifax, Nova Scotia.
Haines, L. 2002. Army Environmental Center. Presentation to the NRC Committee on Source Removal of Contaminants in the Subsurface. August 22, 2002.
Haines, L. 2003. Army Environmental Center. Presentation to the Committee on Source Removal of Contaminants in the Subsurface. April 14, 2003.
Jurgens, B., and W. Linn. 2004. Drycleaner Site Assessment & Remediation—A Technology Snapshot (2003). State Coalition for the Remediation of Drycleaners. http://drycleancoalition.org/download/2003surveypaper.pdf
Interstate Technology Regulatory Council (ITRC). 2000. Dense Non-Aqueous Phase Liquids (DNAPLs): Review of Emerging Characterization and Remediation Technologies Technology Overview. Washington, DC: Interstate Technology and Regulatory Cooperation Work Group.
ITRC. 2002. DNAPL Source Reduction: Facing the Challenge. Regulatory Overview. Washington, DC: Interstate Technology and Regulatory Council.
MacKay, D., and J. A. Cherry. 1989. Groundwater Contamination: Pump and Treat Remediation. Environ. Sci. Technol. 23:630–636.

Mercer, J. W., and R. M. Cohen. 1990. A Review of Immiscible Fluids in the Subsurface: properties, models, characterization and remediation. J. Contam. Hydrol. 6:107–163

National Research Council (NRC). 1994. Alternatives for Ground Water Cleanup. Washington, DC: National Academies Press.

NRC. 1997. Innovations in Ground Water and Soil Cleanup: From Concept to Commercialization. Washington, DC: National Academy Press.

NRC. 1999a. Groundwater and Soil Cleanup: Improving Management of Persistent Contaminants. Washington, DC: National Academy Press.

NRC. 1999b. Environmental Cleanup at Navy Facilities: Risk-Based Methods. Washington, DC: National Academy Press.

NRC. 2003. Environmental Cleanup at Navy Facilities: Adaptive Site Management. Washington, DC: National Academies Press.

Pankow, J. F., and J. A. Cherry. 1995. Dense Chlorinated Solvents and other DNAPLs in Ground-water. Waterloo, Ontario: Waterloo Press. 522 p.

Quinton, G. E., R. J. Buchanon, D. E. Ellis, and S. H. Shoemaker. 1997. A Method to Compare Groundwater Cleanup Technologies. Remediation 8:7–16.

Strategic Environmental Research and Development Program (SERDP). 2002. SERDP/ESTCP Expert Panel Workshop on Research and Development Needs for Cleanup of Chlorinated Solvent Sites. Washington, DC: SERDP/ESTCP.

Stroo, H. 2003. Retec. Presentation to the Committee on Source Removal of Contaminants in the Subsurface. January 30, 2003.

Villaume, J. F. 1985. Investigations at sites contaminated with DNAPLs. Ground Water Monitoring Review 5(2):60–74.

# 2

# Source Zones

Understanding the characteristics of subsurface source zones provides a foundation for addressing source characterization, technology options, and decision making. Following the definition given in Chapter 1, source zones are volumes that have been in contact with separate phase contaminants and that act as reservoirs that sustain a contaminant plume in groundwater, surface water, or air or that act as a source for direct exposure. Within these subsurface regions, non-aqueous, sorbed, and dissolved phase contaminants in hydraulically stagnant zones can provide persistent loading of contaminants to groundwater passing through them. First, the five hydrogeologic settings that typify most hazardous waste sites are described. The chapter then turns to an examination of contaminant releases and subsequent transport, storage, and fate, describing the many processes that act on contaminants in the subsurface and how this is manifested in the field-scale distribution of contaminants. The architecture of source zones is then considered for the five hydrogeologic settings. Although many of the processes that control contaminant fate and transport in the subsurface are the same for either chlorinated solvents or chemical explosives, these contaminants are discussed separately because their release mechanisms can be significantly different from one another.

## HYDROGEOLOGIC SETTINGS

Subsurface settings are a product of a set of diverse geological processes that produce an abundance of variations. Common sedimentary systems include wind-blown (eolian) sands, beach sands, alluvial fans, river sequences, glacial outwash, deltaic sequences, and lake-deposited (lacustrine) clays. Common rock systems include limestone, dolomite, sandstone, shale, interbedded sandstone and shale, extrusive volcanic flow sequences, intrusive granitic bodies, and metamorphic systems of crystalline rock. To varying degrees these systems can be fractured, cemented, and/or opened by dissolution (karst). This diversity makes it challenging to develop general statements regarding the characteristics of source zones, the efficacy of remedial technologies, and what endpoints are attainable. For example, the flow of groundwater or remedial fluids (such as surfactants) is substantially different in beach sand than in karst systems, and the tools required to characterize alluvium are substantially different than tools used to characterize rock.

Five general hydrogeologic settings that are broadly representative of the common conditions of concern are illustrated in Figure 2-1. The differentiating features between the five settings are the spatial variations in permeability and porosity (see Box 2-1, which describes the terminology relevant to the following discussion). These parameters control the mechanisms by which contaminants are stored and released from source zones under natural and engineered conditions. The scale (size) of the representative hydrogeologic settings is envisioned

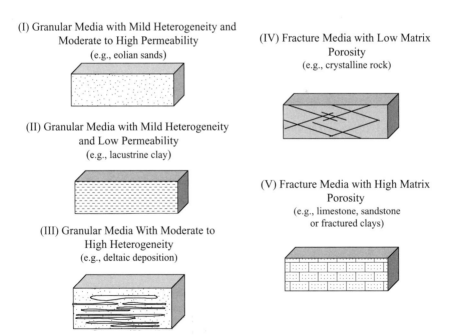

FIGURE 2-1 Five General Hydrogeologic Settings.

## BOX 2-1
## Terminology for Hydrogeologic Settings

The following terms help distinguish among the five hydrogeologic settings discussed in this chapter.

**Consolidated vs. Unconsolidated Media:** Geologic media that are cohesive as a body, firm, or secure are described as consolidated (e.g., most rock formations). Geologic media that are not cohesive as a body, are loosely arranged, and that readily separate into granular components, are described as unconsolidated. Most alluvial deposits (e.g., beach sand) are unconsolidated. Both terms are geotechnical, in that different tools are used to probe consolidated vs. unconsolidated media. The term unconsolidated may not apply to all clays, although all clays are granular. Thus, unconsolidated is a more restrictive term than granular and is used sparingly in this report.

**Grain Size:** From Press and Siever (1974), common labels describing the sizes of granular media are: Clay < $^1/_{256}$ mm < Very Fine Sand < $^1/_{16}$ mm < Fine Sand < $^1/_8$ mm < Medium Sand < $^1/_2$ mm < Coarse Sand < 1 mm < Very Coarse Sand < 2 mm < Granule < 4mm < Pebble < 8 mm < Cobble < 256 mm < Boulder. Grain size and the degree of mixing of different grain sizes (sorting) are primary factors that control the permeability of granular porous media.

**Permeability:** Permeability ($k$) is a property of a porous medium that describes its capacity to transmit fluid. Permeability is independent of the fluid or fluids present in the porous medium and has the units length squared (e.g., $m^2$). Permeability is used in this report as the primary metric for the capacity to transmit fluid because more than one fluid (e.g., air, water, and NAPL) can coexist in the pore space of the medium of interest. Low permeability media are considered herein to be < $10^{-14}$ $m^2$. High permeability media are considered to be > $10^{-10}$ $m^2$. Between $10^{-14}$ and $10^{-10}$ $m^2$ is referred to as moderate permeability media.

**Secondary Permeability:** Secondary permeability refers to the portion of the permeability of a porous medium that can be attributed to secondary (post-emplacement) features of the matrix. Examples of secondary features include fractures, animal burrows, root casts, and solution features. In some media, such as fractured clays or crystalline rock, the dominant factor controlling fluid transmission is commonly secondary permeability.

**Effective Porosity:** Porosity is defined as the volume of void space in the medium divided by the total volume of the medium. In hydrogeology the more important term is the effective porosity of a porous medium, $\phi$, which is a unitless parameter defined as the volume of the *interconnected* void space in the medium divided by the total volume of the medium. Throughout this report, when the term porosity is used, effective porosity is assumed.

$$\phi = \frac{V_{\text{Interconnected Voids}}}{V_{\text{Total}}}$$

For fractured media, the components of porosity include the matrix porosity, $\phi_{matrix}$, and the fracture porosity, $\phi_{fracture}$:

$$\phi_{matrix} = \frac{V_{Matrix\ Voids}}{V_{total}} \qquad \phi_{fractures} = \frac{V_{Fracture\ Voids}}{V_{total}}$$

**Hydraulic Conductivity:** Within the field of groundwater hydrology, the term hydraulic conductivity ($K$) describes a porous medium's capacity to transmit water. In contrast to permeability, conductivity is dependent on the properties of the porous medium *and* the fluid in the porous medium and has the units of length divided by time. Hydraulic conductivity is described as:

$$K = \frac{k\,g\,\rho_{water}}{\mu_{water}}$$

where $k$ is permeability, $g$ is the gravitational constant, $\rho_{water}$ is the density of water, and $\mu_{water}$ is the viscosity of water. $K$ values are included at relevant points in this report for those more familiar with hydraulic conductivity. The relationship between permeability and hydraulic conductivity, and their values for common geologic media, are described in Figure 2-2. In all cases these values are based on the assumption that water is the fluid of interest and that water fully saturates the porous media. Low hydraulic conductivity is considered to be less than $10^{-7}$ m/sec, high is considered to be greater than $10^{-3}$ m/sec, and moderate would fall in between those two values.

**Heterogeneity:** Heterogeneity is used to describe spatial variations in permeability. Heterogeneity can exist over a variety of scales and can be reflected in abrupt changes in permeability at discrete interfaces (caused, for example, by low-permeability inclusions) or by continuous variations in permeability over some length scale (caused, for example, by periodic gradations in grains size). Hetero-geneity is of interest down to the scale of centimeters. In terms of the extent of heterogeneity, media with spatial variations in permeability of less than three orders of magnitude are referred to as *mildly heterogeneous*. This builds on (1) the classification of the Borden Aquifer (Canadian Forces Base Borden, Ontario) as "mildly heterogeneous" (Domenico and Schwartz, 1998) and (2) the observation of nearly three orders of magnitude variation in the permeability in the Borden Aquifer (Sudicky, 1986). Media with greater than three orders of magnitude spatial varia-tion in permeability are described as having either moderate or high heterogeneity. *Anisotropy* refers to the condition in which the permeability of a geologic formation varies with the direction of measurement about a point. This commonly occurs in layered sedimentary deposits where vertical permeability is often less than $^{1}/_{10}$th of the horizontal permeability. This anisotropy tends to foster lateral spreading and horizontal flow.

*continued*

## BOX 2-1 Continued

**Transmissivity:** Transmissivity describes the bulk capacity of a vertical interval of geologic media to transmit water. Transmissivity is the product of the average hydraulic conductivity of an interval and the thickness of the unit. The units of transmissivity are length squared divided by time.

**Layered:** This term refers to horizontal beds of material with different permeability and porosity that are commonly encountered in natural geologic media. Individual layers typically reflect changes in the mode of deposition (e.g., flowing or stagnant conditions in water). The thickness and lateral extent of layers depends on the mode of deposition.

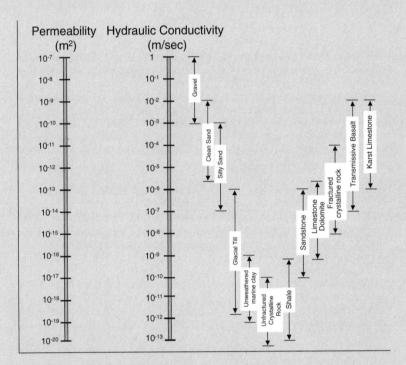

FIGURE 2-2 Permeability and hydraulic conductivity for common geologic media. SOURCE: Adapted from Freeze and Cherry (1979).

in the range of a few meters, whereas the size of an entire source zone can be on the order of tens of meters. Source zones can occur within a single hydrogeologic setting (e.g., a sand dune deposit) or can include multiple hydrogeologic settings (e.g., alluvium overlying fractured crystalline rock). The latter case can be challenging in that the mechanism of contaminant storage and release can be substantially different in adjacent portions of a single source zone.

The following section describes the hydraulic characteristics (primarily permeability and porosity) of the five general settings. The likely distribution of contaminants in each of these settings is developed in subsequent sections. Although they do not entirely capture the diversity of hydrogeologic systems, these five settings are useful for highlighting major differences in how source zones store and release contaminants. The taxonomy used in this chapter is purposefully general and could easily be expanded to more rigorously reflect the range of hydrogeologic conditions that exist.

### Type I – Granular Media with Mild Heterogeneity and Moderate to High Permeability

Type I media include systems with porosities that are consistent with typical granular media (e.g., 5 percent to 40 percent), with permeabilities that are consistent with sand or gravel deposits ($>10^{-14}$ m$^2$ or hydraulic conductivity $>10^{-7}$ m/s), and mild heterogeneity (less than three orders of magnitude). As conceptualized here, this material is about as uniform as it can be in nature and thus is relatively uncommon. Deposits of this nature are encountered in association with windblown sands and beach deposits. Examples include beach sands at the Canadian Forces Base Borden, Canada, and dune deposits at Great Sand Dunes National Park, Colorado (Figure 2-3). Due to its mild heterogeneity and moderate to high permeability, all portions of this media type can transmit groundwater.

### Type II – Granular Media with Low Heterogeneity and Low Permeability

Type II settings have porosities that are consistent with typical granular media (e.g., 5 percent to 40 percent), low spatial variation in permeability (less than three orders of magnitude), low permeability consistent with silt or clay deposits ($k < 10^{-14}$ m$^2$), and low hydraulic conductivity ($K < 10^{-7}$ m/s). An example is a clay deposit with no significant secondary permeability features (such as fractures, root holes, animal borrows, or slickenslides). These systems are somewhat uncommon (especially in the near-surface environment where releases typically occur), although some examples include TCE-contaminated clays at the Department of Energy's Savannah River Site in South Carolina. More typically, low-permeability materials contain significant secondary permeability features and thus fit better into the Type V setting description (see below).

FIGURE 2-3  Example of Type I media from Great Sand Dunes National Monument.
SOURCE: http://www.nps.gov/grsa

## Type III – Granular Media with Moderate to High Heterogeneity

Type III encompasses systems with moderate to large variations in permeability (greater than three orders of magnitude) and porosities that are consistent with granular media (e.g., 5 percent to 40 percent). Given large spatial variations in permeability (at the scale of centimeters to meters), portions of the zone are comparatively transmissive while others contain mostly stagnant fluids. As an example, an interbedded sandstone and shale is shown in Figure 2-4. For the purpose of this report, the more transmissive zones in Type III media have a permeability greater than $10^{-14}$ m$^2$ ($K > 10^{-7}$ m/s). Near-surface deposits of this nature are common due to the abundance of alluvium with large spatial variations in permeability and are encountered in either rock or alluvium associated with deltaic, fluvial, alluvial fan, and glacial deposits. Examples include the Garber-Wellington Aquifer in central Oklahoma, the Chicot Aquifer in Texas and Louisiana, and varved sediments near Searchmont, Ontario (Figure 2-5).

## Type IV – Fractured Media with Low Matrix Porosity

Fractured media with low matrix porosity are common in crystalline rock including granite, gneiss, and schist. Examples include bedrock in the Piedmont and Blue Ridge Mountain region of the southeastern United States and plutonic cores of mountain ranges in the western United States (see Figure 2-6 for an example). The primary transmissive feature in Type IV settings is secondary permeability caused by fractures, because little to no void space exists in the unfractured matrix. The permeability of the unfractured matrix is considered to

FIGURE 2-4 Interbedded sandstone and shale, shown as an example of Type III media. SOURCE: Reprinted, with permission, from http://geology.about.com. © 2004 About.com.

be less than $10^{-17}$ m$^2$ ($K < 10^{-10}$ m/s). However, the bulk permeability of the media is dependent on the frequency, aperture size, and degree of interconnection of the fractures, such that the anticipated range of bulk permeability values is $10^{-15}$–$10^{-11}$ m$^2$ ($K = 10^{-8}$–$10^{-4}$ m/s). The porosity of both the matrix and the fractures is typically small—less than 1 percent. However, in regions where crystalline rock has been extensively weathered (e.g., at the top of bedrock), the bulk media can behave more like a porous medium than what would be expected from a fractured rock type setting. A primary feature that differentiates Type IV from Type I is that contaminants in Type IV will occur in a sparse network of rock fractures that may or may not be hydraulically interconnected. In general, sources zones in fractured media with low matrix porosity are less commonly encountered than sources zones in Type III and Type V settings. This reflects the fact that many surface releases never reach bedrock and, in the United States, crystalline bedrock occurs less frequently than sedimentary bedrock (Back et al., 1988).

## Type V – Fractured Media with High Matrix Porosity

This setting includes systems where fractures (secondary permeability) are the primary transmissive feature *and* there is large void space in the matrix. The permeability of the unfractured matrix is considered to be less than $10^{-17}$ m$^2$ ($K < 10^{-10}$ m/s). The anticipated range of bulk permeability values is $10^{-16}$–$10^{-13}$ m$^2$ ($K = 10^{-9}$–$10^{-6}$ m/s). The porosity of the fractures relative to the total unit

FIGURE 2-5 Interbedded sand and silt layers associated with annual depositional cycles from the Varved Sediments, near Searchmont, Ontario, shown as an example of Type III media. SOURCE: Reprinted, with permission, from http://geology.lssu.edu/NS102/images/varves.html. © 2004 Department of Geology and Physics, Lake Superior State University.

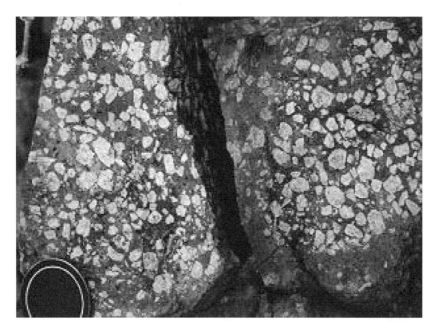

FIGURE 2-6 Fractured crystalline rock shown as an example of Type IV media. Photo taken near Kitt Peak Observatory, Arizona. SOURCE: Reprinted, with permission, from http://geology.asu.edu/~reynolds/glg103/rock_textures_crystalline.htm. © 2004 Department of Geological Sciences, Arizona State University.

volume is small (e.g., <1 percent). However, unlike Type IV, in Type V hydrogeologic settings the porosity of the unfractured matrix is anticipated to fall in the range of 1 percent to 40 percent. Fractured media with high matrix porosity are commonly encountered in sedimentary rock (e.g., limestone, dolomite, shale, and sandstone) and fractured clays. Examples include the Niagara Escarpment in the vicinity of the Great Lakes (see Figure 2-7) and fractured lake-deposited clay in Sarnia, Ontario, Canada.

An important variant of the Type V setting is karst, which is common in carbonates (e.g., limestone or dolomite). In this scenario, transmissive zones include sinkholes, caves, and other solution openings that vary widely in aperture and have the potential to store and transport significant contaminant mass (see Figure 2-8). Permeability in karst terrains varies over tens of orders of magnitude from low permeabilities between fractures to open channel flow in channels and caves (Teutsch and Sauter, 1991; White, 1998, 2002). Karst is characterized by both rapid transport along sparse dissolution features and a high ratio of stagnant to transmissive zones. As such, it is one of the most challenging hydrogeologic settings to characterize and manage.

FIGURE 2-7 Fractured limestone, Door County, Wisconsin, shown as an example of Type V media. SOURCE: Reprinted, with permission, from http://www.uwgb.edu/dutchs/ GeoPhotoWis/WI-PZ-NE/BayshorePark/bayshcp3.jpg. © 2004 Natural and Applied Sciences, University of Wisconsin-Green Bay.

## Relating the Hydrogeologic Settings to Specific Sites

The five hydrogeologic settings defined above represent distinct members in the continuum of settings observed at actual sites. Type I, with mild heterogeneity and moderate to high permeability, grades gradually into Type III as heterogeneity increases. With an increasing clay fraction, Type III grades to Type II. Natural systems range from clean sands to clayey sands to sandy clays to clays in a continuum. In a similar manner, the degree of importance of fractures may vary from insignificant in Type III to dominant in Type V. Because of these gradients, the presence of stagnant zones and the degree of diffusion and sorption vary continually.

Source zones, especially those above a certain size, may also encompass more than one hydrogeologic setting. This commonly occurs in the instance of shallow alluvium over bedrock. For example, in the Piedmont of the southeastern United States, one can find fluvial deposits (Type III) and saprolite (Type V) overlying fractured crystalline rock (Type IV) (Figure 2-9). Selecting characterization tools and source management technologies is challenging under these conditions, because although contamination may exist throughout, the appropriate

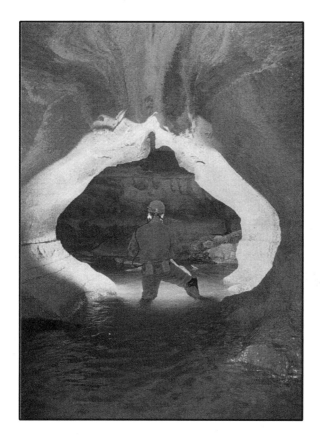

FIGURE 2-8 Large- and small-scale solution features in karst limestone, Redstone Arsenal. SOURCE: Courtesy of De la Paz and Zondlo, Shaw E&I (2003).

tools for one hydrogeologic setting may not work in the adjacent hydrogeologic setting.

In the face of these complexities, the Army was nonetheless asked to estimate the percentage of its DNAPL sites that exist in the five hydrogeologic settings, for the purposes of providing context to the above discussion. Of the Army's 43 active and BRAC (base realignment and closure) installations potentially thought to have DNAPLs (out of a total of 120), 26 percent are located in hydrogeologic setting I, 16 percent are located in hydrogeologic setting II, 16 percent are located in hydrogeologic setting III, 14 percent are located in hydrogeologic setting IV, and 28 percent are located in hydrogeologic setting V (Laurie Haines, Army Environmental Center, personal communication).

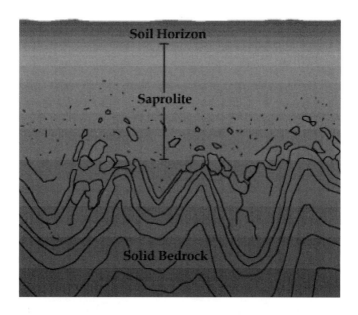

FIGURE 2-9 Mixed hydrologic settings of highly weathered saprolite overlying crystalline bedrock. SOURCE: Reprinted, with permission, from http://web.wm.edu/geology/virginia. © 2004 The Geology of Virginia, Department of Geology, College of William and Mary.

## DNAPLS

How chlorinated solvents are distributed in the subsurface depends on the particular hydrogeologic setting, described above; on the chemical and physical properties of the solvents, the amount, mode, and timing of initial release of the solvents, and their fate and transport processes in the subsurface; and on human activities that may subsequently alter the source zone architecture (such as excavation). Table 2-1 lists several properties of chemicals commonly found in DNAPLs (including both chlorinated solvents and other organic compounds). As discussed below, these properties play important roles in shaping the migration capacity and eventual distribution of DNAPLs, and they can be used to predict the potential for chemical transformation or phase transition processes.

### Contaminant Releases

Contaminants are typically released to the subsurface as constituents of a liquid phase, such as a dilute aqueous solution, a concentrated aqueous solution

TABLE 2-1 Properties of Organic Chemicals Found in DNAPLs

| | CAS # | Aqueous Solubility (mg/L) | Density (g/cm$^3$) | Vapor Pressure (mm Hg) | Absolute Viscosity (cP) | Log $K_{ow}$ | Henry's Law Coefficient ($10^3$ atm m$^3$/mol) |
|---|---|---|---|---|---|---|---|
| **Chlorinated Solvents** | | | | | | | |
| Tetrachloroethene (perchloroethylene, PCE) | 127-18-4 | 150 | 1.62 | 20 | 0.89 | 2.88 | 15 |
| Trichloroethene (trichloroethylene, TCE) | 79-01-6 | 1,100 | 1.46 | 74 | 0.57 | 2.53 | 9.1 |
| cis-1,2-Dichloroethene (cis-1,2-DCE) | 156-59-2 | 3,500 | 1.28 | 200 | 0.48 | 1.86 | 3.4 |
| 1,1-Dichloroethene (1,1-DCE) | 75-35-4 | 400 | 1.21 | 590 | 0.36 | 2.18 | 15 |
| 1,1,1-Trichloroethane (1,1,1-TCA) | 71-55-6 | 1,300 | 1.34 | 120 | 0.87 | 2.48 | 15 |
| 1,2-Dichloroethane (1,2-DCA) | 107-06-2 | 8,300 | 1.23 | 87 | 0.80 | 1.48 | 1.1 |
| Tetrachloromethane (carbon tetrachloride, CT) | 56-23-5 | 800 | 1.59 | 120 | 0.97 | 2.83 | 30 |
| Trichloromethane (chloroform, CF) | 67-66-3 | 8,200 | 1.49 | 200 | 0.58 | 1.95 | 3.4 |
| Dichloromethane, DCM (methylene chloride) | 75-09-2 | 13,000 | 1.33 | 440 | 0.43 | 1.25 | 2.0 |
| **Other Hydrocarbons** | | | | | | | |
| Naphthalene | 91-20-3 | 32 | 1.14 | $8.2 \times 10^{-2}$ | solid | 3.36 | 0.46 |
| Benzo[a]pyrene | 50-32-8 | 0.004 | 1.35 | $5.6 \times 10^{-9}$ | solid | 6.04 | $3.4 \times 10^{-4}$ |
| Aroclor 1254 (PCB mixture) | 1097-69-1 | 51 | 1.50 | $7.7 \times 10^{-5}$ | 700 | 6.5 | 2.7 |
| Aroclor 1260 (PCB mixture) | 1096-82-5 | 80 | 1.56 | $4.0 \times 10^{-7}$ | resin | 6.9 | 0.34 |

Units and abbreviations:
CAS # is a unique identifier used by the American Chemical Society to identify chemicals in databases.
Aqueous solubility in milligrams per liter at 25°C;
Density in grams per cubic centimeter at 20°C;
Vapor pressure in millimeters of mercury at 25°C;
Absolute viscosity in centipoise at 20°C;
Log $K_{ow}$ = log of the octanol–water partition coefficient (unitless);
Henry's Law coefficient is in atmospheres meters cubed per mole at 25°C.

SOURCE: Montgomery (2000), except as noted. Mackay et al. (1993) for properties of cis-1,2-DCE; Cohen et al. (1993) for absolute viscosity.

(leachate), or an organic liquid (nonaqueous phase liquid or NAPL) that is not miscible with water. The characteristics of the released liquid will greatly influence the migration pathways, extent of travel, and persistence of the released contaminants. For example, a considerable mass of a contaminant that is only slightly soluble in water may travel large distances in the subsurface as a constituent of an organic liquid.

Sources of NAPLs are numerous and include surface spills, leaking drums, pipelines, storage tanks, and liquid waste disposal operations. Many NAPLs encountered in the subsurface tend to be mixtures of several different chemical compounds, due to their use history prior to release, the sequence of chemical releases, and biotic or abiotic reactions within the subsurface environment subsequent to release. A NAPL consisting of more than one compound is a multicomponent NAPL. Volatile organic compounds (VOCs), polychlorinated biphenyls (PCBs), and polycyclic aromatic hydrocarbons (PAHs) are often found as components of NAPLs. The two most prevalent classes of NAPL components are chlorinated solvents and petroleum hydrocarbons (including gasoline and fuel oils) (Mercer and Cohen, 1990). Other common NAPLs include coal tars, transformer oils (primary carriers of PCBs), and creosote (a primary carrier of PAHs and phenols).

NAPLs that have a density greater than that of water are commonly known as dense NAPLs (DNAPLs). Of the classes of compounds cited above, chlorinated solvents are the most common DNAPL components (e.g., PCE, TCE, *cis*-1,2-DCE, 1,1-DCE, 1,1,1-TCA, 1,2-DCA, carbon tetrachloride, chloroform, and methylene chloride). PCBs, PAHs, creosote, and coal tar often form or are constituents of DNAPLs. Petroleum hydrocarbons and transformer oils tend to form light NAPLs (LNAPLs), with densities less than that of water. It should be noted that it is not uncommon for multicomponent DNAPLs to contain significant concentrations of petroleum hydrocarbons such as benzene, toluene, ethylbenzene, and xylene (BTEX). Thus, it is important to make a distinction between a chemical component of a DNAPL, which may be denser or lighter than water in its pure liquid form, and the DNAPL itself, which is a separate phase organic liquid composed of various chemical species.

## DNAPL Fate and Transport in the Subsurface

*Immiscible Flow*

Bulk NAPL migration within a formation is governed by properties of the organic fluid, including density, viscosity, and interfacial tension with the pore water, as well as properties of the formation solids, including texture and surface characteristics. As a NAPL enters the subsurface, it will first encounter the unsaturated zone, which contains natural pore water and air. In this zone, NAPL migration (in which the NAPL displaces air) is driven primarily by gravity and thus will tend to be vertically downward. Depending upon its solubility and volatility, the NAPL may dissolve into the pore water and/or volatilize into the pore air as it moves within the unsaturated zone. As the NAPL continues its downward migration, it will eventually encounter the saturated zone in which water completely fills the pore space. Given its density, a DNAPL will tend to continue to migrate vertically, displacing the groundwater until a less permeable stratum is reached or the volume of DNAPL is depleted or a sufficient vertical hydraulic gradient is met. Alternatively, a LNAPL will tend to spread preferentially within the capillary fringe zone at the top of the saturated domain, migrating primarily in the direction of natural groundwater flow.

Representative specific gravities (densities) of DNAPL constituents, which play a role in their tendency to migrate downward through the saturated zone, are given in Table 2-1 (along with several other parameters of interest). Note that chlorinated solvents are characterized by densities of 1.2 g/cm$^3$ or more and hence exhibit substantial propensity for vertical flow (Pankow and Cherry, 1996). Creosote, coal tar, and other PAH-based DNAPLs, however, tend to have densities (e.g., 1.1 g/cm$^3$) much closer to that of water and thus may not experience as large a driving force for vertical migration.

While the density of a NAPL influences its propensity to migrate vertically within the saturated groundwater zone, its viscosity influences its *rate* of migration. In general, fluids of lower viscosity tend to migrate more rapidly due to reduced resistance to flow. Table 2-1 reveals that many DNAPLs, including chlorinated solvents, have viscosities smaller than that of water (1 cP). Thus, these fluids will tend to flow readily. Creosotes and coal tars, however, generally have a much higher viscosity and tend to migrate more slowly under similar hydraulic gradients.

A third phenomenon that influences the migration of a DNAPL in the subsurface is known as capillarity. Capillarity is the physical manifestation of the interfacial forces that occur between phases—either liquid/liquid or solid/liquid. The nature and extent of these interfacial forces exert primary control on migration pathways, the extent of DNAPL spreading, and DNAPL entrapment in a saturated formation. Capillarity is controlled by both by the geometry of the pores and two interfacial properties: interfacial tension and wettability. Interfacial tension is a property of the aqueous–DNAPL interface and is defined as the energy per unit interfacial area required to create a new surface (Hiemenz and Rajagopalan, 1997). DNAPL–water interfacial tensions are typically in the range of 20–50 dynes/cm (Mercer and Cohen, 1990) for pure phase organic liquids. However, interfacial tension can be significantly affected by co-contaminants or additives in the DNAPL phase, including organic acids and bases and surfactants, and by dissolved pore water constituents, such as natural humic substances. Such compounds can behave as surface-active agents, substantially reducing interfacial tension (Adamson and Gast, 1997). A reduction in interfacial tension tends to (1) decrease the spread of a DNAPL transverse (perpendicular) to its primary direction of migration and (2) decrease the force required for the DNAPL to displace water from a saturated pore.

Wettability refers to the tendency of one fluid to spread on or adhere to a solid surface in the presence of another immiscible fluid. It controls the distribution of fluids within the pores. In water/NAPL/solid systems, the liquid having the higher affinity for the solid surface coats the solid and is referred to as the wetting phase, while the other liquid is known as the nonwetting phase. The contact angle, a measure of wettability, is the angle the liquid–liquid interface makes with the solid surface (Hiemenz and Rajagopalan, 1997). For many natural minerals, including quartz and carbonates, water is more strongly attracted to the mineral surface than are common DNAPL constituents. Thus, in such media, water generally is the wetting phase, distributing itself along the solid surfaces and in small-aperture pore regions and fractures. A solid is said to be water-wet if the contact angle is between $0°$ and $60°$, as measured through the aqueous phase. Conversely, as the contact angle approaches $180°$, the surface is said to be strongly NAPL wetting. A surface is termed intermediate-wet if the contact angle ranges from approximately $70°$ to $120°$ (Morrow, 1976). The condition of mixed wettability, in which the larger pores are organic-wetting and the smaller pores are water-wetting, has long been recognized in the petroleum industry (e.g., Salathiel, 1973). The term fractional wettability is generally used to describe media with surfaces of varying wettability (Anderson, 1987). Water-, intermediate-, and organic-wetting conditions can exist simultaneously in the subsurface due to natural variations (Anderson, 1987) or through the interaction of the released NAPLs with the solids. For example, contact with NAPL mixtures containing surface-active constituents can render a porous medium intermediate- to organic-wet (e.g., Powers et al., 1996). These and other studies suggest that variations in wettability may be common in the contaminated subsurface. Such variations may influence NAPL migration and persistence in natural settings.

In water-wet media, the DNAPL, which is the nonwetting phase, tends to be concentrated in the center of the pores and in larger pores and fractures. The wetting phase is able to easily enter new pore spaces, but the nonwetting phase has to overcome capillary forces to do so. The required displacement force is a function of the pore geometry, the interfacial tension, and the contact angle. In a cylindrical pore, the pressure differential required to displace the wetting phase $(\Delta P)$ is given by the Laplace-Young equation:

$$\Delta P = 2 \, \sigma \cos \theta / r \qquad (1)$$

where

$\theta$ = contact angle
$\sigma$ = interfacial tension
$r$ = pore radius.

This required pressure differential can be supplied by an applied pressure on the nonwetting phase or by the weight of the accumulating nonwetting phase above the pore. For a nonwetting DNAPL, the height ($h$) that must accumulate prior to pore penetration is thus related to the pressure differential by:

$$\Delta P = P_{nw} - P_w = \Delta\rho \, h \, g \tag{2}$$

where
  $\Delta\rho$ is the difference in density between the DNAPL and water (the nonwetting and wetting phases)
  $g$ is the gravitational constant
  $P_{nw}$ is the nonwetting liquid pressure
  $P_w$ is the wetting liquid pressure.

Once this pressure differential is achieved, the pore will be invaded by the DNAPL.

Although capillary forces are well understood for simple pore geometry (e.g., cylindrical pores), the complex pore structure of natural porous media makes precise predictions of interface positions difficult. Thus, a macroscopic relation between the liquid pressure differential and the degree of saturation of the wetting fluid, known as a capillary pressure–saturation curve, is typically used in practice to describe the capillary behavior of a particular medium. The critical pressure differential that must be achieved for any of the wetting fluid to be displaced is known as the *entry pressure*. Although the relationship shown in Equation (1) must be modified for pore geometry and surface roughness considerations in natural porous media, the general form of this relationship is still valid. Equation (1) thus suggests that the existence of capillary forces tends to create a (capillary) barrier to the movement of a nonwetting DNAPL into fine pore water-saturated media.

The process of nonwetting DNAPL migration is thus one of vertical migration until a finer-grained layer is encountered, at which point the DNAPL spreads horizontally either until a sufficient thickness of DNAPL accumulates to overcome the entry pressure or until a path with lower entry pressure, perhaps from lithologic variation, is encountered during the horizontal migration. Subtle textural variations that create differences in entry pressure sufficient to affect DNAPL flow can occur even in apparently homogeneous units (Kueper et al., 1993), and DNAPL migration in saturated sandy media has been shown to be sensitive to small-scale variation in permeability and capillary characteristics (Poulsen and Kueper, 1992; Brewster et al., 1995). Such small-scale entry-pressure variations can cause uneven DNAPL penetration of a macroscopically uniform subsurface layer, leading to the formation of narrow vertical preferential flow pathways, commonly known as fingers, that can serve as rapid conduits for DNAPL migration deep into the subsurface. Because small-scale textural variability is not easily quantifiable, these preferential pathways appear to be somewhat random in their distribution (e.g., Rathfelder et al., 2003). This fingering behavior can make it very difficult to locate a DNAPL in situ, due to the small lateral dimensions of the fingers. The propagation of fingers is more common in coarser-textured media and under conditions of low interfacial tension, since capillary forces will tend to oppose the formation of extensive fingers.

The resistance to downward flow by finer-grained layers typically results in horizontal spreading over a distance that is large relative to the lateral extent of the vertical pathways. Thus, DNAPL often follows a highly irregular path, resulting in a source zone that contains narrow vertical pathways connected to thin, laterally extensive horizontal lenses. This will be particularly true of a Type III geologic setting that contains persistent finer-textured horizontal layers of high permeability contrast or of Type IV or V media with extensive horizontal fractures.This could also be characteristic of a Type I setting, if fine-scale layering were present. When the finer-textured or impermeable layers are dipping, the DNAPL will also tend to flow down-dip, even if the hydraulic gradient is in another direction. This combination of horizontal spreading until a more permeable path is found and of down-dip migration may displace a DNAPL a large distance from its point of entry. Variations in the topography of fine-grained layers may trap a portion of the DNAPL, creating isolated pools (as shown in Figure 2-10 for a DNAPL spill in a fractured rock system).

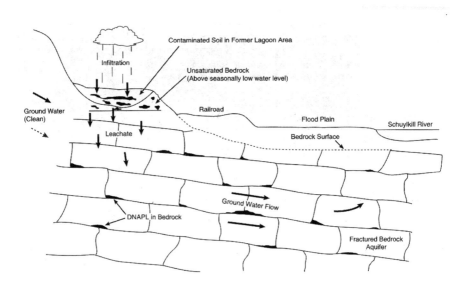

FIGURE 2-10 Schematic of DNAPL pooling in a fractured rock aquifer (Type V) in Pennsylvania. The site is contaminated with 1,2,3-trichloropropane, a DNAPL, and it has moved along bedding plane fractures. SOURCE: EPA (1992).

The DNAPL migration characteristics discussed above are illustrated in Figures 2-11 and 2-12, which contain photos of laboratory sand box experiments exploring PCE release in quartz sand media (representative of Type I media with permeability changes of 1–2 orders of magnitude). Figure 2-11(A) demonstrates the influence of a capillary barrier on PCE migration. Here the PCE was released within a coarse water-saturated sand with a permeability of $1.2 \times 10^{-10}$ m$^2$. It migrated downward under the influence of gravity, until it encountered a layer of finer-textured material with a permeability of $8.2 \times 10^{-12}$ m$^2$. At the textural interface, the PCE spread laterally and was able to cascade around this fine layer before the entry pressure was exceeded, again finding a path through the coarser medium. The PCE then pooled at another textural interface at the bottom of the tank. Inspection of the figure reveals the small thickness of the pool perched on the finer lens and the presence of vertical DNAPL fingers that are barely discernable paths to the lower capillary barrier. As noted above, these are typical characteristics of DNAPL migration behavior. Figure 2-11(B) illustrates the effect of reduced interfacial tension on PCE migration. Here the physical system is identical to that shown in (A), but the resident pore (wetting) fluid includes a surfactant that has lowered the interfacial tension between the PCE and aqueous phases (from 47.8 to 0.5 dynes/cm). Notice that the entry pressure height is now exceeded as the PCE spreads on the finer sand, and the PCE is able to penetrate

**A High Interfacial Tension**          **B Low Interfacial Tension**

 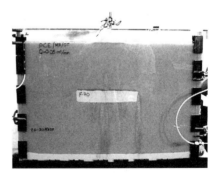

FIGURE 2-11 The influence of interfacial tension (IFT) on DNAPL migration. Photographs of aquifer cell experiments (~60 cm length by ~35 cm height) capture the final DNAPL (dark dye) distribution in (A) water-saturated sand (IFT ~ 48 dyne/cm) and (B) sand saturated with an aqueous solution containing a mixture of surfactants selected to reduce IFT to 0.5 dyne/cm. SOURCE: Reprinted, with permission, from Rathfelder et al. (2003). © 2003 Elsevier Science.

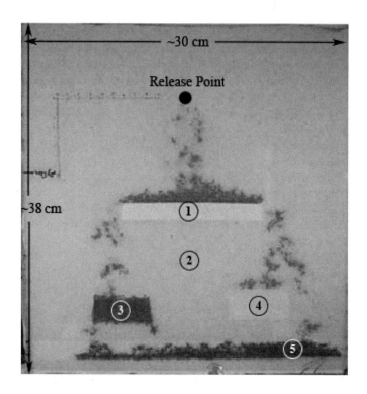

FIGURE 2-12  Effect of wettability on PCE (dark dye) migration in a sand box. Wettability and permeability of sands: (1) water-wet, $4.7 \times 10^{-12}$ m$^2$; (2) water-wet, $4.0 \times 10^{-10}$ m$^2$; (3) organic-wet, $6.4 \times 10^{-11}$ m$^2$; (4) water-wet, $6.4 \times 10^{-11}$ m$^2$; (5) organic-wet, $4 \times 10^{-10}$ m$^2$. SOURCE: Reprinted, with permission, from O'Carroll et al. (2004). © 2004 Elsevier Science.

the finer-textured material. Also note the presence of preferential flow paths through the fine-grained material, attributable to small-scale variations in lens packing that are not easily discernable.

Figure 2-12 illustrates the influence of wettability on DNAPL migration. Here, again, a release of PCE has occurred in a coarse sandy medium. This medium is embedded with finer-textured lenses of varying properties. Those labeled as organic wet sands in the figure have been treated with an octadecyltrichlorosilane coating that has rendered the quartz surface NAPL-wetting. Water is now easily displaced from these treated lenses, since capillary forces act to "pull" the DNAPL into the organic-wet material.

*Residual Entrapment*

As a DNAPL migrates through the subsurface, small globules or ganglia of this organic phase are retained within the pores due to the presence of capillary forces. This entrapped NAPL is frequently quantified as a *residual saturation*—the volumetric ratio of entrapped organic phase to the total pore volume. Entrapment occurs when capillary forces are sufficiently large to overcome the forces exerted by flowing water and gravity. Residual saturations are thus a function of pore geometry, organic phase properties (including interfacial tension, viscosity, and density), flow velocity, and porous medium wettability. The entrapped residual is also a function of the maximum DNAPL saturation reached prior to drainage of the organics and consequently of the release history—that is, higher residuals are found in media that experienced larger release rates or within areas where NAPL had pooled. Furthermore, because larger amounts of an organic compound are entrapped when the release rate is higher, a released volume of DNAPL will tend to migrate further and deeper in a formation if it is released more slowly. Thus, release rate tends to affect the migration patterns of a DNAPL spill. It should be noted that pooled organic can be mobilized by a subsequent change in the groundwater flow field or a breach of the capillary barrier on which the pool is perched. At groundwater velocities commonly encountered under natural or pump-and-treat conditions, however, residual saturations are expected to be essentially independent of velocity (Powers et al., 1992).

Local maximum residual saturations of DNAPLs measured in field-scale and laboratory experiments are typically in the range of 10 percent to 35 percent in saturated, unconsolidated media, with levels as high as 50 percent in materials of low permeability (Conrad et al., 1987; Schwille, 1988). Average residual saturations reported at typical DNAPL field sites, however, are usually much smaller, on the order of 0.1 percent to 1.0 percent (Meinardus et al., 2002) of the affected pore volume. This apparent discrepancy relates to the scale of the saturation measurement, that is, the volume over which the organic mass is averaged. For example, consider the PCE release scenario shown in Figure 2-11(A). Here quantification of the average DNAPL saturation in the tank (the total volume released divided by the tank pore volume) yields a value of 0.1 percent, while saturations measured on selected 2-cm$^3$ subsamples ranged from 0.8 percent to 19 percent (Rathfelder et al., 2003). The first saturation value (0.1%) thus may be more characteristic of an average field-scale value, but it should be recognized that this average value is not representative of the actual distribution of DNAPL within the volume.

NAPL entrapment is illustrated in Figure 2-13(A), in which residual PCE entrapment can be observed in a sand column packed with a coarse quartz sand. Figure 2-13(B) shows a range of entrapped ganglia, polymerized in a similar sand column experiment that entrapped an LNAPL (styrene) in a more graded sand.

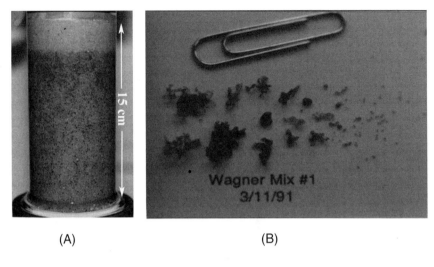

(A)                                                (B)

FIGURE 2-13 PCE entrapment in coarse sand (A) and representative ganglia from a graded sand (B). SOURCE: Reprinted, with permission, from Powers et al. (1992). © (1992) American Geophysical Union.


*Other Processes Influencing DNAPL Persistence*

The sections above have described the processes controlling the migration and entrapment of DNAPL in geologic materials to form a DNAPL source. Once the DNAPL is emplaced, the long-term persistence of this source zone will be controlled by the rates of dissolution of the DNAPL mass to the flowing pore water and by any alternations in the DNAPL properties due to chemical weathering or microbial transformations.

Local DNAPL dissolution has been the subject of much investigation. Researchers have found that the rates of dissolution are controlled by a number of variables, including the solubility of the DNAPL constituents, the local groundwater velocity, the textural heterogeneity of the geologic media, the wettability characteristics of the solid, and the saturation of the DNAPL (e.g., Powers et al., 1992, 1994; Bradford et al., 1999). In general, mass transfer rates tend to increase with groundwater velocity, grain size uniformity, and DNAPL solubility. Because DNAPL solubilities (see Table 2-1) and groundwater velocities are typically very small, dissolution rates tend to be very slow in DNAPL source zones under natural flow conditions. DNAPLs at residual saturation can thus persist for decades (Lemke et al., 2004). As would be anticipated from thermodynamic equilibrium considerations, for DNAPL mixtures the more soluble constituents tend to dissolve faster, resulting in a change in DNAPL composition over time.

Such preferential dissolution can potentially alter DNAPL solubility, viscosity, density, interfacial tension, and toxicity.

In zones of stagnation or very slow flow, dissolution is further limited by the diffusion of the organic solute away from the DNAPL interface (diffusive flux). According to Fick's law of diffusion, diffusive flux is the product of the concentration gradient and the diffusivity of the DNAPL constituent in water. Typical aqueous diffusivities of DNAPL constituents are on the order of $10^{-5}$ cm$^2$/s. This small value, coupled with the low solubility of these compounds, makes diffusion an extremely slow process. Thus, if the DNAPL is able to enter a stagnant zone of a formation or if water flow bypasses the DNAPL region (as will occur for high-saturation pools), dissolution rates tend to be extremely small, and the DNAPL may persist for centuries under natural conditions.

The process of molecular diffusion will also tend to spread dissolved contaminant mass into stagnant regions of a formation. Thus, even if a capillary barrier or low-permeability zone has prevented the downward propagation of a DNAPL, the low-permeability layer underlying a DNAPL pool may become a reservoir for dissolved organic mass. If the DNAPL pool is subsequently removed, the direction of the concentration gradient will be reversed, and this dissolved organic mass will slowly diffuse out of the low-permeability zone over time (reverse diffusion), serving as a persistent source of contamination to flowing groundwater in the overlying stratum.

The potential reservoir of organic mass for reverse diffusion from stagnant regions of a formation is also strongly influenced by the sorptive capacity of the solids for the DNAPL constituents. This sorbed portion of the contaminant mass is not generally quantifiable from groundwater sampling alone and can only be assessed through aquifer coring. Sorbed contamination is the organic mass that is associated with the solid matrix material. For organic compounds, the sorption capacity is generally related to the fraction and character of the solid phase organic carbon, the surface area of the solids, and the compound's octanol–water partition coefficient ($K_{ow}$) (see Table 2-1). In general, larger sorptive capacity is associated with a higher fraction of organic carbon, a higher surface area, and a larger octanol–water partition coefficient. The extent of diagenesis of the organic solid fraction will tend to affect its sorption rates and sorption reversibility (e.g., Huang and Weber, 1998; Weber et al., 1999).

Under equilibrium conditions, there is a quantifiable relationship between sorbed organic and aqueous phase organic concentrations. This relationship is known as an equilibrium isotherm. In the vicinity of a DNAPL, the persistent high aqueous concentrations, coupled with diffusion, tend to promote sorption. Conversely, if the DNAPL is removed, desorption will serve as a continuing source of contaminant mass, feeding the reverse diffusion process. Thus, in heterogeneous formations, reverse diffusion from stagnant zones that contain aqueous phase or sorbed mass can sustain plume concentrations after the DNAPL has been depleted either through remediation efforts or natural processes.

Prediction and assessment of contaminant storage and elution from low-permeability zones is a complex task because groundwater flow velocities vary by orders of magnitude within different interbeds of natural sediments (e.g., sand and silt). If groundwater flow is relatively slow, as is typical under natural flow conditions, the sorption is typically considered as an equilibrium process; that is, these solid–water exchanges are assumed to occur instantaneously as the water flows through the solids. However, when flow is relatively fast, such as may occur when groundwater is extracted by pumping, these exchanges may lag behind the flow and lead to (1) earlier-than-expected contaminant arrival or (2) later-than-expected contaminant extraction, when a plume is arriving at or passing an observation point, respectively. Biological and chemical contaminant transformation in the aqueous phase may similarly be limited by desorption, especially where those transformations are rapid. To date, the process of reverse diffusion has been studied in only simple geologic scenarios (Sudicky et al., 1985; Parker et al., 1994, 1997; Ball et al., 1997; Liu and Ball, 1998; Mackay et al., 2000).

Until recently, it was believed that biotransformation processes could not occur near a chlorinated solvent source due to the toxicity of high contaminant concentrations associated with the presence of NAPL (Robertson and Alexander, 1996). A number of studies, however, have recently documented microbial activity at concentrations at or near the aqueous solubility of PCE (e.g., Yang and McCarty, 2000; Cope and Hughes, 2001; Sung et al., 2003). Recent studies have suggested that microbial activity can enhance DNAPL dissolution by a factor of 5 or more in the laboratory by increasing local concentration gradients and enhancing aqueous solubility (Yang and McCarty, 2002; Cope and Hughes, 2001), although enhanced dissolution has not yet been well documented in the field. Under natural subsurface conditions, the absence of electron donors typically limits microbial activity.

## Field-Scale Distribution of Contaminants in Source Zones

In view of the many processes affecting DNAPL distribution discussed above, it is not surprising that only a small fraction of the subsurface volume at a contaminated site actually contains DNAPL. Within this volume, the DNAPL will be irregularly distributed both horizontally and vertically, with the distribution not easily predicted by the position of the release point. The DNAPL mass will be distributed within both residual ganglia and more saturated pools. During or immediately after a release, DNAPL will be the largest component of contaminant mass in the source zone. With time, diffusive mass transfer to the aqueous, gas, and solid phases depletes the DNAPL. As this occurs, dissolved mass in stagnant aqueous zones (generally in the rock matrix) and mass sorbed to solids can become the dominant fractions of the source mass, depending on the hydrogeologic setting. Figure 2-14 shows a conceptual distribution of contaminant mass in these various phases.

The tendency for the source to exist as DNAPL, sorbed, or dissolved phase mass is discussed below for each of the five hydrogeologic settings introduced earlier.

*Type I Settings*

In instances of granular media with mild heterogeneity and moderate to high permeability, all of the source zone can be viewed as transmissive. Where the media has a low sorptive capacity (e.g., where organic matter content is low), little mass will be retained on the solids. Thus, there are typically no stagnant zones or persistent elution areas in Type I settings, and the most likely source of the dissolved phase plume is DNAPL. Note, however, that extensive DNAPL pooling may occur in Type I settings at boundaries with other less-permeable units (see Figure 2-12), and such pools may be difficult to treat.

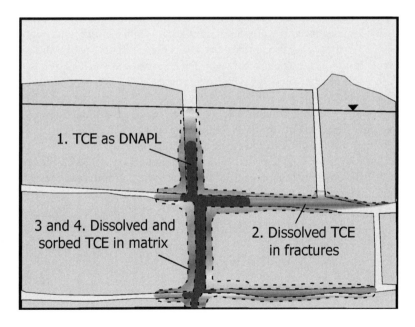

FIGURE 2-14 Conceptual diagram of contaminant mass in the subsurface showing (1) DNAPL, (2) mass dissolved in a transmissive zone (and thus part of the plume as discussed later), and (3) and (4) mass sorbed to the solid matrix or existing as a dissolved phase in stagnant zones within the matrix.

*Type II Settings*

More rare are source zones comprised of granular media with mild heterogeneity and low permeability. Given the absence of secondary permeability features that characterizes this setting, Type II media are difficult for contaminants to invade due to high NAPL displacement pressures and/or low conductivities to fluids. An exception to this is where DNAPL preferentially wets the fine-grained media and is drawn into the material by capillary forces. An example of this preferential wetting behavior in a lower-permeability material is shown in Figure 2-12. Here the finer-grained lens on the left near the bottom of the tank "absorbs" the DNAPL, while the lens on the right side is not penetrated. Such preferential wetting in low-permeability media can be the result of differences in mineralogy or, more commonly, the result of previous contact with groundwater containing co-contaminants that are surface-active, such as organic acids and bases.

*Type III Settings*

This setting involves interbedded transmissive and stagnant zones with high porosity, which is common in alluvium where low-permeability interbeds are often horizontally oriented and laterally extensive. Initial migration of DNAPL through this type of setting is along the pathways in the transmissive layers due to relatively low displacement pressures. Low-permeability layers, with high displacement pressures, act as capillary barriers. Often, DNAPL within the source zone will reside at the contact between transmissive and stagnant zones (e.g., DNAPL pools perched on clay layers). With time the DNAPL will partition into the aqueous phase where it will either be flushed out of the source zone (through transmission in flowing groundwater) or driven into the stagnant zones of low-permeability layers via diffusion. A large fraction of contaminant mass may also be adsorbed to the solids in the low-permeability layers (Parker et al., 1994). Once all of the NAPL is depleted (either by natural or engineered processes), reverse diffusion of contaminants from stagnant zones can sustain concentrations of contaminants in the transmissive layer for extended periods. This can be a primary factor driving contaminant rebound after application of source control measures that largely deplete contaminants in transmissive zones (e.g., Sudicky et al., 1985; Parker et al., 1997; Liu and Ball, 2002; Sale et al., 2004). Schematic contaminant fluxes about a DNAPL in a simplified Type III setting are illustrated in Figure 2-15.

*Type IV Settings*

Type IV settings involve fractured media with low matrix porosity, such that the primary space in which contaminants can be stored are fractures. Critical attributes of fracture networks are that they typically represent a small fraction of

A) With DNAPL Pool

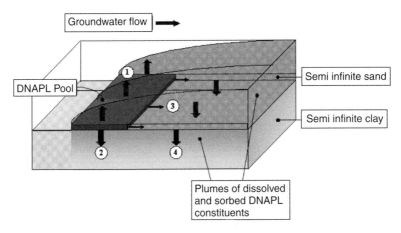

B) Post DNAPL Pool

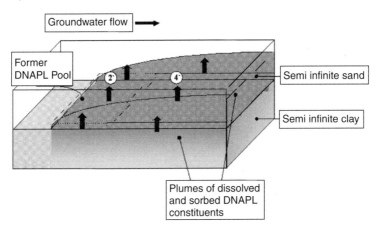

FIGURE 2-15 Contaminant fluxes in a Type III setting. (A) Contaminant fluxes (1–4) and distribution about a DNAPL pool at a contact between a sand and a clay layer, which typifies Type III settings, and (B) reverse diffusion fluxes and contaminant distribution following DNAPL depletion. Fluxes in (A) include (1) diffusion into the overlying transmissive layer, (2) diffusion into the underlying stagnant zones (3) advection through the DNAPL pool, and (4) diffusion into the stagnant zone downgradient of the DNAPL pool. In (B), the 2`and 4` fluxes are back diffusion into the transmissive zone post DNAPL dissolution. SOURCE: Sale et al. (2004).

the total rock matrix volume, and the fractures can be either well or only poorly interconnected (Parker et al., 1996). Because of low overall porosity (including the fracture zones) of Type IV settings, contaminants released into these settings tend to create a relatively large source zone. In addition, the rates of aqueous phase contaminant transport can be large due to small cross-sectional areas of flow and an absence of contaminant attenuation via diffusion into stagnant zones. Because matrix porosity in this hydrogeologic setting is low, little if any contaminant is stored as sorbed or stagnant dissolved phase mass in the source zone. However, in instances where the fracture networks are poorly connected, a subset of the fractures may behave as stagnant zones, and DNAPLs in dead-end fractures may act as persistent sources of dissolved phase contaminants that are difficult to remediate. Lastly, characterization of this type of system can be difficult due to the potentially sparse network of fractures and the limitations of characterization tools in crystalline rock.

*Type V Settings*

Type V settings involve fractured media (rock or low-permeability alluvium) with high matrix porosity. Thus, unlike Type IV settings, stagnant zones in Type V settings have the potential to store contaminants in the dissolved and sorbed phases and can represent a large fraction of the source mass. DNAPL itself tends to flow through the fracture networks (along paths of low displacement pressure) and is typically precluded from the matrix blocks (due to high displacement pressure) (Kueper and McWhorter, 1991). Given a finite release of moderate- to high-solubility DNAPL, complete transfer of DNAPL from the transmissive fractures into stagnant zones through dissolution and sorption is possible (Parker et al., 1994). The challenges of managing contamination in this hydrogeologic setting include describing the extent of the source zone, characterizing the fracture network, delivering remedial solutions to the targeted areas in some cases, and understanding the potential for reverse diffusion to sustain contaminant concentrations in the transmissive fractures after depletion of DNAPL.

\* \* \*

Figure 2-16 is a conceptual diagram of a DNAPL in a system of clay and sand layers that depicts the hypothetical DNAPL distribution given the known processes described above. Depending on the thickness of the beds and the size of the spill, this site can be conceptualized as either a Type II setting or as a combination Type I (sand)–Type V (fractured clay) setting. In either case, the DNAPL migrates vertically through the upper sand aquifer to the upper clay layer. Based on field studies (Poulsen and Kueper, 1992; Kueper et al., 1993) and numerical simulations (Kueper et al., 1991a,b), the DNAPL in the sand layer is most likely

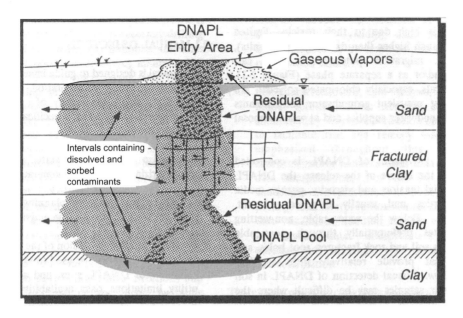

FIGURE 2-16 Hypothetical DNAPL source zone. In addition to residual and pooled DNAPL, the figure depicts a vapor plume in the unsaturated zone and a halo of dissolved and sorbed contamination in the saturated zone about the DNAPL. The plumes of dissolved and sorbed contaminants extend downgradient of the DNAPL in the sand layers, into the stagnant clay beds about the DNAPL, and into the clay beds about the plumes in the sand. Note that the residual DNAPL is more likely to occur in sparse pools and fingers, rather than in the massive bodies inferred in the picture. SOURCE: Adapted from Cohen et al. (1993).

to occur in sparse horizontal lenses and vertical fingers that have drained to near residual saturations. Upon reaching the upper fractured clay layer, the DNAPL continues to migrate downward through secondary permeability features. Large aqueous phase concentration gradients would likely drive contaminants into the clay matrix via diffusion (Parker et al., 1994, 1997). In the second sand layer, the DNAPL forms sparse horizontal lenses and vertical fingers that have drained to near residual saturations. Vertical migration ceases at the lower clay layer, where substantial pooling occurs. The pathway leaves residual DNAPL in the unsaturated soil zone and in sand and clay layers, sorbed and dissolved contaminants on the aquifer sediments that may serve as a long-term source, a pool of DNAPL deep in the subsurface, and two contaminant plumes containing dissolved compounds.

## The Contaminant Plume

Given the definition of "source zone" in Chapter 1, it is important to differentiate between source zones and the contaminant plume. Contaminant plumes develop downgradient of the source material in cases where the DNAPLs are moderately soluble in water and are resistant to natural biodegradation. For example, chlorinated solvents are not easily broken down by oxidative microbial processes (being more likely to be biodegraded, at least partially, through reductive dechlorination or cometabolic oxidative processes), and they sorb weakly on aquifer materials. Thus, chlorinated solvents can form extensive plumes downgradient of the source. Depending on the extent of various biogeochemical processes (discussed previously), the chemical composition and concentration range of the contaminant suite in the plume may be quite different than in the source zone.

In general, groundwater plumes tend to have larger spatial extents and to be more continuous in nature in comparison with contaminant mass distributions within source zones. For a finite, uniform, aqueous phase contaminant release, the textbook plume shape is generally a three-dimensional ellipsoid characteristic of idealized homogeneous sandy aquifers. This plume shape is created under uniform, unidirectional flow conditions, assuming three-dimensional Gaussian mixing or dispersion. Such a regular plume concentration distribution, however, is only possible in the most idealized cases. Closer to the source zone, plumes will tend to take on spatial characteristics that mimic the irregular saturation distributions upgradient. The existence of DNAPL pools and the small vertical mixing under natural flow conditions tend to create highly irregular and stratified plumes (see Box 4-1). Furthermore, in more complex hydrogeologic regimes [e.g., highly heterogeneous granular settings (Type III), fractured bedrock or karst systems (Types IV and V)], plume shape may be dominated by macroscopic velocity variations, non-Darcian (fracture-based) flow, and matrix diffusion, and may bear little resemblance to the textbook Gaussian shapes.

The shape and extent of a groundwater plume may vary with time if the groundwater flow conditions are subject to changes, such as may stem from seasonally fluctuating infiltration. However, under reasonably steady flow conditions, and given sufficient time to satisfy the sorption capacity of the aquifer solids, groundwater plumes will appear to be fairly stationary. Under such conditions, the plume is fed by the source zone, which is losing mass at a sufficiently slow rate to appear constant over the time period of observation (perhaps several years). On the plume periphery, mass is lost or, more accurately, attenuated to below detection limits by groundwater dilution processes (advection and dispersion) or by transformation processes such as biodegradation.

It should be noted that sorption or diffusion of contaminants *from the plume* onto aquifer solids (and subsequent reverse diffusion) is common in many hydrogeologic settings. This phenomenon, which is shown schematically in Figure 2-17, may support the long-term contamination of groundwater. Indeed, it is thought

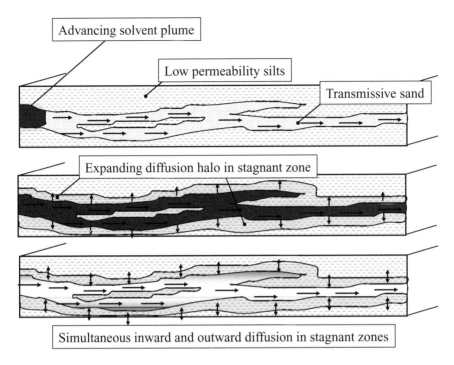

FIGURE 2-17 Conceptualization of reverse diffusion of contaminants from stagnant to transmissive zones after contact with a solvent-containing plume. SOURCE: Modified from Sale et al. (2004).

that the chronic low-level concentrations of chlorinated solvents produced by pump-and-treat systems can have more to do with reverse diffusion of this dispersed non-DNAPL mass than with discrete DNAPL or mass in the stagnant areas of the source zone (Liu and Ball, 2002). Although this sorbed or stagnant-zone mass can be a chronic supply of aqueous phase DNAPL constituents, it does not constitute a source zone as defined in Chapter 1, because it does not exist where the NAPL once was. Therefore, not all groundwater contaminant plumes imply the presence of a source zone.

## CHEMICAL EXPLOSIVES

The chemical explosives that have caused the greatest environmental impact, and which are of greatest concern to the Army, are the organic explosives 2,4,6-trinitrotoluene (TNT), 2,4-dinitrotoluene (DNT), hexahydro-1,3,5-trinitro-1,3,5-triazine (RDX), and octahydro-1,3,5,7-tetranitro-1,3,5,7-tetrazocine (HMX),

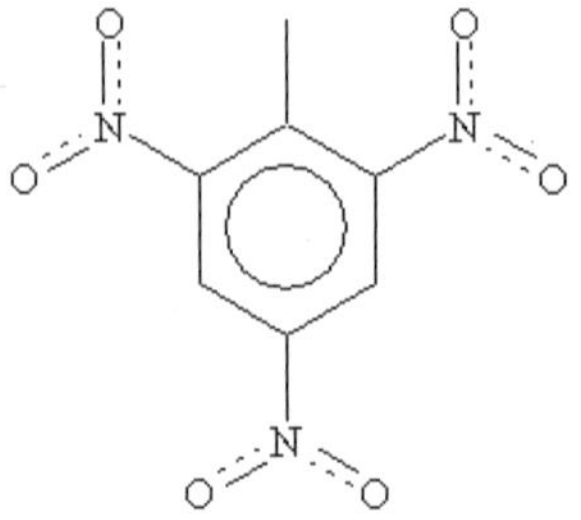

TNT
2,4,6-Trinitrotoluene

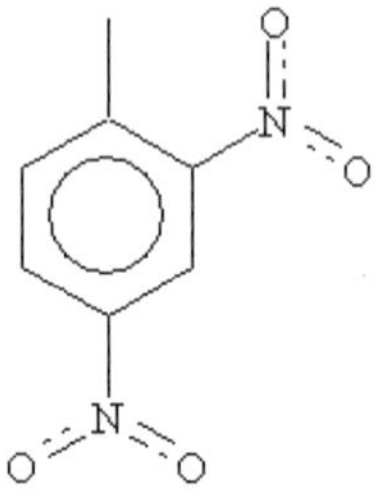

DNT
2,4-Dinitrotoluene

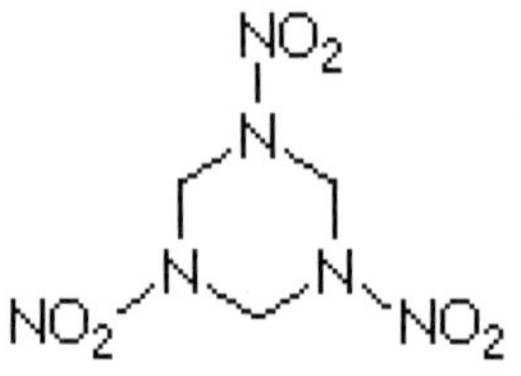

RDX
hexahydro-1,3,5-trinitro-1,3,5-triazine

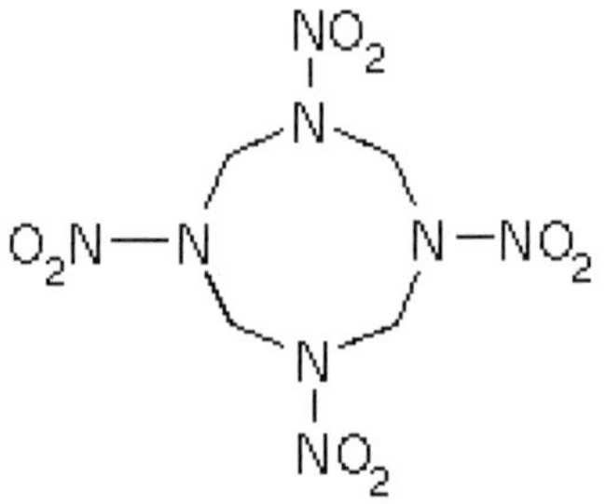

HMX
octahydro-1,3,5,7-tetranitro 1,3,5,7-tetrazocine

FIGURE 2-18  Structural diagrams of TNT, DNT, RDX, and HMX.

whose structures are shown in Figure 2-18.[1] As with DNAPLs, the extent of subsurface contamination by chemical explosives is governed by the nature of the contaminant release and the physics of soil–chemical interactions in the subsurface. However, much less is known about how chemical explosive source material interacts with soil compared to chlorinated solvents. Hence, this section primarily describes the nature of the contaminant releases, highlighting several unique features that make sites contaminated with these compounds particularly challenging to remediate.

Table 2-2 shows the key properties of the predominant chemical explosives found as environmental pollutants. These materials are characterized by low

_____

[1]This report does not consider unexploded ordinance (UXO).

TABLE 2-2 Properties of Common Chemical Explosives

| Explosive | CAS # | Aqueous Solubility @ 20°C (mg/L) | Melting Point | Log $K_{ow}$ |
|---|---|---|---|---|
| TNT | 118-96-7 | 108 | 80°C | 1.86 |
| 2,4-DNT | 121-14-2 | 170 | 72°C | 1.98 |
| RDX | 121-82-4 | 45 | 204°C | 0.86 |
| HMX | 2691-41-0 | 3 | 286°C | 0.06 |

SOURCE: Rosenblatt et al. (1991).

aqueous solubilities, and they exist as solids at environmental temperatures. While the extent of sorption of explosives onto aquifer solids is dependent on many factors, the $K_{ow}$ values for each compound and the fraction and type of organic carbon present are particularly important and are used below to illustrate how the extent of sorption varies for the explosive compounds highlighted in this report. Given the $K_{ow}$ values in Table 2-2, sorption controls the fate and transport of TNT and DNT to a much greater degree than it does RDX and HMX, as observed at Department of Defense (DoD) sites that contain multiple contaminants (e.g., Louisiana AAP and Cornhusker AAP) in which groundwater plumes are most extensive for HMX, followed by RDX and then DNT and TNT. HMX is included in this discussion for completeness, although it rarely drives cleanup efforts due to its low toxicity, its less strict drinking water advisory levels, and its typically low concentrations found in the groundwater.

### Explosives Releases

Environmental contamination by organic explosives at Army sites can be categorized by three generalized release mechanisms: (1) production process discharges, (2) manufacturing process discharges, and (3) military training and testing operations. More explosives problems are derived from historic production process discharges and from manufacturing processes (milling/machining or in demilitarization of ordnance) than from end-use detonations. This differentiates explosives sites from chlorinated solvent sites, most of which have arisen from spills or discharges of the final product during end-use applications.

*Production Process Discharges*

Explosives production operations have been performed at numerous facilities to support wartime needs and to create national security stockpiles. The explosives produced in the highest volume include TNT, RDX, and DNT. Within the Army, the principal production facilities for TNT were Volunteer AAP, Joliet

**BOX 2-2**
**DNT Production at Badger Army Ammunition Plant**

The Badger Army Ammunition Plant, built in 1942, was the principal DNT production facility in the United States until it changed to standby status in 1977. At this facility, waste pits were used to burn organic solvents, propellant wastes, and lumber. Three waste pits received up to 500 gallons (1893 L) per day of DNTs, solvents, and other constituents. Investigation of the waste pits showed concentrations of DNTs up to 28 percent. Interim remedial actions led to the excavation and incineration of the top 13–20 ft (4–6 m) of material from each waste pit. In addition, six soil vapor extraction wells installed at each waste pit removed 1,600 pounds (726 kg) of solvents. However, subsurface soils 15–25 ft (4.6–7.6 m) below the bottom of the waste pits still contained DNT well over 1 percent. Source treatment continues today at Badger with some success using in situ bioremediation along with in situ wetting to induce solid phase DNT mass transfer to soil pore water; however, delivery of nutrients and management of pH and nitrite are necessary to optimize field-scale biodegradation (Fortner et al., 2003).

AAP, Radford AAP, Louisiana AAP, Longhorn AAP, Cornhusker AAP, and Iowa AAP. DNT was produced primarily at Badger AAP (see Box 2-2), while RDX was made at Holston AAP.

**Trinitrotoluene (TNT).** TNT is the most prevalent explosive used in military ordnance. TNT production in the United State occurs solely at military arsenals and peaked at 65 tons per day during World War II (Kaye, 1980). In a refined form, TNT is one of the most stable explosives and can be stored for long periods of time.

Commercial TNT production begins with a batch process through the sequential nitration of toluene. The first unit process produces mononitrotoluene (MNT or mono-oil) by the addition of nitric and sulfuric acids to toluene under heated conditions. The mono-oil is converted to dinitrotoluene (DNT or bi-oil) using nitric acid-fortified waste acid. In the final conversion to TNT, bi-oil was heated with oleum[2]-fortified sulfuric and nitric acids. In each step, the batch process produces a mixture of compounds. Thus, mono-oil and bi-oil are terms that include MNT and DNT, respectively, but also other manufacturing byproducts. Pre-product TNT is melted and washed with soda ash solution and then washed with sodium sulfite to separate 2,4,6-TNT from the other less desir-

[2]Oleum is a heavy oily liquid mixture of sulfur trioxide in sulfuric acid.

able isomers. Waste acids from each process are routed for use in the previous batch step. Wash water from the final purification steps is routed to a "redwater" (meaning explosives-contaminated water) treatment area by a network of flumes (Urbanski, 1967a). In 1968, continuous manufacturing of TNT began at Radford AAP, where the nitric acid and oleum were introduced countercurrent in a six-stage process (Kaye, 1980). Both batch and continuous TNT production use water-filled drown tanks at each production house to stop the process if an out-of-control reaction begins. Leaks in material transfers between production houses and from storage in holding tanks, other spills, and discharges of material to the drown tanks could be significant sources of explosives contamination in the subsurface.

Historic literature uses the term "nitrobody" to represent the wide assortment of nitroaromatic molecules present in the production process prior to completion of the final product (TNT). The composition of the nitrobody in each stage of TNT production varies widely, as shown in Table 2-3 for the Radford AAP continuous production line.

The nitrobody production materials discharged to the drown tanks contained mono-oil, bi-oil, and TNT in mixtures of nitric and sulfuric acids and residual toluene, often at elevated temperatures. Evaluations have shown that the drown-tank material from stage 1 of the Joliet AAP batch process and from stages 1 and 2 of the Radford AAP continuous process were liquids containing 75 percent to 85 percent MNT (mostly 2-MNT and 4-MNT) (Persurance, 1974) that could possibly behave like DNAPLs. However, the specific gravity of MNTs ranges from 1.155 to 1.160 g/cm$^3$ depending on the isomer, which is much less dense than chlorinated solvents (see Table 2-1). The nitrobody material from later stages in both processes was found to be a solid at ambient temperature. Unfortunately, there is no information on the frequency with which drown tanks were used among the batch or continuous operations nor on the intervals at which

TABLE 2-3 Composition of Nitrobody during Continuous TNT Production (percent).

| Process | Composition of Nitrobody, % | | | Temp (°C) |
|---|---|---|---|---|
| | MNT | DNT | TNT | |
| 1 | 77 | 18 | 4 | 50-55 |
| 2 | 0 | 71 | 29 | 70 |
| 3 | 0 | 30 | 69 | 80-85 |
| 4 | 0 | 10 | 90 | 90 |
| 5 | 0 | 2 | 98 | 95 |
| 6 | 0 | 0 | 100 | 100 |

SOURCE: Kaye (1980).

drown-tank materials were removed and destroyed. The physical–chemical properties of the TNT production-process discharge materials are further complicated by the high sulfuric acid content, which alters the density of the material and the solubility of the explosives compounds. The density of 100 percent sulfuric acid is 1.84 $g/cm^3$ and 78 percent sulfuric acid has a density of 1.71 $g/cm^3$. MNT isomers have a solubility of 34 percent in 90 percent sulfuric acid at 50°C; DNT isomer solubility is 20 percent in 90 percent sulfuric acid at 70°C; and TNT isomer solubility is 10 percent in 90 percent sulfuric acid at 80°C (Urbanski, 1967a). All these complexities make it very difficult to understand the miscibility of nitrobody production materials and to determine whether they consist of emulsions of separate nonaqueous phase liquids.

Thus, the physical–chemical properties of explosives material that might be released at TNT production facilities vary depending on the process stage and on the mixed acid content of the material, the completeness of the reactions in that stage, and the extent of dilution in drown tanks or waste lagoons. In some situations, a separate phase NAPL containing mostly MNTs may be present, and in others, a dense miscible phase liquid (DMPL) containing very high concentrations of MNT, DNT, and TNT may be present. At most explosives sites, there is limited information on these factors, making it difficult to assess the distribution of explosives in various hydrogeologic settings.

**Dinitrotoluene (DNT).** In general, the production of DNT mimics that of TNT, but the process stops after the second nitration. Thus, the factors that control the physical–chemical properties of any release material from DNT production are similar to those for TNT production. Nitration of MNT isomers produces various DNT isomers. For example, nitration of o-nitrotoluene produces 2,4-DNT and 2,6-DNT, while nitration of p-nitrotoluene produces only 2,4-DNT. Nitration of m-nitrotoluene produces 3,4-DNT, 2,3-DNT, and 3,6-DNT, all of which are undesirable in the production of 2,4,6-TNT. DNT production in the Army and the resulting subsurface contamination are discussed in Box 2-2.

**RDX (Royal Demolition eXplosive/Research Demolition eXplosive).** Cyclotrimethylenetrinitramine, hexahydro-1,3,5-trinitro-sym-triazine, and Cyclonite are all synonyms for RDX, which is a white crystalline solid with a nitrogen content of 37.84 percent. RDX is usually used in mixtures with other explosives, oils, or waxes. It has a high degree of stability in storage and is considered one of the most brisant[3] of the military high explosives. Pure RDX is used in press-loaded projectiles. Cast loading is accomplished by blending RDX with a relatively low melting point substance. RDX is also used as a base charge in detonators and in blasting caps.

_____

[3]Brisant is defined as of or relating to the power (the shattering effect) of an explosive.

Just prior to and during World War II, numerous methods to synthesize RDX emerged. The first method was through the direct nitration of hexamine with nitric acid. The yield on this process was low, and in 1941 Americans and Germans simultaneously developed a method where hexamine dinitrate is reacted with ammonium dinitrate in the presence of acetic anhydride (Urbanski, 1967b). This process is the principal one used in the United States today, and it contains a constant impurity of 8 percent to 12 percent HMX (see below).

RDX was principally produced at the Holston Ordnance Works in Kingsport, Tennessee. Initial characterization efforts at this location have shown RDX in the groundwater, but only two sites located below production buildings have concentrations that imply a subsurface source zone (~ 2 percent to 4 percent of RDX's aqueous solubility) (USACHPPM, 2003). RDX groundwater contamination is much more prevalent at other facilities where manufacturing process discharges occurred (as described below).

**HMX (High Melt eXplosive or Her Majesty's eXplosive).** Cyclotetramethylenetrinitramine, octahydro-1,3,5,7-tetranitro-1,3,5,7-tetrazocine, and Octagen are all synonyms for HMX. HMX is used in military ordnance where the greatest explosive power per mass is needed. HMX is formed by the nitration of hexamine (or hexamethylenetetramine) in the presence of glacial acetic acid, acetic anhydride, ammonium nitrate, and nitric acid. The reaction produces a mixture of RDX and HMX, and the RDX is selectively destroyed by base hydrolysis. HMX is also a byproduct of production of RDX and has been produced at the Holston Ordnance Works. The groundwater investigations at Holston have found HMX in the same wells that RDX was found in, but at much lower concentrations and none over drinking water health advisory limits (USACHPPM, 2003).

*Manufacturing Process Discharges*

Manufacturing processes are defined here as post-production operations that operate with the solid phase explosive material. Examples include load, assemble, and pack facilities and demilitarization operations that remove the explosive fill from expired munitions. The most prevalent explosive fill material contains TNT and Comp B (60% RDX/40% TNT). Both operations typically use hot water or steam as a washdown or washout material. Spent cleaning solutions are typically filtered with coarse fabric to remove the suspended particulates and then are discharged via pipelines, open flumes, or ditches to infiltration or evaporation ponds. Continuous releases of aqueous solutions containing explosives as solutes can infiltrate into soils, causing large areas of contamination.

The Umatilla Army Depot provides an example, where from about 1955 to 1965 a munitions washout facility was operated where hot water and steam were used to remove explosives from munitions bodies. An estimated 85 million gallons (322 million L) of wash water containing TNT and RDX was discharged to two surface impoundments that covered about half an acre (0.2 hectares). Source control measures began by excavating the top 20 feet (6 m) of soil for treatment by composting. The remaining 30 feet (9 m) of soil above the water table is being treated with in situ flushing using groundwater pump and granular activated carbon treatment.

*Military Training and Testing Operations*

Military training and testing operations relevant to the release of explosive materials include live fire exercises that use military ordnance. Detonation of military ordnance typically consumes the majority of the explosive fill. Experimental characterization of detonations of 81- and 120-mm mortars and 105-mm artillery showed only trace residues of TNT, RDX, and HMX present on the ground surface (Jenkins et al., 2001). However, a low-order detonation can distribute solid phase energetic material onto and into near-surface soil (DeLaney et al., 2003).

The low-order detonation process is very ill-defined, but it is generally characterized by initiation of detonation, which recedes before consumption of all of the explosive fill. Rupture of the case distributes large chunks to small particulates, and it can occur above the soil surface or in ground after impact. In addition, Explosive Ordnance Disposal methods (e.g., open detonation) to safely dispose of unexploded ordnance can also cause incomplete detonation of target objects (Lewis et al., 2003). These detonation methods have not been fully characterized for energetic material releases.

## Chemical Explosives Fate in the Subsurface

The above sections have noted the various materials that might be released from military operations involving chemical explosives. For production and manufacturing process discharges, aqueous solutions are the most common type of waste. Under circumstances where the environmental temperatures are significantly lower than the discharge water temperatures, explosives may precipitate out of solution and create a separate solid phase material in the soil. For example, the aqueous solubility of TNT drops from 250 mg/L at 40°C to 110 mg/L at 20°C. Similarly, the solubility of RDX is 115 mg/L at 40°C and 45 mg/L at 20°C. Precipitation of explosive solids following production and manufacturing process discharges most likely occurs in the near soil surface (< 20 feet or 6 m), such that surface excavation and treatment (e.g., via incineration or composting) are effective remediation strategies. It is possible that leaks or spills from pipelines and flumes, and infiltration from unlined ditches, could lead to contaminant accumulation in unsaturated soil pores and fracture matrices and thus provide a long-term reservoir of contamination to groundwater.

As mentioned previously, some highly concentrated wastes in production process discharges might act like DNAPLs or DMPLs. However, even heated and concentrated production materials that might initially behave like a DNAPL are likely to undergo significant change once introduced into soil, where environmental conditions would tend to decrease both temperature and acidity, promoting the creation of a separate solid phase material. Of course, recharge of the subsurface during rain events can lead to dissolution of solid phase explosives and subsequent transfer of explosives mass to soil pore water and perhaps groundwater. Subsequent dilution of the more soluble, and/or degradation of the more labile, compounds could over long periods of time (decades) result in a mixture with a very different signature than the original release materials.

Rainfall can dissolve the solid phase energetic material from the detonation of military ordnance and transfer the mass to soil pore water. The potential contaminant threat from detonation of military ordnance is determined by the depth to groundwater and the various contributions of recharge, dissolution, sorption, and degradation rates. The nature and the impact of this type of source are only beginning to be understood, and current programs are in progress to improve scientific understanding and develop mitigation approaches (SERDP, 2003). Management of these source areas is compounded by the presence of unexploded ordnance and continued operations.

Once in the subsurface, biotic and abiotic redox reactions are the principal processes that degrade the chemical explosives discussed above, although there are significant differences within this group. TNT is rapidly transformed to mono- and then di-aminonitrotoluenes by naturally occurring microorganisms and soil minerals under both aerobic and anaerobic conditions (Ahmad and Hughes, 2000). TNT biodegradation appears to be mostly cometabolic. Reduction rates decrease with the successive reduction of each nitro group due to the destabilization of the aromatic ring and a decrease in the electrophilic nature of the remaining nitro groups. A portion of the aminonitrotoluenes can continue to participate in reactions with soil organic matter, becoming covalently bound in a multistep humification process (Thorne and Leggett, 1999). 2,4-DNT and the 2,6-DNT manufacturing impurity can be biodegraded under aerobic conditions where specific bacteria use these materials for carbon, nitrogen, and energy sources (Fortner et al., 2003). Environmental reactions with RDX are strongest under reducing conditions that sequentially reduce the nitro groups to mono-, di-, and trinitroso products of RDX, followed by ring cleavage to produce a variety of short-chain compounds (Hawari, 2000). Very little work has been done to understand the fate of HMX in the subsurface, although a similar sequential reduction of the nitro groups followed by ring cleavage, as with RDX, is hypothesized.

Natural attenuation mechanisms favor the loss of TNT>DNT>RDX>HMX in the environment. TNT sorption and aerobic degradation provides for continuous elimination reactions. However, DNT appears to require nutrients and nitrite byproduct elimination to support significant biodegradation. RDX and HMX both sorb poorly to soils and require strongly reducing conditions for natural elimination reactions. For this reason, and considering the toxicity of RDX, RDX has become one of the more challenging organic explosives for environmental remediation.

### Field-Scale Distribution of Explosives in Source Zones

At this time, characterization of chemical explosive source zones is immature compared to the knowledge base for DNAPLs, primarily because when the major production operations ceased 20–30 years ago, the knowledge of past waste management practices and the physicochemical properties of the explosive production mixtures faded. Distinct source zones of explosives generally contain solid phase material, although for TNT production, the presence of mono-oil and bi-oil material as a significant source material has been speculated. Explosive debris distributed on or in surface soils by detonations is emerging as a potential source material at military training and testing ranges.

How chemical explosives might be distributed in the five hydrogeologic settings described earlier is difficult to determine at this time. The presence of reprecipitated solid phase explosive compounds in granular media is suspected, based on phase partitioning laws and maximum aqueous phase limits. But the dynamics of explosive material reprecipitation, dissolution, and transport that would define source zone architecture are not well understood. In addition, no work has been performed to understand the miscible/immiscible flow characteristics of production process wastes in drown tanks or disposed of in surface impoundments.

### SUMMARY

This chapter has outlined important physical and chemical features of contaminated sites that should be understood (or at a minimum discussed) prior to any site remediation. First and foremost, it is imperative to be able to categorize the hydrogeologic setting of a site, as this plays a significant role in determining the overall subsurface distribution of contamination. Furthermore, the existing hydrogeologic setting limits both the types of tools that can be used to characterize the source zone and the technologies that might achieve reductions in source mass. Given the combination of heterogeneity in hydrogeology and in physical-chemical properties, complex sites are the norm rather than the exception.

Chlorinated solvents that exist as DNAPLs in the subsurface are the primary concern of the Army and many other potentially responsible parties, and thus constitute the major focus of this report. Among the many distinguishing features of DNAPL sites is the fact that the distribution of DNAPL in the subsurface is typically sparse and highly heterogeneous (depending on the site hydrogeology). Furthermore, depending on a site's porosity, permeability, and sorption capacity, a substantial portion of the contaminant mass that might have been released to the subsurface as a DNAPL may transition into stagnant zones as either sorbed or dissolved phase contamination. These sources have the potential to be a chronic supply of contamination to groundwater plumes.

In comparison to DNAPLs, the state of the art for explosives source zone characterization is quite immature. Most explosives are released to the environment as aqueous mixtures, from which chemical explosives precipitate out. The source zone architecture created by production process discharges, manufacturing process discharges, and military training and testing operations requires scientific investigation before remediation technologies can be considered, designed, and deployed with confidence. In addition, an important constraint not found with DNAPLs is explosives safety. Management of detonation hazards, especially the drilling and handling of source material, will require additional resources and technologies.

Even though five hydrogeologic settings are discussed in the chapter, there are many more than five typical contaminant distributions. A site's contaminant distribution will be influenced by transformation and transport processes, by the nature of the contaminant release, and by the hydrogeologic setting (or combination of settings). Thus, this chapter should not be viewed as a cookbook for how to categorize sites and determine their contaminant distribution. Rather, source characterization is necessary, as discussed in the next chapter.

## REFERENCES

Adamson, A. W., and A. P. Gast. 1997. Physical Chemistry of Surfaces, 6th ed. New York: Wiley.

Ahmad, F, and J. B. Hughes. 2000. Anaerobic Transformation of TNT by *Clostridium*. *In*: Biodegradation of Nitroaromatic Compounds and Explosives. J. C. Spain, J. B. Hughes, and H. Knackmuss (eds.). Boca Raton, FL: Lewis Publishers.

Anderson, W. G. 1987. Wettability literature survey—part 4: effects of wettability on capillary pressure. J. Pet. Technol. 39:1283–1300.

Back, W., J. S. Rosenshein, and P. R. Seaber. 1988. Hydrogeology—The Geology of North America Volume O-2. Boulder, CO: Geological Society of America.

Ball, W. P., C. Liu, G. Xia, and D. F. Young. 1997. A diffusion-based interpretation of tetrachloroethene and trichloroethene concentration profiles in a groundwater aquitard. Water Resources Research 33(12):2741–2758.

Bradford, S. A., R. Vendlinski, and L. M. Abriola. 1999. The entrapment and long-term dissolution of tetrachloroethylene in fractional wettability porous media. Water Resources Research 35(10):2955–2964.

Brewster, M. L., A. P. Annan, J. P. Grenhouse, B. H. Kueper, G. R. Olhoeft, J. D. Redman, and K. A. Sander. 1995. Observed migration of a controlled DNAPL release by geophysical methods. Groundwater 33(6):977–987.

Cohen, R. M., J. W. Mercer, and J. Matthews. 1993. DNAPL Site Evaluation. C. K. Smoley (ed.). Boca Raton, FL: CRC Press.

Conrad, S. H., E. F. Hagan, and J. L. Wilson. 1987. Why are residual saturation of organic liquids different above and below the water table? *In*: Proceedings—Petroleum Hydrocarbons and Organic Chemicals in Groundwater: Prevention, Detection and Restoration. Worthington, OH: National Water Well Association.

Cope, N., and J. B. Hughes. 2001. Biologically-enhanced removal of PCE from NAPL source zones. Environ. Sci. Tech. 35(10):2014–2021.

De la Paz, T., and T. Zondlo. 2003. DNAPLs in Karst the Redstone Experience. Presentation to the Committee on Source Removal of Contaminants in the Subsurface. January 30, 2003, San Antonio, Texas.

DeLaney, J. E., M. Hollander, H. Q. Dinh, W. Davis, J. C. Pennington, S. Taylor, and C. A. Hayes. 2003. Characterization of explosives residues from controlled detonations: low-order detonations. Chapter 5, Distribution and fate of energetics on DoD test and training ranges: Report 3, ERDC TR-03-2. Vicksburg, MS: U. S. Army Engineer Research and Development Center, Environmental Laboratory.

Domenico, P. A., and F. W. Schwartz. 1998. Physical and Chemical Hydrogeology, 2nd ed. New York: John Wiley & Sons.

Environmental Protection Agency (EPA). 1992. Evaluation of Ground-Water Extraction Remedies: Phase II, Volume I—Summary Report. Publication 9355.4-05. Washington, DC: EPA Office of Emergency and Remedial Response.

Fortner, J. D, C. Zhang, J. C. Spain, and J. B. Hughes. 2003. Soil Column Evaluation of Factors Controlling Biodegradation of DNT in the Vadose Zone. Environ. Sci. Technol. 37(15):3382–3391.

Freeze, R. A., and J. A. Cherry. 1979. Groundwater. New Jersey: Prentice-Hall.

Hawari, J. 2000. Biodegradation of RDX and HMX: from basic research to field application. In: Biodegradation of Nitroaromatic Compounds and Explosives. J. C. Spain, J. B. Hughes, and H. Knackmuss (eds.). Boca Raton, FL: Lewis Publishers.

Hiemenz, P. C., and R. Rajagopalan. 1997. Principles of Colloid and Surface Chemistry, 3rd ed. New York: Marcel Dekker.

Huang, W. L., and W. J. Weber. 1998. A distributed reactivity model for sorption by soils and sediments. 11. Slow concentration dependent sorption rates. Environ. Sci Technol. 32(22):3549–3555.

Jenkins, T. F., J. C. Pennington, T. A. Ranney, T. E. Berry, P. H. Miyares, M. E. Walsh, A. D. Hewitt, N. M. Perron, L. V. Parker, C. A. Hayes, and E. G. Wahlgren. 2001. Characterization of Explosives Contamination at Military Firing Ranges. ERDC TR-01-5. Vicksburg, MS: U. S. Army Engineer Research and Development Center.

Kaye, S. M. 1980. Encyclopedia of Explosives and Related Items. PATR 2700, Volume 9. Dover, New Jersey: U.S. Army Armament Research and Development Command, Large Caliber, Weapon Systems Laboratory.

Kueper, B. H., and D. B. McWhorter. 1991. The behavior of dense nonaqueous phase liquids in fractured clay and rock. Journal of Ground Water 29(5):716–728.

Kueper, B. H., D. Redman, R. C. Starr, S. Reitsma, and M. Mah. 1993. A field experiment to study the behavior of tetrachloroethylene below the water table: spatial distribution of pooled DNAPL. Groundwater 31:756–766.

Kueper, B. H., and E. O. Frind. 1991a. Two phase flow in heterogeneous porous media. 1. Model development. Water Resources Research 27(6):1049–1057.

Kueper, B. H., and E. O. Frind. 1991b. Two phase flow in heterogeneous porous media. 2. Model application. Water Resources Research 27(6):1058–1070.

Lemke, L. D., L. M. Abriola, and J. R. Lang. 2004. DNAPL source zone remediation: Influence of hydraulic property correlation on predicted source zone architecture, DNAPL recovery, and contaminant mass flux. Water Resources Research. In press.

Lewis, J., S. Thiboutot, G. Ampleman, S. Brochu, P. Brousseau, J. C. Pennington, and T. A. Ranney. 2003. Open detonation of military munitions on snow: An investigation of energetic material residues produced. Chapter 4, Distribution and fate of energetics on DoD test and training ranges: Report 3, ERDC TR-03-2. Vicksburg, MS: U.S. Army Engineer Research and Development Center, Environmental Laboratory.

Liu, C., and W. P. Ball. 1998. Analytical modeling of diffusion-limited contamination and decontamination in a two-layer porous medium. Advances in Water Resources 24(4):297–313.

Liu, C., and W. P. Ball. 2002. Back diffusion of chlorinated solvents from a natural aquitard to a remediated aquifer under well-controlled field conditions: predictions and measurements. Journal of Groundwater 40(2):175–184.

Mackay, D. M., W. Y. Shiu, and K. C. Ma. 1993. Illustrated Handbook of Physical-Chemical Properties and Environmental Fate for Organic Chemicals, Vol. III. Chelsea, MI: Lewis Publishers.

Mackay, D. M., R. D. Wilson, M. P. Brown, W. P. Ball, G. Xia, and D. P. Durfee. 2000. A controlled field evaluation of continuous versus pulsed pump-and-treat remediation of a VOC-contaminated aquifer: site characterization, experimental setup, and overview of results. Journal of Contaminant Hydrology 41:81–131.

Meinardus, H. W., V. Dwarakanath, J. Ewing, G. J. Hirasaki, R. E. Jackson, M. Jin, J. S. Ginn, J. T. Londergan, C. A. Miller, and G. A. Pope. 2002. Performance assessment of NAPL remediation in heterogeneous alluvium. Journal of Contaminant Hydrology 54:173–193.

Mercer, J. W., and R. M. Cohen. 1990. A review of immiscible fluids in the subsurface: properties, models, characterization and remediation. Journal of Contaminant Hydrology 6:107–163.

Montgomery, J. H. 2000. Groundwater Chemicals, 3rd ed. Boca Raton, FL: Lewis Publishers and CRC Press.

Morrow, N. R. 1976. Capillary-pressure correlations for uniformly wetted porous media. Journal of Canadian Petroleum Technology 15(4):49–69.

O'Carroll, D. M., S. A. Bradford, and L. M. Abriola. 2004. Infiltration of PCE in a system containing spatial wettability variations. Journal of Contaminant Hydrology 73:39-69.

Pankow, J. F., and J. A. Cherry (eds.). 1996. Dense Chlorinated Solvents and other DNAPLs in Groundwater. Portland, OR: Waterloo Press.

Parker, B. L., D. B, McWhorter, and J. A. Cherry. 1997. Diffusive loss of non-aqueous phase organic solvents from idealized fracture networks in geologic media. Ground Water 35(6):1077–1088.

Parker, B. L., J. A. Cherry, and R. W. Gillham. 1996. The effect of molecular diffusion on dnapl behavior in fractured porous media. Chapter 12 In: Dense Chlorinated Solvents and Other DNAPLs in Groundwater. J. F. Pankow and J. A. Cherry (eds.). Portland, OR: Waterloo Press.

Parker, B. L., R. W. Gillham, and J. A. Cherry. 1994. Diffusive disappearance of immiscible–phase organic liquids in fractured geologic media. Journal of Groundwater 32(5):805–820.

Persurance, R. 1974. Explosion Hazard Classification of Drowning Tank Material from TNT Manufacturing Process. Picatinny Arsenal Technical Report 4613. Dover, NJ.

Poulsen, M. M., and B. H. Kueper. 1992. A field experiment to study the behavior of tetrachloroethylene in unsaturated porous media. Environ. Sci. Technol. 26(5):889–895.

Powers, S. E., L. M. Abriola, and W. J. Weber, Jr. 1994. An experimental investigation of NAPL dissolution in saturated subsurface systems: transient mass transfer rates. Water Resources Research 30(2):321–332.

Powers, S. E., W. H. Anckner, and T. F. Seacord. 1996. Wettability of NAPL-contaminated sands. Journal of Environmental Engineering 122:889–896.

Powers, S. E., L. M. Abriola, and W. J. Weber, Jr. 1992. An experimental investigation of NAPL dissolution in saturated subsurface systems: steady-state mass transfer rates. Water Resources Research 28(10):2691–2705.

Press, F., and R. Siever. 1974. Earth. San Francisco, CA: W. H. Freeman and Company.

Rathfelder, K. M., L. M. Abriola, M. A. Singletary, and K. D. Pennell. 2003. Influence of surfactant-facilitated interfacial tension reduction on organic liquid migration in porous media: observations and numerical simulation. J. Contaminant Hydrology 64(3-4):227–252.

Robertson, B. K., and M. Alexander. 1996. Mitigating toxicity to permit bioremediation of constituents of nonaqueous-phase liquids. Environ. Sci. Tech. 30:2066–2070.

Rosenblatt, D. H., E. P. Burrows, W. K. Mitchell, and D. L. Parmer. 1991. Organic Explosives and Related Compounds. *In* The Handbook of Environmental Chemistry, Vol 3. G. O. Hutzinger (ed.). Berlin and Heidelberg: Springer-Verlag.

Salathiel, R. A.. 1973. Oil recovery by surface film drainage in mixed-wettability rocks. J. Pet. Technol. 25(OCT):1216–1224.

Sale, T., T. Illangasekare, F. Marinelli, B. Wilkins, D. Rodriguez and B. Twitchell. 2004. AFCEE Source Zone Initiative—Year One Progress Report. Colorado State University and Colorado School of Mines, Prepared for the Air Force Center for Environmental Excellence.

Schwille, F. 1988. Dense Chlorinated Solvents in Porous and Fractured Media. Translated by J. F. Pankow. Chelsea, MI: Lewis Publishers.

SERDP. 2003. Annual Report to Congress—Fiscal Year 2002, from the Strategic Environmental Research and Development Program. Arlington, VA: SERDP Program Office.

Sudicky, E. A. 1986. A natural gradient experiment on solute transport in a sand aquifer: spatial variability of hydraulic conductivity and its role in the dispersion process. Water Resources Research 22(13):2069–2082.

Sudicky, E. A., R. W. Gillham, and E. O. Frind. 1985. Experimental investigations of solute transport in stratified porous media: (1) the non reactive case. Water Resource Research 21(7):1035–1041.

Sung, Y., K. M. Ritalahti, R. A. Sanford, J. W. Urbance, S. J. Flynn, J. M. Tiedje, and F. E. Loffler. 2003. Characterization of two tetrachloroethene (PCE)-reducing, acetate-oxidizing anaerobic bacteria, and their description as *Desulfuromonas michiganensis* sp. nov. Applied and Environmental Microbiology 69:2694–2974.

Teutsch, G., and M. Sauter. 1991. Groundwater modeling in karst terranes: scale effects, data acquisition and field validation. Pp. 17–35 *In:* Proceedings of the Third Conference on Hydrogeology, Ecology, Monitoring, and Management of Ground Water in Karst Terranes, Nashville, TN.

Thorne, P. G., and D. C. Leggett. 1999. Investigations of explosives and their conjugated transformation products in biotreatment matrices. Special Report 99-3. U.S. Army Corps of Engineers, Cold Regions Research and Engineering Laboratory. Hanover, NH: Army Corps of Engineers.

Urbanski, T. 1967a. Chemistry and Technology of Explosives, Vol. 1. Oxford: Pergamon Press.

Urbanski, T. 1967b. Chemistry and Technology of Explosives, Vol. 3. Oxford: Pergamon Press.

USACHPPM. 2003. Interim Measures Report, Site-Wide Groundwater, Area B (Explosives Production Area), 28 May through 13 June 2003, Holston Army Ammunition Plant, Kingsport, Tennessee. Aberdeen Proving Ground, MD: U.S. Army Center for Health Promotion and Preventative Medicine, Ground Water and Solid Waste Program.

Weber, W. J., W. L. Huang, and E. J. LeBoeuf. 1999. Geosorbent organic matter and its relationship to the binding and sequestration of organic contaminants. Colloids and Surfaces A-Physicochemical and Engineering Aspects 151 (1–2):167–17.

White, W. B. 2002. Karst hydrology: recent developments and open questions. Engineering Geology 65(2–3):85–105

White, W. B. 1998. Groundwater flow in karstic aquifers. Pp. 18–36 *In:* The Handbook of Groundwater Engineering. J. W. Delleur (ed.). Boca Raton, FL: CRC Press.

Yang, Y., and P. L. McCarty. 2000. Biologically enhanced dissolution of tetrachloroethene DNAPL. Environ. Sci. Tech. 34(14):2979–2984.

Yang, Y., and P. L. McCarty. 2002. Comparison between donor substrates for biologically enhanced tetrachloroethene DNAPL dissolution. Environ. Sci. Tech. 36:3400–3404.

# 3

# Source Zone Characterization

One of the goals of this study was to explain the importance of characterization to the effectiveness of source remediation, including a discussion of tools or methods used to delineate sources of organics contamination in the subsurface. The environs of a hazardous waste site described in Chapter 2—that is, the hydrogeologic environment and the distribution of contaminants—are revealed through site characterization. Site characterization is a continuous, dynamic process of building and revising a site conceptual model that captures relevant aspects of a hazardous waste site, including the source zone. The site conceptual model represents current understanding of the site in terms of the relevant subsurface materials and processes, serves as the basis for more sophisticated site characterization, and will ultimately support the evaluation of various remedial alternatives. Because of the inherent scarcity of available data at field sites, the site conceptual model can only provide an approximation to the real world. Indeed, at the early stages of site conceptual model development, it is possible that several realizations will be tenable. However, as more monitoring and other data become available, the various plausible site conceptual models should gradually converge into a single picture encompassing all salient fluid flow and material transport and transformation processes. Site conceptual models are continually refined, possibly using computer models, to address site-specific complexities involving spatial and temporal variations in flow, transport, and transformation processes.

Although it is impossible to prescribe a specific step-by-step source zone characterization process because of differing conditions from site to site, there are four broad categories of information that are critical for characterizing all source zones:

1. **Understanding source presence and nature**. What are the components of the source, whether a DNAPL or explosive material, and what is the expected behavior of the individual components based on known information?

2. **Characterizing hydrogeology**. What are the lithology of the subsurface and groundwater flow characteristics as they pertain to the source zone? Are there multiple aquifers at the site, and how are they connected? What are the properties and connectedness of the low permeability layers or zones? Can the flow system be described at the specific site and at a larger scale? Can the groundwater velocity and direction (and the spatial and temporal variation in both) be measured?

3. **Determining source zone geometry, distribution, migration, and dissolution rate**. Where is the source with respect to lithology? Is it present as pooled DNAPL, distributed as residual saturation, or both? Is it crystalline explosive material, or is it sorbed? What is the current vertical and lateral extent of the source material, and what is the potential for future migration based on the hydrogeologic characteristics of the site? How fast is the source dissolving?

4. **Understanding the biogeochemistry**. What roles do transport and transformation processes play in attenuating the source zone and the downgradient plume? How will possible remediation strategies affect the geochemical environment (e.g., by releasing other toxic substances, or by adding or removing substances upon which microbial activity and contaminant degradation depend)?

There may be an overall work plan directing that the source characterization activities described above be conducted in a particular order. However, each of the activities is related to the others, and a good deal of iteration between the general categories is not only desirable but critical to the process. Furthermore, iteration between source zone characterization and other site conceptual model building blocks should be employed to constantly reassess site understanding and integrate new data from all facets of the characterization.

This chapter addresses several aspects of source zone characterization, beginning by examining some potential ramifications of inadequate source zone characterization. A subsequent section discusses the four primary categories of information important to source zone characterization and outlines a broad array of characterization methods and tools. General methods for site characterization have been described elsewhere (ITRC, 2003; EPA, 2003a; Thiboutot et al., 2003) and will not be detailed here. Specific source characterization methods for explosives are not well developed and are also not addressed in detail in this chapter. The chapter closes by discussing (1) the importance of source zone characterization to determining cleanup objectives, (2) scale issues, and (3) coping with uncertainty during the process.

A recurring theme in this chapter is that source zone characterization should be carried out in a manner that best informs the entire source remediation process. Decisions regarding the objectives of remediation and the remediation technologies selected will have a strong impact on the source zone characterization strategy and vice versa. These subjects are addressed in Chapters 4 and 5, respectively, and the reader is encouraged to keep the interrelationship between these three key topics in mind.

## KEY PARAMETERS OF SOURCE ZONE CHARACTERIZATION AND THE TOOLS TO MEASURE THEM

The four categories of information important for source zone characterization are (1) the nature and presence of the source material, whether it be a DNAPL or chemical explosives, (2) the hydrogeologic setting, (3) source zone delineation, including geometry, distribution, migration, and dissolution rate in the subsurface, and (4) the biogeochemical environment of the site. These categories of information, and the tools necessary to measure certain parameters, are discussed in detail below.

Because of the variation and complexity of the subsurface environment and the various human activities performed at different sites, no two DNAPL or explosives-contaminated sites are the same. Therefore, there is not a standard suite of tools that can be prescribed for source characterization. Each site must be characterized in a manner that addresses its particular set of constraints and challenges. Before the necessary source zone characterization tools are chosen, it is important that the capabilities and limitations of the tools and the uncertainty of the data generated be generally understood. Many tools are appropriate for both source zone and general site characterization and can provide useful information that spans several of the four categories listed above.

The impact of cost, regulator acceptance, and other nontechnical factors should also be considered in decisions on appropriate characterization tools. For example, drilling and core analysis to assess DNAPL distribution and saturation is an inexpensive method that is accepted by the regulatory community. Partitioning interwell tracer tests (PITT), on the other hand, have been less widely used (primarily for cost reasons), even though they have advantages over drilling and coring in terms of determining the volume of residual DNAPL. Costs and regulatory requirements pertaining to the handling and disposal of investigation-derived wastes can be high for both DNAPLs and explosives.

Safety issues will also vary depending on the source material involved. When performing field characterization of suspected explosives source areas, field teams must be vigilant because of the risk of detonation (EPA, 1993). For example, soils contaminated with ~12 percent to 15 percent TNT or RDX could propagate a detonation after initiation by flame and shock (Kristoff et al., 1987). For this reason, the Army considers explosives in soil at greater than 10 percent to constitute a detonation risk. Thus, geophysical methods are often used to safely get information on site hydrogeology before drilling is commenced on production and training ranges (see Thiboutot et al., 2003).

Table 3-1 summarizes various characterization methods and tools and their applicability for addressing the four categories relevant to source zone characterization. More detailed information about each tool is presented in Table 3-2, including a brief description of the tool and what it measures, the general application of the tool, and the general limitations of the tool for source zone characterization. A variety of methods and tools are presented here including noninvasive characterization approaches ranging from collecting historical information to certain geophysical techniques, invasive sampling tools, methods for laboratory analysis, and tools that represent a combination of the above. The tools found in Tables 3-1 and 3-2 are not equivalent, as some are approaches to removing contaminant samples from the subsurface, some measure specific chemicals either in situ or following sample extraction, some perform both functions, and some do neither. Furthermore, some of the tool categories are much broader than others, and some may overlap slightly. The tables are meant to be inclusive and provide a broad overview of the array of tools and methods often used in source zone characterization.

There are a number of references that provide additional information on the applications and limitations of these techniques. For example, Cohen et al. (1993) and ITRC (2002, 2003) provide details on many of these techniques. NRC (2003)

TABLE 3-1  Various Characterization Methods and Their Potential for Providing Source Zone Information

| Method/Tool | Source Material | Hydrogeology | Source Zone Delineation | Biogeo-chemistry |
|---|---|---|---|---|
| Historical Data | Maybe | Maybe | Maybe | No |
| Regional Geology | No | Yes | No | Maybe |
| Geophysical Tools | No | Yes | No | No |
| Direct Push | Maybe | Yes | Yes | Yes |
| Core Analysis | Maybe | Yes | Maybe | Yes |
| Downhole Methods | Maybe | Yes | No | No |
| Piezometers | No | Yes | No | No |
| Pump Tests | No | Yes | No | No |
| Groundwater Analysis | Maybe | No | Maybe | Yes |
| Solid (Matrix) Characterization | No | No | No | Yes |
| Microbial Analyses | No | No | No | Yes |
| Soil Vapor Analysis | Maybe | No | Maybe | No |
| DNAPL Analysis | Yes | No | No | No |
| Partitioning Tracer Tests | No | Maybe | Yes | No |
| Ribbon NAPL Samplers | Yes | No | Yes | No |
| Dyes | Maybe | No | Maybe | No |

NOTE: The term "maybe" indicates that in some situations the method/tool could provide information relevant to the category.

TABLE 3-2 Summary of Various Methods and Tools and Their Application to Source Zone Characterization

| Method/Tool | Tool Description and What Is Measured | Application/ Relevance to Source Zones | Limitations |
|---|---|---|---|
| **Historical Data** | Information about types and amounts of chemicals used and practices for chemical handling and disposal. | Provides understanding of DNAPL composition and source location. | Subsurface solvent migration unknown. Chemical composition changes with time. |
| **Regional Geology** | Information about fractures, sink-holes, springs, and discharge points. | Used for site conceptual model and determining hydrogeologic setting. | Site-specific details difficult to infer from this information. |
| **Geophysical Methods** | Methods include: | | |
| | a) *Seismic refraction and reflection.* Seismographs measure the subsurface transmission of sound from a point source. | Provides 3-D stratigraphic map. Useful for defining geologic heterogeneities. | Not specific for DNAPL detection. |
| | b) *Electrical resistivity* measures bulk electrical resistance during transmission of current between subsurface and ground surface electrodes. | Used to determine site stratigraphy, water table depth, buried waste, and conductive contaminant plumes. | Not applicable for DNAPL detection. |
| | c) *Electrical conductivity* measures bulk electrical conductance by recording changes in the magnitude of electromagnetic currents induced in the ground. | Used for determining lateral stratigraphic variations and the presence of conductive contaminants, buried wastes, and utilities. | Not applicable for DNAPL detection. |

*continued*

TABLE 3-2 Continued

| Method/Tool | Tool Description and What Is Measured | Application/ Relevance to Source Zones | Limitations |
|---|---|---|---|
| **Geophysical Methods (Continued)** | d) *Ground-penetrating radar* measures changes in dielectric properties of materials by transmitting high-frequency electromagnetic waves and continuously monitoring their reflection from interfaces between materials with different dielectric properties. | Used to determine site stratigraphy, and the location of buried wastes and utilities. | Cannot penetrate clay layers. Not specific for DNAPL. |
|  | e) *Magnetic techniques* measure perturbations to the earth's magnetic field caused by buried ferrous metal objects. | Used for finding steel drums at landfill sites. | Limited to ferrous metal detection. |
| **Direct Push** | Direct push techniques are used both for retrieving subsurface samples and for performing in situ analyses of physical and chemical parameters. Two major techniques include cone penetrometer (CPT) and rotary hammer methods. They are similar in their principles of operation but differ in scale and in some of their applications. CPT systems, which are used mainly for in situ measurements, make use of sensors that measure soil and sediment resistance. CPT is often used in conjunction with aqueous phase (drive point) sampling and probes [e.g., laser-induced fluorescence, neutron probe, membrane interface probe (MIP)]. | Used for gaining information about the physical properties of soils, stratigraphy, depth to the water table, pore pressure, and hydraulic conductivity. Extracted aqueous phase samples may be analyzed quantitatively ex situ. MIP provides semiquantitative subsurface aqueous volatile organic compound (VOC) concentration data, while laser-induced fluorescence detects fluorescing compounds. | Direct push techniques are generally quicker and more mobile than traditional drill rigs, and there is no drilling waste. However, they are not applicable in bedrock, boulders, and tight clays. They are limited to unconsolidated aquifers and to depths of less than 100 ft (30 m). They require calibration with borehole data for accurate interpretation of stratigraphy. Chlorinated solvents do not fluoresce. |

*continued*

## TABLE 3-2 Continued

| Method/Tool | Tool Description and What Is Measured | Application/ Relevance to Source Zones | Limitations |
|---|---|---|---|
| **Core Retrieval and Analysis** | A variety of drilling techniques (rotosonic technologies, flight augers, hollowstem augers, rotary drilling, and cable tool drilling) coupled with different sampling tubes (hollow stem or piston tubes) can be used to collect cores from unconsolidated or consolidated media. | Provides direct information regarding porous media, geology, and stratigraphy. The samples can be tested for contaminants or other biogeochemical species. | Provides a point measurement of spatially variable parameters. Collection methods may alter the physical–chemical properties of the core. Expensive at radioactive sites. |
| **Downhole Methods** | a) Downhole video (e.g., GeoVIS) illuminates soil in contact with a sapphire window and images it with a miniature color camera. | Provides visual imaging of borehole. NAPL possibly visible as discrete globules. | Conditions for effectiveness not well defined. |
| | b) Downhole flow metering impeller or thermal flowmeters measure groundwater inflow rate. | Identifies zones of preferential flow. | Calibration with other flow metering techniques necessary to ensure accuracy. |
| | c) Caliper logging tool follows borehole wall and measures hole diameter. | Identifies cavities or fractures in borehole. | Provides only point measurements. |
| | d) Specific conductance probe determines fluid conductivity with depth. | Can identify inflow zones and contamination zones. | Limited to contaminants that change fluid conductivity (i.e., not DNAPL). |
| | e) Natural gamma logging measures emissions from isotopes preferentially sorbed in clay and shale layers. | Reveals presence of shale or clay layers. | |
| | f) Gamma-gamma log measures media response to gamma radiation. | Provides information about formation density. | |

*continued*

TABLE 3-2 Continued

| Method/Tool | Tool Description and What Is Measured | Application/ Relevance to Source Zones | Limitations |
|---|---|---|---|
| Downhole Methods (Continued) | g) Neutron logging measures media response to neutron radiation. | Measures moisture content and porosity. | |
| | h) Electrical resistance or conductance devices measure these properties of formation fluids and media. | Enables identification of lithology, stratigraphy, or high ionic strength-contaminated water. | Typically used in conjunction with core analysis or other borehole data. |
| Piezometers | Primarily used to determine pressure head spatially and temporally on the site. | Used for potentiometric mapping to understand groundwater flow. Screen length is important. | May not provide the detail needed within the source zone. Because head distribution changes over time, sampling can be required over an extended time. |
| Pump Tests | Pumping groundwater and then monitoring the drawdown cone and rebound can be used to estimate permeability, hydraulic conductivity, the radius of influence, and flow boundaries. Standard tracer tests (e.g., bromide or iodide) are frequently used during pumping to confirm flow models and optimize flow. | This information is necessary for site conceptual model development and remedial activities. | Provides a spatially averaged estimate. Not specific for locating preferential paths or highly permeable zones. |
| Ground-water Analysis | Discrete water samples can be collected with various pumps, bailers, or samplers and then analyzed for different contaminants and groundwater constituents of interest. Multilevel sampling allows water sampling at various depths within a single well. | Helps delineate source areas on site and document preremediation conditions in order to later evaluate whether remedial objectives have been met. | Good understanding of groundwater flow, biogeochemistry, and DNAPL composition is needed for proper interpretation. |

*continued*

TABLE 3-2 Continued

| Method/Tool | Tool Description and What Is Measured | Application/ Relevance to Source Zones | Limitations |
|---|---|---|---|
| **Solid (Matrix) Characterization** | Includes analysis of organic matter content, percent clay and clay type, silt content, mineral composition, and wetting behavior. | Improves understanding of the source zone and the impact of the subsurface environment on remedial actions (e.g., oxidation). | Difficult to quantitatively relate bulk soil measurements to contaminant behavior. |
| **Microbial Analyses** | Microbial community composition and functional potential can be measured for extracted subsurface samples using a combination of molecular techniques and tools based upon conventional culturing and microcosm approaches. | Identifies organisms and genes that are present within the subsurface community to evaluate potential activity and quantify functional activity associated with the active microbial population. | Difficult to extrapolate laboratory activity measurements and rates to in situ field activity. |
| **Soil Vapor Analysis** | Soil probes or passive soil-gas collectors are used to withdraw soil gas. A variety of analytical techniques are used to measure the actual contaminants (e.g., GC-MS). | May be used for indicating "hot spots" of contamination. | Provides point measurements. Understanding of partitioning and NAPL composition is necessary for interpretation. Reflects DNAPL distribution in the vadose zone only. |
| **DNAPL Analysis** | A variety of analytical techniques are used to determine DNAPL chemical composition (e.g., GC-MS) and chemical-physical properties such as viscosity (e.g., viscometer), interfacial tension (e.g., pendant drop method), and density (e.g., densitometer). | Used to better interpret groundwater and vapor sample measurements and to enhance site conceptual model (SCM) and modeling efforts. | DNAPL samples are difficult to obtain and may be variable across the site and with time. |

*continued*

TABLE 3-2 Continued

| Method/Tool | Tool Description and What Is Measured | Application/ Relevance to Source Zones | Limitations |
| --- | --- | --- | --- |
| **Partitioning Tracer Tests** | Hydrophobic chemicals such as higher-weight alcohols (partitioning tracer) are injected through a contaminated zone with a conservative tracer. The reactive tracers partition into DNAPL and experience a delay in breakthrough as compared to the conservative tracer. The retardation and partition coefficients are used to determine NAPL saturation. | Provides in situ estimates of DNAPL saturation. Can provide information on DNAPL distribution when coupled with multilevel sampling. | Expensive. Limited to media with sufficiently high permeability. |
| **Ribbon NAPL Samplers** | Material is placed on a core or in bore holes that reacts with NAPL. Flexible Liner Underground Technologies Everting (FLUTe) is an example. | Provides continuous record of DNAPL distribution in borehole. | Only indicates presence (not amount) of DNAPL. Not proven to be responsive in all cases; thus, negative results are not conclusive. Time in borehole may be important. |
| **Hydrophobic Dyes** (such as Sudan IV) | Hydrophobic dye shake test for detecting DNAPL in soil samples. | Onsite screening tool for locating DNAPL. | Can only indicate presence (not amount) of DNAPL. |

provides information on various sensors and analytical techniques and their appropriate applications. An expert panel report to the U.S. Environmental Protection Agency (EPA) on DNAPL source depletion presents a summary of characterization tools (EPA, 2003a), Kram et al. (2001, 2002) provide a comparison of various analytical techniques with cone penetrometers, and Griffin and Watson (2002) provide a comparison of field techniques to confirm DNAPLs. A large amount of information on sampling technology can be obtained from the Department of Energy's (DOE) Environmental Management Science Program, EPA's Technology Innovation Office (http://fate.clu-in.org), and EPA's Environmental

Technology Verification program (http://www.epa.gov/etv). Sampling in fractured rocks is discussed in Shapiro (2002). Thiboutot et al. (2003) provides extensive information on characterization of explosives soil sites, primarily on military training and testing ranges, including the risks of detonation and appropriate sampling strategies and chemical analysis methods.

## Source Presence and Nature

Before extensive source zone characterization methods are undertaken, an effort should be made to first determine the nature of the source. Determining the composition of the DNAPL or explosive material is useful for a variety of site management activities. Knowing the components of the source and being able to predict the expected behavior of the individual components based on known information is important for performing risk assessment and surmising appropriate remedial actions for the site. For DNAPLs, the physical–chemical properties such as solubility, density, specific gravity, viscosity, interfacial tension, wettability, contact angle, and the tendency to partition between sediment and water should be determined if possible (see Cohen et al., 1993, for analytical methods relevant to characterizing DNAPLs). The concentrations analyzed in sediment and water can be related to health-based standards, and estimates of the human and ecosystem exposure to the contaminants can be predicted. This information is necessary to guide subsequent phases of site characterization.

Kram et al. (2001) provide an excellent summary of field techniques and information for determining DNAPL source material information based both on direct and indirect evidence. Direct detection of a DNAPL source can be accomplished via various analyses of soil, rock, or water cores and samples. These range from such simple techniques as visual observation (such as with downhole video) and soil shake tests with hydrophobic dyes, to measurements of UV fluorescence in situ or within extracted samples or cores, or to ribbon NAPL samplers used either ex situ to test extracted cores or in situ within boreholes. These various techniques are generally used to determine the presence or absence of DNAPL and not necessarily total mass or chemical composition.

Understanding the presence and nature of the source material can be a challenge at sites where a DNAPL or solid phase explosive sample cannot be isolated from the source zone. In such cases, indirect methods such as measuring high aqueous or vapor contaminant concentrations relative to saturated aqueous or vapor concentrations, or measuring high contaminant concentrations in soil cores, are used for inferring the presence of a separate phase (see Box 3-1 for an example relevant to chemical explosives). For example, aqueous concentrations in excess of 1 percent of DNAPL solubility (Mackay et al., 1991; Cohen et al., 1993) or soil concentrations greater than about 10,000 mg/kg (EPA, 1992) are generally considered to be indicative of DNAPL presence. Caution should be taken when using this technique to infer DNAPL presence because of the highly

## BOX 3-1
## Inferring Pure Solid Phase Explosives from a Soil Sample

One method to infer the potential presence of a separate solid phase is through evaluation of phase partitioning equilibria (Jury et al., 1991; Phelan and Barnett, 2001). Jury et al. (1991) shows that the total concentration $(C_T)$ of a chemical constituent in soil can be partitioned among the soil-sorbed, soil-water, and soil-air phases:

$$C_T = \rho_b K_d C_l + \theta C_l + a K_H C_l$$

where $\rho_b$ is the soil bulk density (g/cm$^3$), $K_d$ is the linear soil-water partitioning coefficient (cm$^3$/g), $C_l$ is the concentration in the soil water (g/cm$^3$), $\theta$ is the volumetric soil moisture content (cm$^3$/cm$^3$), $a$ is the air-filled porosity (cm$^3$/cm$^3$), and $K_H$ is the Henry's constant (unitless). At a specific temperature, the maximum concentration of a solute in water is specified by the solubility limit. If the temperature dependent solubility limit is used with estimates for $\rho_b$, $K_d$, $\theta$, $a$, and $K_H$ in the equation above, $C_T$ is the maximum total concentration, as determined from a field soil sample, before a separate chemical solid phase must appear.

Figure 3-1 is the result of an analysis that shows the maximum total soil concentrations of RDX (a solid phase energetic material) that can be partitioned in soil before a separate solid phase must exist. At 20°C, the maximum solubility of RDX in water is about 45 mg/L. Due to the low vapor pressure and air-water partitioning coefficient $(K_H)$, the soil-water partitioning coefficient $(K_d)$ is the principal factor influencing the maximum total soil residue. Figure 3-1 shows that total soil residue concentrations above 30 to 70 $\mu$g/g (sum of liquid, sorbed to soil, and vapor phases) indicate the potential presence of solid phase material in the soil system. The lower the soil-water partitioning coefficient $(K_d)$, the lower the maximum total soil residue because of the smaller sorption capacity of the soil. This case is for nearly saturated soil ($S_l = 99$ percent, where $S_l$ is defined as the volumetric soil moisture content $\theta$ divided by the soil porosity).

heterogeneous distribution of DNAPL. Aqueous concentrations may vary by an order of magnitude over small vertical and horizontal distances, such that sampling from specific depth intervals (using multilevel monitor wells, drive point sampling at specific depths, or short well screens) may be advisable for providing more resolution on potential sources than conventional monitor wells with large screened intervals. In addition, the solubility of individual components in a multi-component DNAPL is lower than that of pure compounds. Empirical evidence clearly indicates that the lack of observations above these numerical limits does not preclude the presence of DNAPL (Frind et al., 1999; Dela Barre et al., 2002). It should be kept in mind that groundwater and soil sampling provide general composition information only for the soluble components of the source material

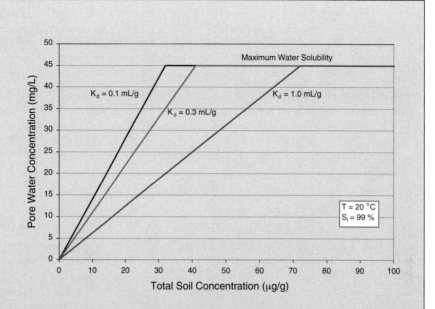

FIGURE 3-1 Effect of soil–water partitioning coefficient ($K_d$) on maximum soil residue for RDX. SOURCE: Phelan et al. (2003).

(unless one is fortunate enough to collect a core sample that contains DNAPL in its pores, something that is unusual due to the highly inhomogeneous DNAPL distribution). Furthermore, solids can interfere with the extractability of certain components. Historical records may be able to provide some information on the age and identity of the compounds and how they were used on the site. Yet, short of collecting actual chemical samples such as a DNAPL sample, it may not be possible to fully understand the nature of the source material (in terms of the key physical and chemical parameters mentioned above).

Table 3-3 lists the likelihood of a source zone being present at a site, given the occurrence of certain events. For example, if there was a known or probable historic release of a DNAPL, then there is high certainty that there is a source.

TABLE 3-3  The Degree of Uncertainty Regarding the Presence of a Source
Zone at a Site Based on the Occurrence of Various Events.

| Event | DNAPL Source |
|---|---|
| Known or probable historical release of DNAPL | High certainty |
| Process or waste practice suggests probable DNAPL release | High certainty |
| DNAPL visually detected in subsurface, monitoring wells, etc. | High certainty |
| Chemical analyses indicate DNAPL presence ($\geq$ saturation) | High certainty |
| DNAPL chemicals used in appreciable quantities at site | Likely; some uncertainty |
| Chemical analysis suggests possible source zone | Likely, some uncertainty |

SOURCE: Modified from Cohen et al. (1993) diagram, Figure 7-1.

The wide range of chemicals used and disposal methods practiced can result
in great variability in the chemical mass and composition across the site. Therefore, finding and analyzing one DNAPL sample may not provide a representative
assessment of the overall contamination. It is also possible that the source area is
not necessarily stable over time. Source material can change as the more soluble,
biodegradable, or volatile components are removed, altering the material's composition and, in turn, affecting its chemical and physical properties. In addition,
the source itself, as in the case of DNAPLs, can continue to move due to natural
changes in the subsurface and to various characterization and remediation activities
performed at the site.

Identification of explosives as subsurface source material will most likely
occur via chemical analysis of soil samples. Field-screening methods are an ideal
method to determine high concentrations during field campaigns. These methods
require extraction of a 2- to 20-g soil sample with acetone or methanol. Then,
either a colorimetric or immunoassay detection method is used for obtaining
quantitative or semiquantitative results (Crockett et al., 1998). These methods are
applicable to TNT, DNT, RDX, and HMX as well as to other explosive compounds, with detection limits for most methods at ~1 ppm. When employing
these methods, users must be aware of possible interferences with the colorimetric
methods and of cross-reactivity for immunoassay methods. Other emerging
chromatography-based field methods may provide more quantitative results
(Hewitt and Jenkins, 1999).

## Characterizing the Hydrogeology

An accurate depiction of the subsurface and its flow characteristics is critical
to the overall site conceptual model and necessary for developing successful
remediation strategies. The primary difficulty in characterizing the subsurface is
in determining the heterogeneity of the geologic material and understanding how
this affects groundwater and contaminant transport. Nonetheless, it is often

possible to develop a general description of a site, including the surface topography, the regional geology such as whether the subsurface material is consolidated or unconsolidated, and the regional groundwater patterns (e.g., flow from areas of recharge—rainfall, surface impoundments, or ponds—to areas of discharge— surface water features or pumping wells), without large amounts of site-specific data.

Characterization of subsurface hydrogeology involves a variety of hydraulic and tracer tests as well as detailed hydrostratigraphic determination. Data pertaining to lithology from soil borings and hydraulic head distributions will aid in the development of a more detailed picture of the site layering in terms of relatively permeable and impermeable units, as well as groundwater flow direction. When and if sufficient data are collected, a flow model can be useful to describe water movement through permeable media and around impermeable media to discharge points. In fractured rock systems, an understanding of the flow system is necessary for understanding where the DNAPL has migrated or where it has the potential to migrate.

Determining permeability and effective porosity can be difficult at the scale that is needed for accurately portraying flow at a given site. In heterogeneous aquifers, permeability can range over 13 orders of magnitude. Pump tests are often used for determining permeability in the field; however, they provide a spatially averaged value that cannot yield insight into the heterogeneity often observed in Type III and V settings, which are characterized by low-permeability zones. In some fractured rock aquifers, such as karst aquifers, obtaining adequate data on the hydraulic properties and determining flow paths are even more difficult, if not impossible. Understanding the flow system and being able to determine groundwater velocities are important in order for mass flux reduction to be used as a metric for source remediation (see Chapter 4).

A number of noninvasive characterization techniques developed over the years for delineating subsurface heterogeneities and anomalies are useful for source zone geologic characterization (although they have little applicability to locating sources and defining their distribution within the subsurface). Geological mapping and interpretation from outcrops and other geomorphic features can provide information on aquifer characteristics and possible zones of infiltration near a source zone. Other techniques (see Table 3-2), such as seismic refraction and reflection methods, electrical resistivity, electrical conductivity, ground-penetrating radar (GPR), and magnetic techniques, involve the input of some type of energy into the subsurface and an appropriate detector that captures the transmitted energy. For example, electrical resistivity involves the transmission of a current between the subsurface and implanted electrodes and measurement of the bulk electrical resistance of the media though which the current passes.

The successful application of noninvasive technologies is often dependent on the field experience of the user as well as on a lack of interference from structures and power lines. The latter can be very difficult to ensure in a manufacturing setting or on a site where there is significant historic activity such as filling. Generally, the user must have detailed site knowledge or must independently verify the results with other techniques. Data quality for noninvasive techniques can be very good, but the data are typically calibrated with coring data. More detailed information on such characterization tools can be found in Cohen et al. (1993) and ITRC (2002).

Beyond noninvasive tools, direct push techniques are widely used for characterizing site and source zone hydrogeology. As the name implies, rods equipped with probes, samplers, or other tools are directly pushed into the subsurface using a hydraulic ram, hammer, or vibratory method. Cone penetrometers (CPT) typically use a hydraulic ram for pushing the tool string into the ground, while other drive point methods often use a rotary hammer. Drive point methods are used for conducting continuous and single-point groundwater sampling, for piezometer and monitoring well installation, and for vapor sampling. Cone penetrometers outfitted with a sensor in the cone tip can measure lithostatic pressure, hydrostatic pressure, electrical resistivity, and pore pressure, and they can provide centimeter scale vertical resolution of lithology. In addition, numerous analytical methods have been developed for use within the cone. These include fluorescence spectroscopy, Raman spectroscopy, and UV absorption, which can aid in determining contaminant presence and concentration (see Kram et al., 2001, 2002). Since direct push methods are generally less expensive and less intrusive than conventional drill-based technologies, they are increasingly used whenever site conditions allow (direct push tools cannot penetrate bedrock and may be refused by cobbles or very dense layers).

Downhole methods are increasingly useful for characterizing source zone hydrogeology. Typically used in conjunction with other characterization tools such as soil cores, downhole methods can provide information about stratigraphy, formation density, porosity, fractures, flow paths, and moisture content, depending on the method used. For example, various downhole imaging techniques can provide information about the location of fractures and their orientations, if the fractures intersect the boreholes. It should be noted that the fracture density has little correlation with permeability within a fractured rock aquifer (Paillet, 1998). Therefore, fracture connections in such settings are generally elucidated using conventional aquifer tests with hydraulic stress being applied to a single centrally located borehole and with drawdown being measured in observation boreholes (Tiedeman and Hsieh, 2001). Because most hydraulic testing produces large quantities of contaminated water for disposal and may cause spreading of contaminants, these methods should not be used in source zones without a good understanding of the site. Methods that produce less water, such as cross-borehole flowmeter pulse tests that use short stress times, may be useful for defining subsurface connections between discrete fractures (Williams and Paillet, 2002).

## Source Zone Delineation

Source zone delineation refers to determining the location of a source in the subsurface (both horizontally and vertically), its strength,[1] and how it moves among and between phases. For example, is the DNAPL pooled at the bedrock, is it pooled at the top of a confining layer in a multilayer sequence of high- and low-permeable units, is it distributed in sediment, is it locally above residual saturation or primarily at or below residual saturation? Without this knowledge, there will be considerable uncertainty about the effectiveness of the chosen remedy and the time required to meet cleanup goals.

An important first step in delineating a source zone is to gather and analyze historical chemical usage and disposal information about the site. Information that can suggest where to look for DNAPLs or explosives is valuable and can save time and money during the remediation process. Hydrogeologic information from previous or ongoing investigations, either on a site scale or a regional scale, can provide valuable information on possible vertical and horizontal flow paths. However, in some cases, such as complex sites that have multiple interconnected layers and fractured bedrock, such information can be difficult to obtain.

Source delineation is most commonly accomplished by analyzing sediment cores, by measuring dissolved concentrations of specific compounds and, in rare cases where possible, by analyzing the free product found in wells or cores. Obtaining soil and rock cores and dissolved contaminant concentrations is generally accomplished using invasive technologies such as core retrieval via drilling and direct push techniques, as previously described. It should be noted that there is a real danger of cross-connecting water-bearing zones when using drilling techniques. Collecting soil cores allows visual and chemical characterization of the source area, and constructing wells allows the application of a variety of pumping-based assessment techniques. In both cases, data are needed in three dimensions to map the source and its dissolved constituents. Because DNAPL distribution may be extremely heterogeneous on a scale of tens of centimeters vertically and a few meters horizontally, cores would have to be taken and analyzed on a similar scale in order to construct a detailed map of DNAPL distribution. However such detailed characterization is generally unnecessary. As mentioned earlier, ribbon NAPL samplers and hydrophobic dyes can provide onsite detection of DNAPLs within the cores or the borehole itself.

Partitioning interwell tracer tests (PITT) offer a method for estimating mass and distribution of the DNAPL over volumes much larger than is possible with soil sampling. This method has been used at more than 50 DNAPL sites with

---

[1]"Source strength" is a loosely defined term that refers to the mass of contaminant present in the source zone and more so to its flux from the source zone. Thus, it conveys information not only about the longevity of the source, but also about the size of the plume that would be created (due to the reference to flux).

good success (Jin et al., 1997; Annable et al., 1998; Mariner et al., 1999; Londergan et al., 2000; Rao et al., 2000; Kram et al., 2001, 2002; Meinardus et al., 2002; Jayanti, 2003). Meinardus et al. (2002) reported one of the most complete comparisons ever made between DNAPL volumes measured by a PITT and by hundreds of soil samples from cores; the total DNAPL volume from both methods was in excellent agreement despite significant heterogeneity at the site. The advances in field analytical techniques can shorten the time of analysis and augment laboratory analyses. In addition, the use of multilevel samplers during the PITT can provide better estimates of the spatial distribution of the DNAPL. The technique relies on sufficient contact between partitioning tracers swept through the source zone and the resident DNAPL. The presence of large heterogeneities and a highly variable DNAPL distribution can affect this contact and greatly reduce the accuracy of the test. The application of PITT to complex sites, therefore, may require special design methods such as hydraulic control wells to confine the tracer both laterally and vertically to the source zone. Jayanti (2003) recently analyzed the impact of heterogeneities under a wide range of field conditions.

It is particularly difficult to locate source zones in fractured rock aquifers. Collecting water samples for chemical analysis in fractured rocks can yield highly variable results depending on the aquifer properties—the transmissivity, storativity, and hydraulic head of fractures intersecting the borehole. In situations where it is important to characterize the spatial variability of the groundwater chemistry in a bedrock aquifer, it is advantageous to collect water samples from a discrete interval that hydraulically isolates, with packers or liners, a single fracture or group of closely spaced fractures in the boreholes. In situations where water samples cannot be taken from hydraulically isolated intervals and where water samples are collected from an open borehole, the effect of the water volume in the borehole must be considered in the design of the collection system (Shapiro, 2002).

Although sources can be delineated by monitoring contaminant concentrations at sampling points over time, measurements of contaminant mass flux (the amount of contaminant mass migrating through a cross section of the aquifer orthogonal to groundwater flow within a given time) are increasingly being considered as a more accurate way to determine the effectiveness of remediation (see Chapter 4, Feenstra et al., 1996). To calculate flux, it is necessary to have a good understanding of the flow system and the vertical distribution of concentrations of the source components in at least one transverse transect of the plume prior to commencing remediation. Although the documentation is limited, methods for measuring mass flux are becoming more common in the literature (as described in Box 3-2).

## BOX 3-2
## Tools for Measuring Mass Flux

Determination of the mass flux of a contaminant has been proposed as a metric to assess the progress of DNAPL source remediation (Feenstra et al., 1996; Einarson and Mackay, 2001; API, 2003; EPA, 2003a). The mass flux is the amount of contaminant migrating though a cross-sectional plane that is perpendicular to the direction of groundwater flow. It is expressed as the mass of contaminant moving across a surface per unit area per unit time. In actual use in groundwater field studies, mass discharge (mass per unit time) is determined rather than mass flux, and it is assumed that the area covers the entire plume width. One method to determine mass flux at a specific location is from contaminant concentration data in a cross-sectional area and specific discharge. Because the contaminant concentration and groundwater discharge vary within a cross section, the mass flux can be estimated in small zones and summed to obtain the total flux. The accuracy of the measurement is related to the sample point density and the determination of hydraulic properties of the aquifer. At complex sites this information is difficult to obtain.

Another method for determining mass flux is the use of aquifer tests in a down-gradient transect of wells and measurement of the mass of contaminant pumped (Bockelmann et al., 2001). If the downgradient wells capture the entire contaminant plume, the mass flux can be calculated from the contaminant concentration and the pumping rate. In many cases the entire plume is not captured, and the method requires knowledge of the pumping well rate, the contaminant data, and other necessary data for use in a flow-and-transport model. All of these requirements have considerable uncertainty. Another concern in using this method is whether pumping will cause additional spreading of the contaminants and the fact that large quantities of water are produced.

A third method, still in the development stage, is to measure mass flux by the use of a sorptive permeable medium that is placed either in a borehole or monitoring well to intercept contaminated groundwater. The medium is spiked with a tracer. By quantifying the mass of tracer released and contaminant sorbed, the groundwater velocity and contaminant mass flux can be calculated (Hatfield et al., 2002). This requires knowledge of the partitioning characteristics of the medium and of the contaminants. Although this method may be easier to implement because it requires less quantitative information about the flow system, a disadvantage is that measurements are made at specific points, and some flow paths may not be intersected. The use of sorptive permeable media has yet to be field tested.

All of these methods are experimental at present. The reported applications are primarily for site evaluation of natural attenuation and at sites where the contaminants are petroleum hydrocarbons or methyltertbutylether (MTBE), rather than DNAPLs (Borden et al., 1997). Yet, the measurement of mass flux holds promise as a more robust method of quantifying the mass of contaminant loss in cases where information on the groundwater flow system, including hydraulic conductivity and hydraulic gradient, can be obtained.

## Biogeochemical Environment of the Site

Some source remediation technologies can greatly change the redox conditions and the resulting water chemistry and biological activity near the source area and in the downgradient plume. For example, certain treatments, such as use of potassium permanganate, can inhibit microbial populations and stop natural biodegradation processes that were occurring. Other agents, such as surfactants, may actually increase microbial activity by providing a carbon source. Conversely, certain biogeochemical conditions can limit the effectiveness of in situ remediation. For example, where there are high concentrations of organic material in the aquifer, larger amounts of oxidants will be required to ensure reaction with the organic contaminants of interest. To better understand and predict these phenomena, an evaluation of the site geochemistry is essential.

Knowledge of the water chemistry and microbial activity within source zones should be gained prior to commencing a remediation action, particularly when remedies that involve microbial or abiotic chemical transformation are being considered. For example, what minerals are dissolving or precipitating, and what trace elements could be released if the water chemistry changes? It is well known that many trace elements are redox-sensitive and migrate in solution along the hydraulic gradient under certain conditions. An assessment can be made about the potential rates of natural bioremediation and possible enhanced bioremediation by examining the indigenous microbial population in the water/sediment (NRC, 2000; Witt et al., 2002); determining the nature and abundance of electron acceptors in the water, such as oxygen, nitrate, and sulfate (Chapelle et al., 1995; McGuire et al., 2000); determining the amount of extractable iron on the sediments (Bekins et al., 2001); and examining the potential for dissimilatory metal reduction (Lovley and Anderson, 2000).

Both conventional enrichment/isolation-based techniques as well as genetic-based molecular techniques can be used to analyze the microbial community and to evaluate its functional potential. For example, microcosm data can be combined with quantitative polymerase chain reaction (PCR) to evaluate and quantify the presence of *Dehalococcoides* species at a site and confirm its functional potential (Hendrickson et al., 2002; Lendvay et al., 2003). However, care must be taken when extrapolating laboratory results on microbial activity to natural environments since it is virtually impossible to entirely replicate field conditions in laboratory studies.

## Source Characterization at Hydrogeologically Complex Sites

Not all of the characterization tools discussed above are applicable or appropriate in all hydrogeologic settings. Most of the tools have been developed and utilized at sites with porous media (particularly unconsolidated granular geologic environments), and thus do not apply to hydrogeologic settings IV and V or karst (see Chapter 2). Thus, sites that exhibit a range of hydrogeologic conditions (such as the site described in Figure 2-16) require a range of tools in order for source zone characterization to be carried out. A particular problem is that the tools available for use in fractured rock systems often provide limited (i.e., point-specific) information because of the high degree of spatial variability at these sites. Consider, for example, karst systems, which arise from a combination of water-soluble carbonate rock and a well-developed secondary porosity. The structure of the rock (e.g., the bedding planes, joints, faults, and fractures) forms the basis for the creation of the network of interconnected openings common in karst systems. These openings may include large conduits and are often capable of transporting water (and contaminants) rapidly throughout the site. In such settings, it can be extremely difficult to determine where the source is located and which fractures are hydraulically connected. Innovative approaches such as using tomography to determine which fractures are connected are still in a research stage. Thus, source zones in karst and fractured systems create extreme characterization and remediation challenges, as exemplified by the case study in Box 3-3.

## Summary on Characterization Tools

Although it is impossible to prescribe exactly what tools should be used to characterize source zones given their site-specific nature, a typical scenario representing the state of the art can be outlined that utilizes some of the tools described above. Because migration of DNAPL (and hence DNAPL distribution) is to a large extent controlled by permeability (in that low permeability layers exclude DNAPL, while high permeability layers channel it), hydrogeologic characterization is an essential first step during source zone characterization. A combination of historical chemical-use data and analyses of water samples is then used to determine likely areas of DNAPL occurrence. Analyses of core samples from drilling or of samples from direct push or down hole analytical tools can then be used to define the source zone. Due to the extremely heterogeneous distribution of DNAPL, the probability of a core or a push intersecting a small DNAPL source is low; thus, conclusive evidence of DNAPL occurrence is not always found. If deemed necessary, a PITT can be used to confirm the amount of DNAPL present. In general, the objectives of the remediation plan will help determine the level of characterization effort required. This effort will also be constrained by the physical characteristics of the site (i.e., consolidated vs. unconsolidated, fractured vs. nonfractured), since they control how difficult it will be to obtain the required data.

It should be noted that characterization data are collected from a wide range of scales; for example, hydraulic conductivity may be determined from pump tests that average a volume of hundreds of cubic meters or from core samples that average a few cubic centimeters. Furthermore, multiple methods are typically used to measure the same parameter, and making comparisons between the results

## BOX 3-3
## The Redstone Arsenal Case Study

The Redstone Arsenal is a 38,000-acre facility (15,378 hectares) immediately adjacent to Huntsville, Alabama. The underlying geology is a well-developed, mantled intricate network of karst conduits with an ultimate discharge to the Tennessee River to the south. The facility has been in operation since 1941 primarily for rocket propellant research and development. In the mid 1980s to 1996, five solvent degreaser facilities in operation within Operable Unit 10 (OU-10) used TCE and later 1,1,1-TCA. The main contaminants in the groundwater in this area include TCE and perchlorate. The overburden and karstic upper bedrock are intimately interconnected in this area to form a single aquifer with a prevailing upward hydraulic gradient.

A sitewide hydrogeologic investigation of the karst system (Phase I) was performed with the following objectives: establish the significance of karst on groundwater flow and contaminant transport, delineate karst watersheds in order to define potential source–receptor flow paths, identify optimal surface and groundwater monitoring locations for long-term monitoring and possible remediation performance assessment, develop a sitewide conceptual model to support decision making, and evaluate the existing perimeter monitoring well network. The actions and outcomes of this investigation are summarized in the following table.

| Action | Purpose/Outcome |
|---|---|
| Thermal infrared flyover and field reconnaissance | Identify discrete groundwater discharge points (e.g., springs and seeps). Over 100 springs were sampled for VOCs. |
| Air photo stereoscopic evaluation | Catalogue surface (e.g., sinkholes) karst features. |
| Alabama Cave Survey database, review and develop database of boring logs and other drilling info. | Catalogue subsurface (e.g., caves, fractures) karst features; 1,100 top of bedrock elevations obtained and 686 bedrock penetrations; 569 solution cavities identified in 293 locations. |
| Water level measurements from 900 wells | Map potentiometric levels across the site. |
| Borehole flowmeter surveys | Confirm hydraulic gradients both in bedrock and select overburden wells. |
| Continuous water quality data acquisition and review of historical groundwater data | Determine dynamics of groundwater flow, understand the surface water/groundwater interactions, and determine the effect of these on contaminant transport. |
| Surface water data integration | Gain insight regarding karst aquifer (e.g., size, seasonal patterns). |
| Offsite wells inventory | Identify potential receptors, assess existing site perimeter wells. |
| Dye trace studies | Provide indication of potential primary flow paths. |

**Source Delineation:** Additional characterization at OU-10 was performed with the purpose of trying to identify and characterize DNAPL in the subsurface, including its lateral and vertical extent and mass. Understanding the subsurface stratification was also attempted, as this would aid in developing a source zone conceptual model. Actions and outcomes related to these objectives are shown in the table below.

| Action | Purpose/Outcome |
|---|---|
| Seismic reflection | Map structural lows where DNAPLs could accumulate, identify faults. |
| Groundwater screening | Refine limits of DNAPL source area. |
| Drive point (DPT) screening | 245 locations were sampled and analyzed for VOCs, perchlorate. Eight FLUTe surveys were conducted in DPT holes. DPT data were combined with monitoring well screening to identify "hotspots." Data were combined with seismic reflection to guide drilling of deep boreholes. |
| Cored deep bedrock boreholes (up to 275 ft or 84 m) | Open hole geophysical logs were used with natural gamma, fluid temperature, resistivity, and caliper techniques to determine stratigraphy, fractures, conduits, flow. |
| | Hydrophysical surveys were used to obtain hydraulic conductivity, transmissivity, interval-specific yield. $K$ values typically an order of magnitude higher than packer test results. Virtually no flow below 200 ft (61 m). |
| | Digitally recorded optical televiewer was used. Naturally occurring hydrocarbons observed oozing out of layers and down boreholes. |
| | Select boreholes (four) were surveyed using FLUTe reactive ribbon technology. DNAPL detected in only one location. |
| | Selected (77) intervals packer tests were performed (for hydraulic conductivity) and sampled for contaminants. |
| Bedrock cores | Scanned with UV fluorescence—did not identify DNAPLs. |
| Flux characterization | Sampled key springs as function of discharge to establish contaminant loading to surface water from groundwater. |

*continued*

---

**BOX 3-3 Continued**

Additional activities, including dye tracers, long-term pump tests in the deep bedrock, and natural attenuation assessment, are planned for the source area in OU-10. This is an interesting example of the difficulty of defining the source at complex sites, since even after doing numerous studies, the actual source area is still in doubt. The success of any source remediation would be jeopardized without a better understanding of the source zone.

---

is not necessarily simple. For example, aqueous concentrations may be determined in a monitor well screened over several meters, from a drive point well screened over 30 centimeters, or from a fluorescence tool on a direct push unit which measures concentrations over a few square centimeters. Analyses may be from labs using standard methods, from field screening kits, or from downhole tools. In every case both the scale of measurement and the method of analysis will affect the results. In developing the site conceptual model, care must be taken to reconcile the diverse data sets produced by different methods and labs and at different scales.

## APPROACH TO SOURCE ZONE CHARACTERIZATION

Gaining an understanding of the complexity of a site through source zone characterization before engaging in expensive remediation technologies can save resources and help to define reasonable cleanup goals. Beyond the four primary types of source characterization information and the associated tools mentioned above, there are other factors site managers should consider during source characterization, as discussed below. Additional issues faced when managing the source characterization process include how to determine what level of characterization is sufficient, the impact of scale, and how to estimate and manage uncertainty. This section also discusses the need for iterative feedback between source characterization information and the selection both of objectives and technologies for source remediation, a topic further developed in Chapter 6 in the context of a decision framework for source remediation.

### Source Zone Characterization Should Reflect Remediation Decisions

The goal of source zone characterization is to provide the basis for decisions on remediation. Once a source is determined to be present, a series of decisions will arise in developing a remediation strategy, each of which will require specific characterization information. The major decisions include defining remediation objectives, determining if there are one or more potentially effective technologies capable of meeting the chosen objectives, selecting the best technology for the site, designing a remediation project, and evaluating the effectiveness of remediation. Each of these decisions is described in detail in the following chapters (objectives in Chapter 4, remedial technologies in Chapter 5, and the decision process in Chapter 6). Because different technologies and different objectives require different types of characterization, source characterization needs must be reexamined at each decision point. This suggests that source zone characterization is a dynamic and iterative process (e.g., see ASTM, 2003), similar in nature to dynamic work plans and the TRIAD approach to environmental data collection (EPA, 2001, 2003b, 2004; NRC, 2003), and able to make use of real-time data as it becomes available. Indeed, characterization efforts over time should lead to a refined conceptual source submodel that is nested within the overall site conceptual model process. The development of the source zone submodel begins with a suspicion about where and how a contaminant release occurred, and it continues until the uncertainty associated with source size and configuration is acceptable in terms of the cleanup strategy ultimately selected.

### Scale Issues Between the Source and the Plume

It is important for site managers to realize that source characterization will require a much more densely deployed sampling plan than does the associated plume characterization. Plumes exist on a relatively large spatial scale (hundreds of meters to kilometers) while their causative sources exist on a smaller scale (meters to tens of meters). Furthermore, plumes are also much more spatially continuous than DNAPL, and they trace groundwater flow paths, so it is more acceptable to infer their geometry in spite of relatively sparse data.

Attempts to delineate a source zone from plume data will be inaccurate unless the hydrogeology is known with great certainty (Sciortino et al., 2000, 2002). Such certainty is needed because source zones are created by multiphase flow phenomena that are generally less predictable in geometry than are plumes, and it is likely that multiple realizations of the source zone will account for the observed plume characteristics. Therefore, to arrive at a relatively unique delineation of a source zone's geometry, it is necessary to execute 3-D sampling efforts of a high spatial granularity.

Successful site managers gain an adequate understanding about the connection between the plume and its source in the context of their remediation objectives and technologies. Plume-scale observations are needed for the purpose of defining potential exposure pathways. However, it is important to avoid over-delineation of the plume at the expense of more localized source zone characterization efforts. This means that as salient information about site hydrogeology and plumes is gleaned from the larger-scale site characterization efforts, potential source zone configurations should be added to the site conceptual model. The sooner the site conceptual model begins to stabilize, the sooner source remediation objectives and technologies can be identified and critically assessed.

## Recognizing and Managing Uncertainty

Uncertainty is an inescapable part of hazardous waste remediation, particularly when a nonaqueous phase source may be involved. It was in response to this fact that the recent NRC report (2003) coined the term "adaptive site management" to stress the importance of managing sites using adaptive management approaches (Holling, 1978; Walters, 1986, 1997; Walters and Holling, 1990; Lee, 1993, 1999) that are both long-term and empirical, in contrast to the objectives of rapid design, cleanup, and closure that were confidently promulgated in the 1980s and 1990s. Adaptive site management is described in NRC (2003) as "an innovative approach to resource management in which policies are implemented with the express recognition that the response is uncertain, but with the intent that the response will be monitored, interpreted, and used to adjust programs in an iterative manner, leading to ongoing improvements." Of course, such a strategy is not unknown to engineering and the applied sciences. For example, Karl Terzaghi and Ralph Peck, pioneers of geotechnical engineering, advanced the observational approach (Terzaghi and Peck, 1967; Peck, 1969) which, among other features, emphasized that engineers should adapt solutions to new information rather than using ready-made or predetermined solutions, however well conceived. That is, even though Terzaghi and Peck advanced the theory of geotechnical engineering and respected its inherent value, they also recognized its limitations, many of which stem from the complexity of soils in the natural environment.

*Sources of Uncertainty*

The causes of uncertainty that are relevant to source zone remediation can be classified into the following categories. Further details are available in NRC (1999).

**Measurement Error.** Measurement error is associated with the imprecise collection, analysis, and interpretation of samples, and it stems both from user error and from the adequacy of the tools used to collect data. User error can cause samples to be disturbed or contaminated during collection or transportation. Furthermore, users may collect data at the wrong time or place, they may misinterpret data if the spatial or temporal scale of the sample is not understood (Sposito, 1998), or they may use incorrect or oversimplified conceptual models to relate what is directly measured to what needs to be determined (e.g., in pumping or tracer tests, the hydrogeology is usually oversimplified to determine hydrogeologic parameters from head or concentration observations). Quality assurance and quality assessment procedures may reduce the size and frequency of user errors or at least give the professionals working at the site an idea of the errors that may occur. A second important source of measurement error is that measurement devices are inexact and may not be capable of detecting compounds of concern at relevant levels. For example, it may not be possible to identify specific components in a complex mixture of contaminants and humic substances.

**Sampling Error.** Sampling error is defined here as the uncertainty that results from having limited spatial and temporal data on which inferences must be drawn. Often there is a lack of information about the location of a source, its chemical composition, how much contaminant was released and when, and the present distribution of the contaminant. At the same time, geologic formations, source zones, and plumes are highly heterogeneous in ways that are hard to describe with exactness. Depending on the hydrogeologic setting and the contaminants present, the source zone may have an irregular distribution of contamination that reflects the natural geologic variability. Unfortunately, at most sites the groundwater and soil samples taken during characterization are limited in number and are nonuniformly distributed. This is due to the high costs involved in site characterization, the creation of potential new exposure pathways, concerns about remobilization of DNAPLs, and worker-exposure risks associated with extensive drilling and sampling. In such environments, having samples at only a few locations forces one to infer the values of hydrogeologic parameters and concentrations over the whole source zone—an activity that is fraught with error. In an analogous fashion, the significant temporal nonuniformity results in errors when interpolating between samples (Kitanidis, 1999; Houlihan and Lucia, 1999). As might be expected, sampling error can be reduced by increasing the density and frequency of observations.

**Simulation Error.** Simulation error, sometimes referred to as model error, is defined as the error associated with inaccuracies (1) in the underlying conceptual models and how they represent physical, chemical, and biological processes and (2) in the implementation of mathematical models. Any models used to help decide whether to employ a certain technology during source remediation (e.g., the UTChem model of surfactant flooding) can be affected by simulation error. For example, the conceptual model may neglect or misrepresent major processes like adsorption or biodegradation. In some cases, model error stems from natural variability caused by aquifer heterogeneity, since this tends to be poorly captured in physical models of the subsurface. In other cases, models are limited by the accuracy of the data used to validate the models, and thus measurement error can exacerbate simulation error. Finally, all mathematical models are approximations of reality, and even the most realistic of them are computer-based numerical models that suffer from roundoff and truncation errors. More important, mathematical simulation models resolve variability at a scale much coarser than the laboratory scale where our process understanding is the most reliable. That is, even if one understood the processes and knew the values of parameters at the scale of centimeters, it is still nontrivial to determine what equations and parameters to use in a model that effectively averages over meters (Dagan and Neuman, 1997; Sposito, 1998; Rubin et al., 1999). Given that simulation models will used more and more frequently prior to remedy implementation, a greater understanding of simulation error is critical to their success.

The degree of incertitude one encounters in source remediation studies is typically much more than in plume remediation studies because in source zones (1) the distribution of mass is more variable, more dependent on the hydrogeologic setting, and harder to describe than in aqueous plumes, (2) there are typically fewer measurements than in plumes, and (3) there are more physical, chemical, and biological processes involved (such as dissolution and repartitioning).

In many applications, a major cause of uncertainty is sampling error. Fortunately, sampling error can be reduced by increasing the sampling of the source zone, by using statistical techniques, or by a combination of the two. Although infrequently done because of a lack of expertise and upfront financial resources, quantitative uncertainty analysis (with respect to hydrology) should become a more routine part of source characterization, especially at complex sites where uncertainties regarding the source composition, distribution, and strength are very large. The following section discusses three approaches to uncertainty analysis classified as statistical, inverse, and stochastic inverse methods.

*Statistical Methods*

One of the most important goals of uncertainty analysis is to estimate unknown quantities and quantify the estimation error, which enables the representativeness of limited sampling data to be determined. For example, one very simple approach (e.g., Moore and McCabe, 1999) is to use "point" samples of pollutant concentration taken from the source zone to calculate the numerical average, which is then thought of as a representative value of the concentration, while the standard deviation of the data divided by the square root of the number of sample points is considered a measure of variability. However, these simple statistical approaches are usually not applicable because they are based on assumptions (namely, that all samples come from the same distribution and that there is no correlation among samples) that are invalid for data from source zones. In reality, two samples obtained in close proximity are more likely to be similar in value than two samples taken at a large distance from each other; that is, the data exhibit correlation or "spatial continuity" or "structure."

Geostatistical methods (e.g., Rouhani et al., 1990a,b; Kitanidis, 1997; Olea, 1999; ASCE, 2003; Rubin, 2003) are preferable because they explicitly account for spatial continuity and spatial correlation among parameters. For example, in inferring the total mass, geostatistical methods weigh the point measurements in a way that accounts for the structure of the source and the neighborhood of influence of each observation. One of the quantities used in geostatistics is the variogram, which is the mean square difference between two sample values as a function of the length and orientation of the segment that separates the two sampling points. From the shape of the variogram of source concentration, one can see whether the concentration fluctuates in space in a continuous and smooth fashion. If the concentration varies smoothly, two observations next to each other provide essentially the same information: thus, each of them should be given less weight in estimating the total mass than an observation that is taken at an undersampled part of the formation. In another example, consider the problem of estimating values of concentration at grid points, which is a prerequisite for drawing a contour map, from points where samples have been taken. The estimate is a weighted average of the observations, and geostatistics assigns weights that account for the separation (length and orientation) of the grid point from each observation point, as well as the separation between observation points. Thus, all else being equal, if the variability is gradual, an observation point near the grid point should generally be given a higher weight than an observation point far from it. If there is stratification, an observation on the same stratum as the grid point should be given more weight than an observation at the same distance from the grid point but at a different stratum. In the final analysis, geostatistical methods produce more reasonable and intuitively appealing results than simple statistics, which assign equal weights to all observations and thus neglect the effects of nonuniform distribution of sample points in space. Geostatistics is more systematic and rigorous than the ad hoc methods of interpolation and averaging that have been proposed in the past, such as inverse-distance weighting.

A major advantage of geostatistical techniques is that they quantify the uncertainty associated with an estimate in the form a standard error, the availability of which should allow a more informed use of the estimate. The error may suggest that more observations are needed and may even suggest where more measurements must be collected. That is, one may evaluate the effect of an additional observation on standard estimation error before the observation is taken, such that statistical methodologies can guide the selection of sampling strategies. In summary, geostatistical methods have important advantages over ad hoc approaches in that geostatistical methods evaluate appropriate estimates and estimation standard errors. Of course, their purported advantages assume their correct implementation; in particular, appropriate attention must be paid to the selection of a reasonable model of spatial structure (e.g., a variogram). Such methodologies have been used to interpolate hydrogeologic parameters in general and contaminant plumes (e.g., Kitanidis and Shen, 1996; Saito and Goovaerts, 2000; Pannone and Kitanidis, 2001) in particular. They could be used in analyzing data from source zones in order to better estimate contour maps or spatial averages of, say, NAPL in the soil and to provide a better appreciation of the uncertainty associated with these estimates.

It is worth noting that geophysical methods, such as seismic refraction and reflection, ground-penetrating radar, and electric resistivity, in contrast to the more common well or penetrometer samples, provide global rather than point information (Rubin et al., 1999; Hyndman, 1999; Chen et al., 2001; Hubbard et al., 2001; Jarvis and Knight, 2002) such that caution should be used in the geostatistical analysis of data from these tools. Such methods hold promise and have reportedly found some application, but they have uncertainties of their own. First, the data produced are spatial averages. That is, because the signal traverses the geologic formation from a source to a receiver, and because sources and receivers are limited in number due to cost and can be placed only at the surface or in some wells, geophysical properties are measured only approximately. Second, and perhaps more important, the measured geophysical properties are usually not the ones of direct interest in remediation studies, but are only related to them. For example, electric resistivity is correlated with salinity or total dissolved solids in the water. This relation is not exact, and it may be an important source of uncertainty in estimating one quantity from another (e.g., salinity from resistivity or moisture content from dielectric constant). A promising approach is the combined use of geophysical surveys and well data for the development of appropriate correlation functions.

*Inverse Methods*

An alternative approach is to utilize simulation models to infer quantities of interest. Using preliminary estimates of unknown parameters as data, one can predict the values of observed quantities. These predictions are then compared to the actual observations, and the parameters are adjusted judiciously in order to improve the agreement between the predictions and the data. For example, from observations of the pressure or hydraulic head in wells or piezometers within a formation, one can infer the spatial distribution of hydraulic conductivity. This approach is widely used in applied hydrogeology and comes under the rubric of inverse modeling, history matching, model calibration, or just parameter estimation (e.g., Yeh, 1986). In fact, classic well tests (like pumping tests to determine transmissivity and storage coefficient, e.g., Boonstra, 1999) involve inverse modeling using a simple conceptual model. At the early stages of a remedial investigation, the presence of a source is usually inferred only indirectly (e.g., from measurement of solute concentrations that are near the solubility limit or from the persistence of the contamination problem even after many contaminated pore volumes have been treated). This approach uses the principles of inverse modeling, though the degree of sophistication of the model used may vary widely.

To save time in calibrating models, automated inverse methods are increasingly being used (Poeter and Hill, 1997). They search for and use as estimates the values of the parameters that minimize a fitting criterion, such as the sum of the squares of differences between model predictions and observations. The idea is that the model should reproduce the data when the right parameter values are used. Inferring the values of spatially distributed variables (e.g., conductivity, pressure, or concentration, which are functions of the spatial coordinates) from limited observations is particularly challenging because the results may depend on the discretization. Using a finer grid, one may achieve higher resolution, but the results may be much less accurate than if a coarse grid is used. An important part of every methodology for the automated estimation of a spatially distributed variable is its approximate representation, such as through subdivision of the domain into an appropriate number of homogeneous zones or superposition of a manageable number of functions with unknown parameters.

*Stochastic Inverse Methods*

A drawback of deterministic approaches to inverse problems is that they produce a single estimate of the parameters, even though it is understood that there are generally multiple solutions that are equally consistent with the data. For example, even with extensive sampling of a plume showing pollutant concentrations near the solubility limit, it may be impossible to determine whether the cause is a separate DNAPL or explosive material source or the slow desorption from solids over a much larger extent. Even if a source exists with traits that make it amenable to removal (e.g., limited spatial extent), the presently available characterization techniques may be inadequate to support such an action because one cannot identify the source's exact location and strength.

Stochastic inverse methods (see reviews by Yeh, 1986; Ginn and Cushman, 1990; McLaughlin and Townley, 1996), which combine the principles of inverse methods with statistical or geostatistical models and methods to describe spatial structure and quantify errors, explicitly recognize uncertainty. They can produce standard errors and confidence intervals, and not just an estimate. Even better, they can be used to generate several different solutions that are equally consistent with the data (called conditional realizations) by using Monte Carlo techniques (e.g., Robert and Casella, 1999). One can use these solutions to evaluate the chance of success of a proposed management scheme, as illustrated in Bair et al. (1991) for the wellhead protection problem.

Stochastic inverse methods can explicitly take into account information that is in addition to observations, such as, for example, that permeability is within a certain range or that the aqueous concentration at two nearby locations is correlated. Bayesian methods (e.g., Christakos, 1990; Copty et al., 1993; Gelman et al., 1995; Carlin and Louis, 2000; Chen et al., 2001; Kennedy and Woodbury, 2002), which are the subset of statistical inference methods that utilize Bayes theorem, are of particular interest because they are well suited to the analysis of sparse or incomplete data. In Bayesian methods, this additional information is encoded into a "prior probability distribution" that is then combined with the information that comes from the observations to produce a posterior probability.

Stochastic methods, though promising, are at an early stage of development, which limits their practical implementation in source zones. A drawback limiting their practical applicability is their heavy computational cost, because their implementation involves numerous runs of deterministic simulation models. Nevertheless, increased use of stochastic methods is encouraged. They are well suited to the analysis of data from source zones in order to improve understanding of site conditions such as estimates of total mass, to provide a better appreciation of the uncertainty associated with these estimates, and to design monitoring programs.

*Decision Making under Uncertainty*

The prominent role of uncertainty in remediation, particularly when a DNAPL source may be involved, demands that decision makers develop approaches to manage uncertainty. Several studies have considered the issue of decision making under uncertainty (e.g., Massman et al., 1991; Lee and Kitanidis, 1991), including the adaptive site management approach promoted in NRC (2003). Furthermore, geostatistical and other stochastic methods have been used to evaluate the worth of data and thus to guide, often through the use of optimization tools, the design of appropriate sampling networks and the selection of sampling frequencies (e.g., Freeze et al., 1992; James and Freeze, 1993; Christakos and Killam, 1993; James and Gorelick, 1994; Capilla et al., 1998). From these studies one can glean the following common features of strategies that work well under conditions of substantial uncertainty.

**Multiple Scenarios.** The existence of uncertainty essentially means that there is no single scenario or possibility one can design for. For example, instead of a unique shape or extent of a source, there are likely to be many that are consistent with available information. Hence, instead of designing for ideal performance under a certain scenario, one should design for satisfactory performance under a range of plausible possibilities; that is, possibilities that are equally likely given the available information.

**Overdesign, Caution, and Hedging.** Uncertainty results in concrete costs because there are fundamental trade-offs between overdesign and increased risk of failure. For example, if the source boundaries are not known, one must exercise caution by targeting a large area or else hazard missing part of the actual source. The overdesign that usually results can be expressed through a safety factor (e.g., the ratio of the area targeted to the area of the "best estimate"). Best estimates are typically located near the middle of the range of plausible scenarios because they usually are mean or median values of a population of possible values, and using them without safety factors may involve unacceptably high probabilities of failure.

**Feedback, Adaptability, Probing.** One should reduce uncertainty by utilizing all measurements and other information whenever they become available to refine the site conceptual model. For example, consider that a pump-and-treat system is designed for plume containment; if one discovers later that the plume extends over a much smaller area than originally thought possible, the scheme should be adapted in order to reduce the cost of operation. Furthermore, the scheme should be designed in a way that elicits an evaluative response that allows one to reduce costs. For example, the pumping wells may be placed in such a way that over an initial phase of operation, they can provide information about the location of the plume. The concept of using feedback from data collection to inform subsequent efforts is known as a dynamic work plan, and it is central to the current emphasis in EPA's Triad Approach (EPA, 2001, 2003b, 2004). Implementation of this process requires rapid feedback and hence rapid analysis methods such as analytical tools on direct push rigs, field analytical labs, or field chemical screening tests.

**Comprehensive Design of Characterization, Remediation, and Monitoring.** Site characterization and monitoring affect the design of remediation; that is, to compensate for less characterization, a decision may be made in favor of a more conservative remediation design. Consider a hypothetical example in which a fixed amount of money must be allocated between the cost of actual remediation and the cost of characterization. In principle, the optimal allocation is when the marginal benefit of a dollar used for characterization is exactly equal to the marginal benefit of using it for remediation, for a fixed degree of risk (which is likely indicated by concentration targets—see Chapter 4). This is shown conceptually in Figure 3-2. Thus, contrary to the widespread myth that more characterization is always better, one must strike the right balance between monies spent for collecting data and monies spent to perform the actual cleanup. Of course, some remediation methods require more data than others both to determine their potential applicability to a site and later to assess performance, and the cost of these data collection activities should be included when considering the overall cost of the remedial option.

**Robustness.** Depending on the technology chosen, more or less uncertainty can be tolerated; that is, certain technologies are more immune to the effects of uncertainty and are thus termed more robust. A remedy's robustness directly affects the level of effort that needs to be made (with commensurate cost) with respect to source characterization prior to and during remediation. For example, consider a scenario in which there is a relatively high level of uncertainty associated with the nature and geometry of the source zone, while the vertical and horizontal boundaries of the source zone area are relatively certain. Here, a certain level and type of source zone characterization effort may be sufficient for

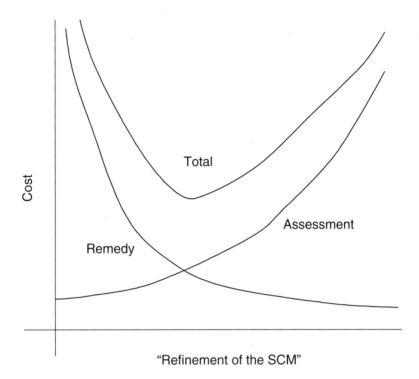

"Refinement of the SCM"

FIGURE 3-2 Illustration of the concept that the allocation of costs between remediation assessment (characterization) depends on how refined the site conceptual model is. The more is spent on assessment, the more precise the SCM. The more precise the SCM is, the less must be spent on the remedy. There is a level of refinement that is optimal for minimizing total cost.

containment or monitored natural attenuation strategies to be used, but would be inadequate for a source depletion technology like surfactant flushing.

\* \* \*

The assessment and removal of a source are challenging tasks that are becoming even more challenging because of the difficulties and high costs associated with collecting measurements. The uncertainty resulting from limited data, hydrogeologic variability, and other factors must be taken into account when evaluating and selecting technologies for source remediation. Most source remediation studies circumvent an objective and systematic analysis of data and evaluation of uncertainty with respect to delineating the source zone and predicting the effectiveness of remediation (although there are frequently used statistical models of human health risk from, e.g., contaminated groundwater). This is unfortunate because such an analysis can provide a better appreciation of the chances of success of a proposed remediation scheme and guidance on how to improve the scheme through additional measurements. Box 3-4 summarizes the progression and intensity of source zone characterization efforts carried out at the relatively small-scale Camp Lejeune, North Carolina. After the development of the overall site conceptual model, which included a preliminary effort focused on defining the source zone and local hydrogeology, this site was selected as a test site for source remediation via surfactant flushing. This project typifies the level of pre- and post-remedial action source zone characterization required to adequately address the uncertainty associated with such an effort.

## REPERCUSSIONS OF INADEQUATE SOURCE CHARACTERIZATION

It is unfortunate that source remediation, especially at complex sites, is usually attempted in the absence of adequate source zone characterization. Impediments to fully understanding source areas are technical, economic, and institutional. For example, the site geology may be such that the technology to delineate a source zone is currently inadequate. Alternatively, the budget allocated for site characterization may be depleted long before the above questions have been conclusively answered. Third, stakeholders and the institutions funding the cleanup may lose patience with the process and call for remedial action to begin. Despite the technical challenges, some level of source zone characterization is indispensable for the effective management of an environmental restoration effort. Severe overestimation of the source size may inflate the cost of remediation efforts to exorbitant levels. Conversely, missing the source material will jeopardize the success of the cleanup and will require additional characterization and restoration work. Source zone assessment must be an integral part of the site characterization process, and failure to adequately characterize a source zone can have major repercussions in terms of future risk.

The following scenarios are intended to illustrate the ramifications of inadequate source zone characterization. All three scenarios are based on a hypothetical DNAPL-contaminated site that has reached a relatively late stage of the characterization process. The hydrogeologic setting is best characterized as a layered Type I and II system (see Chapter 2), where a shallow sand aquifer is separated (to an unknown extent) from an underlying aquifer by a low-permeability clay layer. The site exhibits elevated contaminant concentrations in groundwater near the suspected source area, but data are relatively sparse and variable, and the specific geometry of the source zone is highly uncertain. In one case, the decision is made to forego any additional source zone characterization and begin remediation. In the other two cases, it is decided to dedicate differing amounts of additional time and resources to source zone characterization. The scenarios are depicted in Figure 3-3.

**BOX 3-4**
**Camp Lejeune Case Study**

MCB Camp Lejeune Site 88 is an example of a source depletion action that was accomplished after an adequate amount of source zone characterization was completed (ESTCP, 2001; NFESC, 2001, 2002). The potential PCE source zone area was initially identified at this former dry cleaning facility using chemical usage history and conventional site characterization techniques, such as borehole drilling and groundwater monitoring. Extensive and dense physical sampling efforts in the form of cone penetrometer and soil borings were first undertaken to map the soil lithology in terms of the permeability contrasts and porosity distributions that offer clues as to spatial DNAPL distribution. At the same time, chemical analysis of core samples was used, along with measured organic carbon content, to develop a more direct line of evidence regarding the distribution. As certainty about the DNAPL distribution above and along the clay unit began to increase, source remediation was presented as a potential objective, and wells were installed in and around the source zone. Using these wells, more specific characterization strategies, such as pump tests and partitioning interwell tracer tests (PITT), were undertaken to estimate the integrity of the clay unit as a capillary barrier and to estimate the volume of DNAPL in the source zone. By the end of the pre-removal characterization activity, three injection, six extraction and two hydraulic control wells were situated within the roughly 10-m by 10-m source zone area, which had already been densely probed by means of cone penetrometry and core sampling. This level of characterization was necessary to evaluate the feasibility of the surfactant-enhanced aquifer remediation (SEAR) source remediation technology and prepare for its application.

Evaluation of the SEAR technology at Camp Lejeune involved extensive additional characterization efforts. Pre- and post-SEAR PITTs were used to estimate the amount of DNAPL volume remaining in the source zone, and 60 post-SEAR confirmational core samples were collected from within the small source zone and analyzed. The results from these characterization efforts suggested that while 92 percent to 95 percent of the DNAPL was removed from the permeable portions of the source zone, only 72 percent was removed overall. However, post-SEAR characterization and analyses suggested that the residual DNAPL was as unavailable to dissolution as it had been to the surfactant flushing. Subsequent modeling and flux reduction also indicate this 72 percent mass removal was accompanied by a greater than 90 percent reduction in the dissolution flux emanating from the source zone (Jayanti and Pope, 2004). Follow-up groundwater monitoring efforts aimed at confirming the sufficiency of this reduction in terms of plume size are ongoing. It is important to emphasize that, with the exception of the PITT, conventional characterization procedures were employed to characterize the source zone at Camp Lejeune.

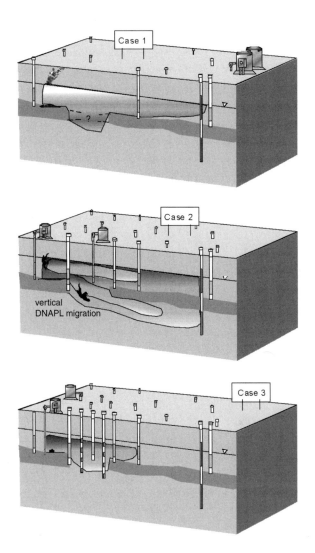

FIGURE 3-3 Illustration of potential repercussions of inadequate source zone character-ization in the case of a poorly defined, discontinuous clay layer. Case 1 represents a prolonged pump-and-treat effort that will contain the source but not deplete the source in the short term. In Case 2, a chemical flushing technology is applied prematurely and results in the migration of a portion of the DNAPL to the lower aquifer. In Case 3, adequate source zone characterization leads to an accurate evaluation of the extent of the source and to more successful execution of the chemical flushing technology, substantial source strength reduction, and subsequent monitored natural attenuation.

*Illustrative Case 1: Minimal Source Zone Characterization.* Despite the uncertainty associated with the existence and nature of DNAPL at the site, the horizontal extremities of the *plume* seem reasonably well defined. Thus, one alternative is to forego any additional source zone characterization and implement a conventional pump-and-treat strategy. Personnel working on the site have experience in the hydraulic capture and treatment of the extracted water. The decision makers are wary of attempting source zone remediation, fearing that it will fail in the face of the uncertainties about the underlying clay layer, source size, and source strength. Thus, source zone remediation is not attempted and a conservative pump-and-treat plan is adopted.

Given their prior decision to halt site characterization, the decision makers have selected a reasonable remedial plan. Negative consequences might have resulted if source zone remediation actions had been undertaken with the inadequate characterization data. First, the costs incurred by this would likely produce limited benefits. Second, additional costs may arise if the source remediation attempt results in transferring portions of the source deeper or into less accessible regions of the subsurface. However, because the source material is not efficiently reduced by the pump-and-treat remedy, the prognosis for this scenario is that there will be a long-term operation and maintenance horizon with costs continuing indefinitely.

*Illustrative Case 2: Insufficient Source Zone Characterization.* In Case 2, the decision makers determine that the potential benefits of source zone remediation (e.g., plume size reduction to within a compliance boundary and reduced time to site closure) make additional source zone characterization worth pursuing. Thus, the decision makers choose to allocate additional resources toward source zone characterization, although their technical staff is relatively less experienced in this aspect of site assessment. A firm time limit of one year is set to ensure that the source zone characterization process does not delay remedial actions for too long. Additional core samples are collected, and monitoring wells are installed in and around the suspected source area. At the end of the one-year period, the corresponding data improve the areal resolution of the source zone, but the vertical extent remains highly uncertain. The decision makers assume that a clay layer identified in the source area is sufficiently continuous to provide a capillary barrier, and the decision is made to design and execute an aggressive chemical flushing technology aimed at mobilizing and extracting the DNAPL (see Chapter 5 for technologies).

In this case, the decision makers have allowed a predetermined time limit to control the quality of the source zone characterization effort. The fact that the clay unit in question is actually not continuous in the area of the source means that the likelihood of success is low. In the end, the costs incurred by the source remediation action are wasted because the mobilization technology will likely drive DNAPL through the discontinuity and deeper into the subsurface.

*Illustrative Case 3: Adequate Source Zone Characterization.* The rationale of the decision makers in Case 3 is similar to that in Case 2 with one exception: in Case 3 they recognize the importance of allowing the source zone characterization to continue until uncertainty regarding the feasibility of partial source depletion is reduced to a manageable level. Outside expertise in the areas of source zone characterization and remediation has been retained to assist the site personnel. As in the previous case, additional core sampling is undertaken, and monitoring wells are installed in and around the suspected source area. At the end of the one-year period, the vertical extent of the source remains highly uncertain, and additional time and resources are allocated to the install and monitor multilevel monitoring wells to better resolve the vertical concentration gradient. At the end of the second year, the source zone and underlying clay unit have been resolved with enough certainty to correctly identify a less aggressive DNAPL solubilization strategy. The decision makers elect to design and execute a chemical flushing technology to reduce the source strength, but they orientate flushing and recovery hydraulics to guard against DNAPL migration through the discontinuity in the clay layer.

In this case the decision makers have expended more time and money during source characterization relative to Cases 1 and 2, respectively. In doing so, they have maximized the likelihood of successful source depletion. The chemical flushing plan is designed, executed, and evaluated, requiring an additional two years. However, according to pre- and postremediation dissolution flux estimates, 95 percent of the original source strength has been depleted. Modeling efforts predict a reduction in source strength, and monitored natural attenuation becomes an acceptable alternative for all stakeholders. Four years have passed since the decision to aggressively characterize the source zone, but operations and maintenance costs for the site are now low compared to those for the previous two cases. (It is not clear, however, whether life cycle costs, which take into account all relevant expenditures over the life of a hazardous waste site, are lower for the third scenario. See Chapter 4 for an in-depth discussion of life cycle analysis.)

* * *

Although these case studies are hypothetical, they serve to illustrate common themes in source zone characterization. For many reasons, responsible parties may be disinclined to commit the considerable funds required for comprehensive source zone characterization. Source identification and delineation are considered technically challenging and expensive, and they may reveal that the contamination problem is more extensive than previously thought, leading to even larger costs. Detailed source zone characterization can be unappealing to the responsible party since it may pave the way to source remediation regarded as complicated, costly, and perhaps of questionable effectiveness. Indeed, it is worthwhile to note that overzealous source zone characterization such as excessive source zone drill-ing without proper precautions can remobilize DNAPL or create worker safety issues related to chemical exposure. However, a nonexistent or halfhearted char-acterization attempt may unnecessarily lead to containment efforts with no fore-seeable end (Case 1) or compromise the effectiveness of any source remediation effort (see Case 2). Avoiding these undesirable consequences requires that the decision process leading to a specific source assessment and mitigation plan be dynamic and iterative. That is, on the basis of preliminary information, funds are allocated for characterization and remediation. As new information is obtained, the goals of the remediation campaign must be revisited and are likely to change. Once it becomes clear that source remediation is a potentially promising approach, one must evaluate, on the basis of the hydrogeochemical conditions and the types of the contaminants, what methods of remediation are appropriate and what levels of characterization they require. While this iterative process may appear time-consuming in the face of stakeholders' desires to expedite cleanup, a rush to judgment on the nature and extent of the contamination can worsen site condi-tions and result in contamination inadvertently remaining behind. The public is more likely to respond positively to an honest acknowledge of the limitations on and uncertainties in the data than to perceived certainty that is later revealed as incorrect.

## CONCLUSIONS AND RECOMMENDATIONS

The committee's review of dozens of case studies of source characterization and remediation (see Chapter 1) suggests that at many DNAPL or explosives-contaminated sites there is inadequate site characterization to support the remedia-tion strategies that were employed and/or to evaluate the actual results of the trial in terms of improvement in water quality, the fraction of the total mass removed, or other appropriate success metrics. In several cases the data were not adequate to determine if there even was a source. This is most likely due to pressure to show progress and meet deadlines, to financial constraints, or to unclear objectives.

Despite these shortcomings, the Army has made substantial progress at improving its source characterization activities, particularly with respect to con-firming that DNAPL is present. The Army case studies reviewed by the committee also suggest that the development of site conceptual models is evolving rapidly in parallel with improved Army characterization efforts. At Fort Lewis, for example, a number of nontraditional characterization technologies were used including CPT-LiF, MIP, GPR, resistivity, dye studies, and multilevel wells. Other sites using innovative technologies included Watervliet, where down-hole geophysics were used to characterize the fractured rock, and Redstone Arsenal, where a major effort was mounted to characterize flow in a complex karst terrain. These efforts are impressive, given that karst, epikarst, and fractured bedrock settings are resistant to detailed source characterization using existing technologies. The Army's realization of the difficulty of site characterization in these hydrogeologic settings has led it to consider applying for technical impracticability waivers (e.g., at Anniston and Letterkenny). Such waivers are appropriate if the objectives of remediation are clearly defined and if sufficient data are obtained to show that the objectives cannot be met by any feasible approach.

The following recommendations regarding source characterization are made.

**Source characterization should be performed iteratively throughout the cleanup process to identify remedial objectives, metrics for success, and remediation techniques.** All sites require some amount of source characterization to support the development and refinement of a site conceptual model. In general, successful source remediation requires information on the nature of the source material, on the site hydrogeology, on the source zone distribution, and on the site biogeochemistry. However, the level of characterization effort required and the tools used at any given site are dependent on site conditions, on the cleanup objectives chosen, and on the technology chosen to achieve those objectives.

**An evaluation of the uncertainties associated with the conceptualization of the source strength and location, with the hydrogeologic characteristics of the subsurface, and with the analytical data from sampling is essential for determining the likelihood of achieving success.** This is often accomplished through the use of statistical, inverse, and stochastic inverse methods. Unfortunately, quantitative uncertainty analysis is rarely practiced at hazardous waste sites. Obtaining a better handle on uncertainty via increased source characterization would allow eventual remediation to be more precise. It is likely that at most sites, there is not an optimum combination of resources and effort expended on source characterization and thus uncertainty reduction vs. remedial action.

## REFERENCES

American Petroleum Institute (API). 2003. Groundwater remediation strategies tool. Publication Number 4730. Washington, DC: API Publishing Services.

American Society of Civil Engineers (ASCE). 2003. Long-term groundwater monitoring: the state of the art. Prepared by the Task Committee on the State of the Art in Long-Term Groundwater Monitoring Design. Reston, VA: American Society of Civil Engineers.

American Society for Testing and Materials (ASTM). 2003. Standard guide for developing conceptual site models for contaminated sites. Document No. ASTM E1689-95(2003)e1. West Conshohocken, PA: ASTM International.

Annable, M. D., P. S. C. Rao, K. Hatfield, W. D. Graham, and C. G. Enfield. 1998. Partitioning tracers for measuring residual NAPLs for field-scale test results. J. Env. Eng 124:498–503.

Bair, S., C. M. Safreed, and E. A. Stasny. 1991. A Monte Carlo-based approach for determining travel time-related capture zones of wells using convex hulls as confidence regions. Ground Water 29 (6):849–855.

Bekins, B. A., I. M. Cozzarelli, E. M. Godsy, E. Warren, H. I. Essaid, and M. E. Tuccillo. 2001. Progression of natural attenuation processes at a crude oil spill site. II. Controls on spatial distribution of microbial populations. J. Contam. Hydrol. 53:387–406.

Bockelmann, A., T. Ptak, and G. Teutsch. 2001. An analytical quantification of mass fluxes and natural attenuation rate constants at a former gasworks site. J. Contaminant Hydrology 53:429–453.

Boonstra, J. 1999. Well hydraulics and aquifer tests. Pp. 8.1–8.34 *In:* The Handbook of Groundwater Engineering. J. W. Delleur (ed.). Boca Raton, FL: CRC Press.

Borden, R. C., R. A. Daniel, L. E. LeBrun IV, and C. W. Davis. 1997. Intrinsic biodegradation of MTBE and BTEX in a gasoline-contaminated aquifer. Water Resources Research 33(5):1105–1115.

Capilla, J. E., J. Rodrigo, and J. J. Gomez-Hernandez. 1998. Worth of secondary data compared to piezometric data for the probabilistic assessment of radionuclide migration. Stoch. Hydrol. Hydraul. 12 (3):171–190.

Carlin, B. P., and T. A. Louis. 2000. Bayes and empirical Bayes methods for data analysis. Boca Raton, FL: Chapman & Hall/CRC.

Chapelle, F. H., P. B. McMahon, N. M. Dubrovsky, R. F. Fujii, E. T. Oaksford, and D. A. Vroblesky. 1995. Deducing the distribution of terminal electron accepting processes in hydrologically diverse groundwater systems. Water Res. Res. 31:359–371.

Chen, J., S. S. Hubbard, and J. Rubin. 2001. Estimating the hydraulic conductivity at the South Oyster site from geophysical tomographic data using Bayesian techniques based on the normal linear regression model. Water Resources Research 37(6):1603–1613.

Christakos, G. 1990. A Bayesian/maximum-entropy view to the spatial estimation problem. Mathematical Geology 22(7):763–777.

Christakos, G., and B. R. Killam. 1993. Sampling design for classifying contaminant level using annealing search algorithms. Water Resources Research 29(12):4063–4076.

Cohen, R. M., J. M Mercer, and J. Matthews. 1993. DNAPL Site Evaluation. C. K. Smoley (ed.) Boca Raton, FL: CRC Press.

Copty, N., Y. Rubin, and G. Mavko. 1993. Geophysical-hydrological identification of field permeabilities through Bayesian updating. Water Resources Research 29(8):2813–2825.

Crockett, A. B., T. F. Jenkins, H. D. Craig, and W. E. Sisk. 1998. Overview of on-site analytical methods for explosives in soil. Special Report 98-4. Hanover, NH: U.S. Army Corps of Engineers, Cold Regions Research and Engineering Laboratory.

Dagan, G., and S. P. Neuman. 1997. Subsurface Flow and Transport: A Stochastic Approach. Cambridge, UK: Cambridge Univ. Press.

Dela Barre, B. K., T. C. Harmon, and C. V. Chrysikopoulos. 2002. Measuring and modeling the dissolution of a nonideally shaped dense nonaqueous phase liquid (DNAPL) Pool in a saturated porous medium. Water Resources Research. 38(8):U143–U156.

Einarson, M. D., and D. M. Mackay. 2001. Predicting impacts of groundwater contamination. Environmental Science and Technology 35(3):66A–73A.

Environmental Protection Agency (EPA). 1992. Estimating potential for occurrence of DNAPL at Superfund sites. Washington, DC: EPA of Solid Waste and Emergency Response.

EPA. 1993. Handbook: Approaches for the remediation of federal facility sites contaminated with explosive or radioactive wastes. EPA/625/R-93/013. Washington, DC: EPA Office of Research and Development.

EPA. 2001. Using the Triad approach to improve the cost-effectiveness of hazardous waste site cleanups. EPA-542-R-01-016. Washington, DC: EPA.

EPA. 2003a. The DNAPL Remediation Challenge: Is There a Case for Source Depletion? EPA 600/R-03/143. Washington, DC: EPA Office of Research and Development.

EPA. 2003b. Using dynamic field activities for on-site decision making: a guide for project managers. EPA 540/R-03/002. Washington, DC: EPA Office of Solid Waste and Emergency Response.

EPA. 2004. Improving sampling, analysis, and data management for site investigation and cleanup. EPA 542-F-04-001a. Washington, DC: EPA Office of Solid Waste and Emergency Response.

Environmental Security Technology Certification Program (ESTCP). 2001. Surfactant enhanced DNAPL removal. ESTCP Cost and Performance Report CU-9714. Washington, DC: ESTCP.

Feenstra, S. J., A. Cherry, and B. L. Parker. 1996. Conceptual models for the behavior or dense non-aqueous phase liquids DNAPLs) in the subsurface. *In:* Dense Chlorinated Solvents and Other DNAPLs in Groundwater. J. F. Pankow and J. A. Cherry (eds.). Portland, OR: Waterloo Press.

Freeze, R. A., B. R. James, J. Massmann, T. Sperling and L. Smith. 1992. Hydrogeological decision analysis: (4) the concept of data worth and its use in the development of site investigation strategies. Ground Water 30(4):574–588.

Frind, E. O., J. W. Molson, M. Schirmer, and N. Guiguer. 1999. Dissolution and mass transfer of multiple organics under field conditions: the Borden emplaced source. Water Resources Research 35(3):683–694.

Gelman, A., J. B. Carlin, H. S. Stern, and D. B. Rubin. 1995. Bayesian Data Analysis. Boca Raton, FL: CRC Press.

Ginn, T. R., and J. H. Cushman. 1990. Inverse methods for subsurface flow: a critical review of stochastic techniques. Stoch. Hydrol. Hydraul. 4:1–26.

Griffin, T. W., and K. W. Watson. 2002. A comparison of field techniques for confirming dense nonaqueous phase liquids. Ground Water Monitoring and Remediation 22:48–59.

Hatfield, K., M. D. Annable, S. Kuhn, P. S. C. Rao, and T. Campbell. 2002. A new method for quantifying contaminant flux at hazardous waste sites. Pp. 25–32 *In:* Groundwater Quality: Natural and Enhanced Restoration of Groundwater Protection. S. F. Thornton and S. E. Oswald (eds.). IAHS Publication No. 275. Oxfordshire, UK: IAHS Press.

Hendrickson, E. R., J. A. Payne, R. M. Young, M. G. Starr, M. P. Perry, S. Fahnestock, D. E. Ellis, and R. C. Ebersole. 2002. Molecular analysis of *Dehalococcoides* 16S ribosomal DNA from chloroethene-contaminated sites throughout North America and Europe. Appl. Environ. Microbiol. 68(2):485–495.

Hewitt, A. D., and T. F. Jenkins. 1999. On-site method for measuring nitroaromatic and nitramine explosives in soil and groundwater using GC-NPD. Special Report 99-9. Hanover, NH: U.S. Army Corps of Engineers, Cold Regions Research and Engineering Laboratory.

Holling, C. S. (ed.). 1978. Adaptive environmental assessment and management. New York: John Wiley & Sons.

Houlihan, M. F., and P. C. Lucia. 1999. Groundwater monitoring. Pp. 24.1-24.40 *In:* The Handbook of Groundwater Engineering. J. W. Delleur (ed.). Boca Raton, FL: CRC Press.

Hubbard, S. S., J. Chen, J. Peterson, E. L. Majer, K. H. Williams, D. J. Swift, B. Mailloux, and J. Rubin. 2001. Hydrogeological characterization of the south oyster bacterial transport site using geophysical data. Water Resources Research 37(10):2431–2456.

Hyndman, D. W. 1999. Geophysical and tracer characterization methods. Pp. 11.1–11.30 *In:* The Handbook of Groundwater Engineering. J. W. Delleur (ed.). Boca Raton, FL: CRC Press.

Interstate Technology and Regulatory Council (ITRC). 2002. DNAPL source reduction: facing the challenge. Regulatory overview. Washington, DC: Interstate Technology and Regulatory Council.

ITRC. 2003. Technology overview: an introduction to characterizing sites contaminated with DNAPLs. Washington, DC: Interstate Technology and Regulatory Council.

James, B. R., and R. A. Freeze. 1993. The worth of data in predicting aquitard continuity in hydrogeological design. Water Resources Research 29(7):2049–2065.

James, B. R., and S. M. Gorelick. 1994. When enough is enough: the worth of monitoring data in aquifer remediation design. Water Resources Research 30(12):3499–3513.

Jarvis, K. D., and R. J. Knight. 2002. Aquifer heterogeneity from SH-wave seismic impedance inversion. Geophysics 67(5):1548–1557.

Jayanti, S. 2003. Modeling tracers and contaminant flux in heterogeneous aquifers. PhD dissertation, University of Texas at Austin, August 2003.

Jayanti, S., and G. A. Pope. 2004. Modeling the benefits of partial mass reduction in DNAPL source zones. Proceedings of the Fourth International Conference on Remediation of Chlorinated and Recalcitrant Compounds, Monterey, CA, May 24–27, 2004.

Jin, M., R. E. Jackson, G. A. Pope, and S. Taffinder. 1997. Development of partitioning tracer tests for characterization of nonaqueous-phase liquid-contaminated aquifers. Pp. 919–930 In: The Proceedings of the SPE 72nd Annual Technical Conference and Exhibition, San Antonio, TX, October 5–8, 1997.

Jury, W. A., W. R. Gardner, and W. H. Gardner. 1991. Soil Physics, Fifth Edition. New York: John Wiley and Sons, Inc.

Kennedy, P. L., and A. D. Woodbury. 2002. Geostatistics and Bayesian updating for transmissivity estimation in a multiaquifer system in Manitoba, Canada. Ground Water 40(3):273–283.

Kitanidis, P. K., and K.-F. Shen. 1996. Geostatistical interpolation of chemical concentration. Adv. Water Resour. 19(6):369–378.

Kitanidis, P. K. 1997. Introduction to Geostatistics. Cambridge, UK: Cambridge University Press. Pp. 249.

Kitanidis, P. K. 1999. Geostatistics: interpolation and inverse problems. Pp. 12.1–12.20 In: The Handbook of Groundwater Engineering. J. W. Delleur (ed.). Boca Raton, FL: CRC Press.

Kram, M. L., A. A. Keller, J. Rossabi, and L. G. Everett. 2001. DNAPL characterization methods and approaches. I. Performance comparisons. Ground Water Monitoring and Remediation 21:109–123.

Kram, M. L., A. A. Keller, J. Rossabi, and L. G. Everett. 2002. DNAPL characterization methods and approaches. II. Cost comparisons. Ground Water Monitoring and Remediation 22:46-61.

Kristoff, F. T., T. W. Ewing, and D. E. Johnson. 1987. Testing to determine relationship between explosive contaminated sludge components and reactivity. Final Report to U.S. Army Toxic and Hazardous Materials Agency, by Hercules Aerospace Company (Radford Army Ammunition Plant) for Arthur D. Little, Inc.

Lee, K. N. 1993. Compass and gyroscope: integrating science and politics for the environment. Covelo, CA: Island Press.

Lee, K. N. 1999. Appraising adaptive management. Conservation Ecology 3(2):3.

Lee, S. -I., and P. K. Kitanidis. 1991. Optimal estimation and scheduling in aquifer remediation with incomplete information. Water Resources Research 27(9):2203–2217.

Lendvay, J. M., F. E. Löffler, M. Dollhopf, M. R. Aiello, G. Daniels, B. Z. Fathepure, M. Gebhard, R. Heine, R. Helton, J. Shi, R. Krajmalnik-Brown, C. L. Major, Jr., M. J. Barcelona, E. Petrovskis, R. Hickey, J. M. Tiedje, and P. Adriaens. 2003. Bioreactive barriers: a comparison of bioaugmentation and biostimulation for chlorinated solvent remediation. Environ. Sci. Technol. 37(7):1422–1431

Londergan, J. T., M. Jin, J. A. K. Silva, and G. A. Pope. 2000. Determination of spatial distribution and volume of chlorinated solvent contamination. In: The Proceedings of the Second International Conference on Remediation of Chlorinated and Recalcitrant Compounds, Monterey, CA, May 22–25, 2000.

Lovley, D. R. and R. T. Anderson. 2000. Influence of dissimilatory metal reduction on fate of organic and metal contaminants in the subsurface. Hydrogeol. J. 8(1):77–88.

Mackay, D. M., W. Y. Shiu, A. Maijanen, and S. Feenstra. 1991. Dissolution of non-aqueous phase liquids in groundwater. J. Contam. Hydrol. 8(1):23–42.

Mariner, P. E., M. Jin, J. E. Studer, and G. A. Pope. 1999. The first vadose zone partitioning interwell tracer test (PITT) for NAPL and water residual. Environ. Sci. Technol. 33(16):2825–2828.

Massmann, J., R. A. Freeze, L. Smith, T. Sperling, and B. James. 1991. Hydrogeological decision analysis: 2. applications to groundwater contamination. Groundwater 29(4):536–548.

McGuire, J. T., E. W. Smith, D. T. Long, D. W. Hyndman, S. K. Haack, M. J. Klug, and M. A. Velbel. 2000. Temporal variations in parameters reflecting terminal electron accepting processes in an aquifer contaminated with waste fuel and chlorinated solvents. Chem. Geol. 169:471–485.

McLaughlin, D., and L. R. Townley. 1996. A reassessment of the groundwater inverse problem. Water Resources Research 32(5):1131–1161.

Meinardus, H. W., V. Dwarakanath, J. Ewing, G. J. Hirasaki, R. E. Jackson, M. Jin, J. S. Ginn, J. T. Londergan, C. A. Miller, and G. A. Pope. 2002. Performance assessment of NAPL remediation in heterogeneous alluvium. J. Contaminant Hydrology 54:173–193.

Moore, D. S., and G. P. McCabe. 1999. Introduction to the Practice of Statistics. New York: Freeman.

Naval Facilities Engineering Service Center (NFESC). 2001. Cost and performance report of surfactant-enhanced DNAPL removal at Site 88, Marine Corps Base Camp Lejeune, North Carolina. Environmental Security Technology Certification Program (ESTCP) Final Report.

NFESC. 2002. Surfactant-enhanced aquifer remediation design manual. Naval Facilities Command Technical Report TR-2206-ENV.

National Research Council (NRC). 1999. Environmental Cleanup at Navy Facilities: Risk-Based Methods. Washington, DC: National Academy Press.

NRC. 2000. Natural Attenuation for Groundwater Remediation. Washington, DC: National Academy Press.

NRC. 2003. Environmental Cleanup at Navy Facilities: Adaptive Site Management. Washington, DC: National Academies Press.

Olea, R. 1999. Geostatistics for Engineers and Earth Scientists. Bingham, MA: Kluwer Academic Publishers.

Paillet, F. L. 1998. Flow modeling and permeability estimation using borehole flow logs in heterogeneous fractured formation. Water Resources Research 34:997–1010.

Pannone, M., and P. K. Kitanidis. 2001. Large-time spatial covariance of concentration of conservative solute and application to the Cape Cod tracer test. Transport in Porous Media 42:109–132.

Peck, R. B. 1969. Advantages and limitations of the observational method in applied soil mechanics. Geotechnique. 44(4):619–636.

Phelan, J., and J. L. Barnett. 2001. Phase partitioning of TNT and DNT in soils. Report SAND2001-0310. Albuquerque, NM: Sandia National Laboratories.

Phelan, J. M., S. W. Webb, J. V. Romero, J. L. Barnett, F. Griffin, and M. Eliassi. 2003. Measurement and modeling of energetic material mass transfer to soil pore water – Project CP-1227 Annual Technical Report. Sandia National Laboratories Report SAND2003-0153. Albuquerque, New Mexico: Sandia.

Poeter, E. P., and M. C. Hill. 1997. Inverse models: a necessary next step in groundwater modeling. Ground Water 35(2):250–260.

Rao, P. S. C., M. D. Annable, and H. Kim. 2000. NAPL source zone characterization and remediation technology performance assessment: recent developments and applications of tracer techniques. J. Contam. Hydrol. 45:63–78.

Robert, C. P., and G. Casella. 1999. Monte Carlo Statistical Methods. New York: Springer. 424 pp.

Rouhani, S., A. P. Georgakakos, P. K. Kitanidis, H. A. Loaiciga, R. A., Olea, and S. R. Yates. 1990a. Geostatistics in geohydrology. I. Basic concepts. ASCE J. of Hydraulic Engineering 116(5):612–632.

Rouhani, S., A. P. Georgakakos, P. K. Kitanidis, H. A. Loaiciga, R. A., Olea, and S. R. Yates. 1990b. Geostatistics in geohydrology. II. Applications. ASCE J. of Hydraulic Engineering 116(5):633–658.

Rubin, Y. 2003. Applied Stochastic Hydrogeology. Oxford, UK: Oxford University Press.

Rubin, Y., S. S. Hubbard, A. Wilson, and M. A. Cushey. 1999. Aquifer characterization. Pp. 10.1–10.68 *In:* The Handbook of Groundwater Engineering. J. W. Delleur (ed.). Boca Raton, FL: CRC Press.

Saito, H., and P. Goovaerts. 2000. Geostatistical interpolation of positively skewed and sensored data in a dioxin-contaminated site. Environ. Sci. Technol. 34:4228–4235.

Sciortino, A., T. C. Harmon, and W. W-G. Yeh. 2000. Inverse modeling for locating dense non-aqueous pools in groundwater under steady flow conditions. Water Resources Research 36(7):1723–1736.

Sciortino, A., T. C. Harmon, and W. W-G. Yeh. 2002. Experimental design and model parameter estimation for locating a dissolving DNAPL pool in groundwater. Water Resources Research 38(5):U290–U298.

Shapiro, A. M. 2002. Cautions and suggestions for geochemical sampling in fractured rock. Ground Water Monitoring and Remediation 22:151–164.

Sposito, G. (Ed.). 1998. Scale Dependence and Scale Invariance in Hydrology. New York: Cambridge University Press.

Terzaghi, K., R. B. and Peck. 1967. Soil Mechanics in Engineering Practice. New York: Wiley.

Thiboutot, S., G. Ampleman, S. Brochu, R. Martel, G. Sunahara, J. Hawari, S. Nicklin, A. Provatas, J. C. Pennington, T. F. Jenkins, and A. Hewitt. 2003. Protocol for Energetic Materials-Contaminated Sites Characterization. The Technical Cooperation Program, Wpn Group - Conventional Weapon Technology, Technical Panel 4, Energetic Materials and Propulsion Technology, Final Report Volume II.

Tiedeman, C. R., and P. A. Hsieh. 2001. Assessing an open-hole aquifer test in fractured crystalline rock. Ground Water 39:68–78.

Walters, C. 1986. Adaptive management of renewable resources. New York: Macmillan.

Walters, C. 1997. Challenges in adaptive management of riparian and coastal ecosystems. Conservation Ecology 1(2):1.

Walters, C. J., and C. S. Holling. 1990. Large-scale management experiments and learning by doing. Ecology 71: 2060–2068.

Williams, J. H., and F. L. Paillet. 2002. Using flowmeter pulse tests to define hydraulic connections in the subsurface: a fractured shale example. Journal of Hydrology 265:100–117.

Witt, M. E., G. M. Klecka, E. J. Lutz, T. A. Ei, N. R. Grasso, and F. H. Chapelle. 2002. Natural attenuation of chlorinated solvents at Area 6, Dover Air Force Base: groundwater biogeochemistry. J. Contam. Hydrol 57:61–80.

Yeh, W. W.-G. 1986. Review of parameter identification procedures in groundwater hydrology: the inverse problem. Water Resources Research 22(1):95–108.

# 4

# Objectives for Source Remediation

The remedial objectives of interest to the Army and other potentially respon-sible range from groundwater restoration and plume shrinkage and containment, to mass removal, risk reduction, and cost minimization. A realistic evaluation of the prospects for, or success of, a source remediation action requires the specifi-cation of these objectives with clarity and precision. The project manager and other stakeholders must know the full range of site remedial objectives, their relative priorities, and how they are defined operationally as specific metrics, in order to determine whether source remedial actions will contribute to meeting objectives for the site. The primary purpose of this chapter is to describe the many objectives possible at sites for which source remediation is a viable option, many of which have been institutionalized within regulatory, risk assessment, and economic frameworks for site cleanup.

Failure to explicitly state remedial objectives appears to be a significant barrier to the use of source remediation. That is, the vagueness with which objec-tives for remedial projects are often specified can preclude effective decision making with regard to source remediation. Too often, either data presented on the effects of source remediation are irrelevant to the stated objectives of the reme-dial project, or the objectives are stated so imprecisely that it is impossible to assess whether source remediation contributes to achieving them.

Evidence supporting the above was received by the committee over the course of its deliberations during numerous briefings on source remediation projects at Department of Defense (DoD) facilities and other sites, supported by extensive documentation on some of these remedial efforts. Other related docu-ments were also reviewed, including many case studies on source remediation efforts available at the U.S. Environmental Protection Agency's (EPA) Technology Innovation Office. In many of the cases reviewed, remedial project managers (RPMs) appeared unable to articulate a clear a priori rationale for undertaking source remediation at a site or to quantify the extent to which source remediation efforts were contributing to accomplishment of remedial objectives. To a significant extent, these interrelated problems appeared to reflect the absence of unambiguously stated remedial objectives for the sites or of clear operational definitions of those objectives. For example, during a brief report on an attempt to use steam recovery to remove a source area from a relatively homogenous unconsolidated aquifer, the project manager expressed considerable frustration with the effort, not only from a technical point of view but, more important, from the point of view that it was consuming significant resources while not contributing to any reduction in human health risk. There was no complete exposure pathway to the contaminated groundwater at the site, nor any expectation of a complete pathway in the near future. This is illustrative of a situation in which an explicit operational statement of site objectives (e.g., a reduction of human health risk as estimated by the procedures specified in EPA's Risk Assessment Guidance for Superfund, RAGS), if made prior to the attempt at source remediation at this site, might have led to a decision not to attempt source remediation.

This widespread problem of vaguely formulated remedial objectives, tenuously linked to performance metrics, is neither specific to the issue of source remediation nor reflective of any unusual failure on the part of the specific project managers with whom the committee interacted. Rather, this ambiguity is embodied in long-standing national policy statements (i.e., the National Contingency Plan) and analytical procedures (as embodied in RAGS). It is compounded by the fact that multiple stakeholders at a site not only may have very different objectives, but may also use very similar language to describe those very different objectives. Moreover, a particular performance metric may potentially correspond to a variety of different objectives and accordingly be viewed quite differently by different stakeholders. Finally, both the DNAPL problem and the effects of source remediation efforts raise temporal issues that are very poorly addressed by conventional analytical frameworks for assessing risks to human health and the environment.

This chapter describes a variety of substantive remedial objectives.[1] It shows how a stated objective can be defined operationally by several different metrics

---

[1]Substantive objectives are concerned with the results of a decision process, in terms of a physical change at the site. In contrast, procedural objectives focus not on the outcome of a remedial effort, but on the process by which a decision is reached (e.g., transparency of the decision process, opportunities for public participation). Procedural objectives are often as important to stakeholders as substantive objectives, but they are not considered here because they are outside the scope of this report.

and how a particular metric may represent the operational definition of several different objectives. This complex relationship between ultimate cleanup objectives and the ways in which they are measured can be the source of considerable uncertainty in the evaluation of remedial alternatives. It can also mask serious differences in stakeholder priorities, which only become apparent when an apparently "successful" remediation fails to satisfy key stakeholders. This is not meant to imply that better specification of objectives and their relationship to metrics of remediation will ensure stakeholder satisfaction. An unambiguous delineation of objectives and metrics will, however, allow the decision on source remediation to be more clearly evaluated. The relevance of source remediation to different stakeholders' objectives can be identified in advance, and progress can be measured.

## FORMULATING OBJECTIVES

One source of ambiguity during site remediation is that various stakeholders may use similar (or identical) language to describe radically different objectives for remediation of a site. Thus, this section provides an approach to clearly describing stakeholder objectives. There are three critical, interdependent elements in the unambiguous specification of a remedial objective for a site: (1) identifying the objective, (2) determining the appropriate metric(s) to measure achievement of the objective, and (3) determining the status of the objective. Although difficulties in specifying each of these elements among the projects reviewed have been noted, the element of *status* is addressed first because it is often overlooked, followed by discussions of common objectives and the selection of appropriate metrics for an objective.

### Status of Remedial Objectives

Status refers to the fact that any identified remedial objective can be seen either as important in itself or as a means to an end. In the former case, the objective is termed *absolute* or *primary,* while in the latter, the objective is *functional.* (For an exposition of the contrast between absolute and functional objectives, see Udo de Haes et al., 1996, and Barnthouse et al., 1997.)

Consider, for example, the objective of reducing contaminant concentrations in groundwater to a specified level at a particular point in time and space. This may be mandated under a particular regulatory framework as a necessary feature of a successful remediation, in which case it represents an absolute objective. Failure to achieve these concentrations represents failure of the remedy. The identical criterion, however, could be selected as a means of ensuring that risks to human health have been reduced to an acceptable level. In this case, the objective is functional, because there may be other objectives that achieve a comparable degree of health protection, such as precluding use of contaminated groundwater.

Confusion about whether an objective is absolute or functional is not uncommon at a wide range of sites, particularly with regard to maximum contaminant levels (MCLs) and how they are viewed by various stakeholders. MCLs are frequently cited as an absolute regulatory objective. Indeed, MCLs (or non-zero maximum contaminant level goals, MCLGs) may be determined to be either an "Applicable" or a "Relevant and Appropriate" requirement under the Comprehensive Environmental Response, Compensation, and Liability Act (CERCLA). However, they can also serve as a functional objective that supports an absolute statutory objective. For example, a state may have determined that all of its groundwater should be protected as a potential source of drinking water. Consistent with state law, a demonstration that concentrations were below MCLs would indicate that the groundwater resource had been adequately protected. The state *might*, however, be open to other indicators that its requirement of resource protection had been met.[2]

There is further complexity with the designation of status. For example, MCLs may serve as a functional objective supporting a *higher-level* functional objective of achieving an acceptable human health risk, which in turn serves the absolute objective of protecting human health. In this case, there are clearly alternative functional objectives, both to meet the risk assessment functional objective and to meet the absolute objective of protecting health. In the first case, it may be that the actual conditions of use would indicate an acceptable level of risk even if the MCL were exceeded. In the second case, there are any number of ways to interrupt the relevant exposure pathway (such as institutional controls or the provision of a public water supply).

The distinction between absolute and functional objectives is important, because trade-offs among different *absolute* objectives cannot be accomplished at a technical level. Rather, they represent social value judgments that must be made among stakeholders.[3] In contrast, trade-offs between *functional* objectives can be made at a technical level, subject to the requirement that equivalence in meeting the corresponding absolute objective can be demonstrated. Thus, functional objectives are fungible.[4] In the above example, the project manager can achieve health protection by precluding use of the contaminated water or by lowering concentrations in groundwater. Similarly, but within the realm of physically specified objectives, if the absolute objective were defined as meeting a concentration at a specified point of compliance (e.g., a fenceline), the project manager could trade off between preventing contaminant migration from the

---

[2]For example, higher concentrations of contaminants that would then be reduced during routine disinfection of raw water sources might be deemed acceptable.

[3]A considerable body of literature on such judgments among qualitatively different environmental objectives has developed in the context of life cycle assessment (e.g., Udo de Haes et al., 1996).

[4]"Fungible" refers to goods or commodities that are freely exchangeable for or replaceable by another of like nature or kind in the satisfaction of an obligation.

source or capturing migrating contaminants before they reached the point of compliance.

It is important to bear in mind that a given functional objective may serve more than one absolute objective, and also that a particular objective may be functional for one stakeholder and absolute for another. For example, limiting the migration of contaminants in groundwater beyond the boundaries of a site may serve the absolute objectives of meeting a state statutory requirement or preventing "chemical trespass" (Gregory, 1993). On the other hand, it may serve the higher-level functional objectives of limiting human health risk to an acceptable level (by reducing exposure potential) or avoiding the effort and uncertainty of applying for an alternative concentration limit under the Resource Conservation and Recovery Act (RCRA). Different stakeholders may all agree that limiting the migration of contaminants in groundwater beyond the boundaries of a site is an important objective, but they would likely have very different responses to any proposals for substituting an alternative objective.

## Metrics for Remedial Objectives

Ultimately, accomplishment of (or progress toward) a remedial objective can only be evaluated if there is a measurable value or metric associated with that objective.[5] Accordingly, any objective that cannot be stated directly in terms of a metric must be assigned one or more subsidiary functional objectives that can be formulated in terms of a metric. This is illustrated in the following simplified example.

The absolute objective is the protection of human health. This is not directly measurable in most cases, where illness has not been recorded in a site-associated population. Accordingly, a common functional objective is the specification of a Hazard Index $< 1.0$ and a cancer risk estimate $< 10^{-4}$ in a quantitative risk assessment. In practical terms, this requires a lower-level functional objective— that "exposure point" concentrations in groundwater be less than a certain value. Two alternatives can be employed to achieve this: (1) change contaminant concentrations at the defined exposure point or (2) change the exposure point, for example, by providing alternate water sources and prohibiting water use near the contaminant source. Each of these functional objectives has an associated metric—a revised concentration in the water at the relevant exposure point.

It is important to bear in mind that not all metrics are as unambiguously specified as is the concentration of a particular chemical at a particular point in

---

[5]Our use of "metric" differs from that of EPA (2003b), which describes classes of metrics that were not in fact measured but were inferred from measured quantities. In our use, such "Type II and Type III" metrics would be considered functional objectives.

time or space, as will become clear in the following section, which discusses common objectives during source remediation and their associated metrics.

## COMMONLY USED OBJECTIVES

Whether defined by the stakeholder as absolute or functional, there are a set of objectives that have been widely used in site remediation. In many cases examined by the committee, the identification of site objectives has been less than clear, such that metrics appropriate to one objective have been employed for a different objective to which they are not applicable. The following sections distinguish between alternative possible objectives (whether absolute or functional) in four areas. There are obviously other kinds of objectives dealt with at sites, including programmatic and societal concerns. Moreover, the list of objectives within each area is merely illustrative and far from exhaustive. The four areas are

1. Objectives related to a physical change at the site
2. Objectives related to risks to human health and the environment
3. Objectives related to life-cycle and other costs
4. Objectives related to the time required to reach particular milestones

### Physical Objectives

There are a number of physical objectives that may drive the design and performance evaluation of source zone treatment methods. These include mass removal, concentration reduction, mass flux reduction, reduction of source migration potential, plume size reduction, and changes in toxicity or mobility of residuals. The specification of metrics for performance evaluation with respect to these objectives is typically easy since the objectives are related to physical, measurable properties. In some cases, however, considerable inference must be interposed between the available metrics and the physical objective. Each of these objectives and their associated metrics are described briefly below.

*Mass Removal*

Removal of contaminant mass from a source zone is a common objective at hazardous waste sites and may be either absolute or functional (depending on the stakeholders, the governing regulations, and other factors). Many of the source zone treatment technologies, particularly those that rely upon fluid flushing of the source area (including surfactant/cosolvent flushing, steam flushing, air sparging, and water flushing), are designed to remove contaminant mass from the swept zone. For these technologies, the injected fluids serve as a carrier medium to transport the contaminant mass to the surface. The mass removed is recovered at the surface through the collection and treatment of these flushing fluids. Other source zone treatment methods, such as chemical oxidation/reduction, soil heating, or enhanced bioremediation, are designed to destroy or convert the form or phase of the contaminant mass in situ.

Common metrics for the mass removal objective are the mass of contaminant recovered or destroyed and the percentage of the total contaminant mass present in the subsurface that was recovered or destroyed. The first metric is relatively straightforward for flushing technologies, for which mass removal is quantified by measuring the contaminant mass recovered as a component of the extracted flushing fluid. For destruction or conversion methods, however, mass removal is less easily quantified, and one must rely on indirect metrics. Under some conditions, the measurement of the concentrations of reaction by-products can facilitate inference of mass removed. Accurate mass balances, however, are typically difficult or impossible to achieve in such situations, since mass conversion or destruction methods do not rely upon injected fluid recovery. Finally, the ability to measure the percentage of the total contaminant mass in the subsurface that is recovered or destroyed depends on estimates of the total mass present, which, depending on site characterization data, may be quite poor.

*Concentration Reduction*

Even more common than the mass removal objective is the objective of reducing contaminant concentration within an affected medium (i.e., soil, sediment, groundwater, etc.) to a desired lower value. The obvious associated metric is contaminant concentration. Like mass removal, concentration reduction can serve as an absolute objective (e.g., meeting MCLs) or as a functional objective for reaching some other absolute objective (e.g., reducing exposure and consequently health risk). The use of reductions in contaminant concentrations as remediation objectives is common because regulations often specify concentration compliance levels.

Concentration is defined as the mass of the target compound per volume (or mass) of the affected medium (pore water, core sample, solid sample). Thus, concentration compliance or target levels can be defined in a number of ways, depending on the sampled medium. For example, groundwater concentration is typically defined as the contaminant mass per volume of pore fluid, while solid-phase concentration is often defined as a contaminant mass per mass of the sampled solid phase. Although concentration is often viewed as a precise metric, it should be noted that it really represents an average value over the volume sampled. If the distribution of contaminant mass within a source zone is highly irregular, local concentrations within a source zone can vary substantially from one another and from the average concentration that would take the entire source volume into consideration. In this report, the term "local" implies that concentrations are sampled over small spatial intervals by extracting small volumes of pore fluid, such that these concentrations are representative of a known physical location within the contaminated zone at the time of sampling.

Remediation technologies seek to reduce concentration levels through mass removal, conversion, or destruction. Thus, it is important to note the connections between mass removal and concentration reduction as remediation objectives. Ideally, if the total mass of the contaminant were removed from the source zone, the concentration in the aqueous phase would be reduced to below detection limits. In practice, however, total mass removal is next to impossible, and may instead be confined to more "treatable" areas. Treatability will depend upon the selected remediation method, but it also depends on permeability (flushing potential), degree of aquifer material affinity for the contaminant, volume and distribution of NAPL present, and composition of the contaminant. For example, aquifer material with high organic content may tend to more strongly sorb contaminants, or the presence of a NAPL pool may limit the degree of contact between the contaminant and the injected flushing fluid. Due to subsurface heterogeneity, treatability typically varies spatially, resulting in a significant spatial variability in the distribution of contaminant mass within the source zone subsequent to treatment. Although some technologies (e.g., steam flushing) may be more robust in their mass removal behavior, it is expected that contaminant concentrations will be locally variable following treatment. Thus, it is very difficult to make generalizations about how removing a certain percentage of mass relates to achieving a certain percent reduction in contaminant concentrations. A number of field studies (e.g., Londergan et al., 2001; Abriola et al., 2003, 2005; EPA, 2003a) have documented that local concentration reductions are achievable with source zone remediation. Such concentration reductions have ranged over one half to two orders of magnitude.

While it is possible that local concentration levels in the treatable zones may be substantially reduced by mass removal, local concentrations in less accessible DNAPL zones will likely remain high (at or near aqueous solubility levels) following treatment. Less accessible zones may also retain significant organic mass on the solid (through sorption) or within stagnant pore fluids (through diffusion). Furthermore, if local groundwater flow rates are low within the source zone under natural conditions, diffusion from less accessible zones may result in increasing contaminant concentration levels in the more accessible areas over time (often termed concentration rebound) once remediation operations have ceased. Thus, the potential effect of mass removal on local contaminant concentrations will be a complex function of source zone properties, including DNAPL distribution, natural groundwater gradients, and the spatial distribution of sorbed mass. Based upon these considerations, the use of local concentration within a source zone as a metric of remedial success is problematic, particularly in the absence of high-resolution sampling. More integrative metrics are discussed below.

## Mass Flux Reduction

While measurements of local concentrations permit the development of a picture of the spatial distribution of contamination within the pore water in the source zone, mass flux quantifies the potential influence of these concentrations on a downstream receptor. Mass flux is typically defined at some cross-sectional (planar) surface selected downstream of the source zone and roughly perpendicular to the direction of flow. The mass flux at a particular location in this transect is defined as the mass of contaminant moving across the surface per unit area per unit time. The total mass flux (or, more accurately, mass flow rate) is then obtained by integrating the mass flux over the plane. The average mass flux can then be obtained by dividing the total mass flow rate by the area of the cross-sectional plane of interest (it should be noted that these approaches may not translate well to fractured flow systems). A related metric, the flux-averaged concentration, is determined by dividing the average mass flux by the average groundwater velocity in the cross section.

Mass flux reduction can be a functional objective that may, for example, support the higher-order objectives of reducing exposure to downstream receptors or preventing the growth of a plume downstream of the source zone. Although conceptually attractive as a remediation objective, mass flux reduction is difficult to quantify in practice, as suggested in Chapter 3. Most existing methods typically involve measurement of contaminant concentration at distinct points in the selected transect. Transformation of these measurements to flux estimates requires application of assumptions about the groundwater velocities at the measurement points. Furthermore, computation of average fluxes from such measurements is subject to a high level of uncertainty. More integrative methods for estimating average mass flux are currently under development. These involve alteration of the flow field through downstream pumping or installation of in situ flow-through devices at selected downstream locations.

The relationship between concentration reduction or mass removal and mass flux reduction is as yet poorly understood. Considerable research is currently being directed at developing information and methodologies for the prediction and quantification of mass flux reduction from data on source zone mass removal and on aquifer and contaminant characteristics. Box 4-1 illustrates some of these efforts. A more extensive numerical investigation of this sort suggests that a two-orders-of-magnitude mass flux reduction may be achievable following partial source zone mass removal in Type I media (Lemke et al., 2004).

## Reduction of Source Migration Potential

Reducing the potential for the source to migrate into clean subsurface areas is a commonly stated objective of the projects reviewed by the committee. Many source zones are characterized by the presence of DNAPL pools, which are

**BOX 4-1**
**Relationship between Partial Source Removal and**
**Mass Flux Reduction**

Recent analytical and numerical modeling efforts provide evidence that partial source zone removal may result in significant (several orders of magnitude) reduction in posttreatment contaminant mass flux (Rao et al., 2002; Lemke and Abriola, 2003; Rao and Jawitz, 2003). Consider two simulated DNAPL cross-sectional saturation profiles shown in Figures 4-1(A) and 4-2(A). Here the modeled formation is based upon an unconfined sandy glacial outwash aquifer located in Oscoda, Michigan, at the site of a former dry cleaning business. Aquifer characterization efforts were conducted in support of a Surfactant Enhanced Aquifer Remediation (SEAR) pilot-scale test (see Chapter 5), designed to solubilize and recover residual tetrachloroethylene (PCE) from a suspected DNAPL source zone at the site (Bachman Road). As part of the SEAR design effort, alternative spatial variability models of the unconfined aquifer were developed from formation core data and were used to generate entrapped PCE distributions using the immiscible fluid flow model MVALOR (Abriola et al., 1992; Rathfelder and Abriola, 1998; Abriola et al., 2002). Further details pertaining to the spatial variability models and simulation conditions can be found in Lemke and Abriola (2003) and Lemke et al. (2004). The simulated spill involved the release of 96 liters of PCE over four grid cells at the top of the model domain at a constant flux of 0.24 liter·day$^{-1}$ for 400 days, with an additional 330 days for subsequent organic infiltration and redistribution.

Examination of the two saturation distributions in Figures 4-1 and 4-2 reveals that while both contain the same total volume of PCE, this mass is distributed more uniformly in Figure 4-1 than in Figure 4-2. Much of the mass in Figure 4-2 is contained within a thin pool, where saturations reach up to 91 percent of the pore space. Alternatively, in Figure 4-1 maximum saturations of PCE do not exceed 31 percent.

The potential influence of source zone mass removal on DNAPL distributions is illustrated in Figures 4-1(B) and 4-2(B), which present saturation mass depletion profiles. Here PCE mass removal was simulated using a lab-validated version of MISER (Taylor et al., 2001; Rathfelder et al., 2001). The initial saturation profiles shown in Figures 4-1(A) and 4-2(A) were flushed with approximately 1.5 pore volumes of surfactant solution and 10 pore volumes of water. Further details pertaining to the simulated surfactant flush can be found in Lemke (2003) and Lemke et al. (2004). Inspection of Figures 4-1(B) and 4-2(B) reveals that different degrees of mass removal are predicted for the different initial mass distributions, even though each was subjected to the same flushing conditions. In Figure 4-1(B), flushing has resulted in 97.8 percent PCE removal, with the remaining PCE distributed in thin of pools of short lateral extent. The mass dissolution for the PCE distribution in Figure 4-2(A) is substantially less, with only 43.2 percent of the mass removed. Here much of the original pooled PCE persists, with maximum concentrations still ranging up to 86 percent (compared to 13 percent in Figure 4-1(B).

Figures 4-1(c) and 4-2(c) illustrate the source zone PCE concentrations evolving from the saturation distributions shown in Figures 4-1(b) and 4-2(b). Notice that despite the substantial mass removal, concentrations are still quite high in the

*continued*

135

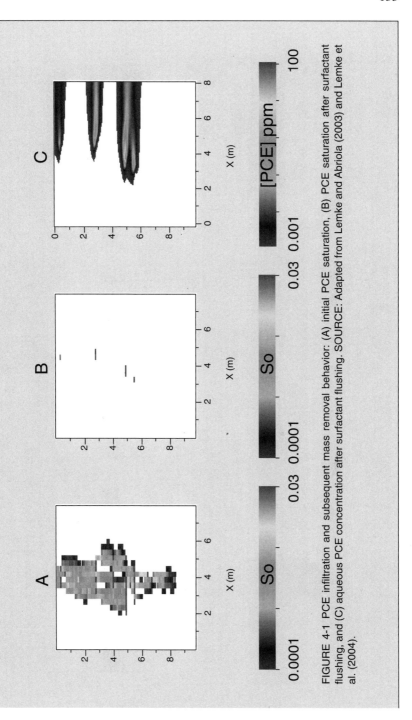

FIGURE 4-1 PCE infiltration and subsequent mass removal behavior: (A) initial PCE saturation, (B) PCE saturation after surfactant flushing, and (C) aqueous PCE concentration after surfactant flushing. SOURCE: Adapted from Lemke and Abriola (2003) and Lemke et al. (2004).

136

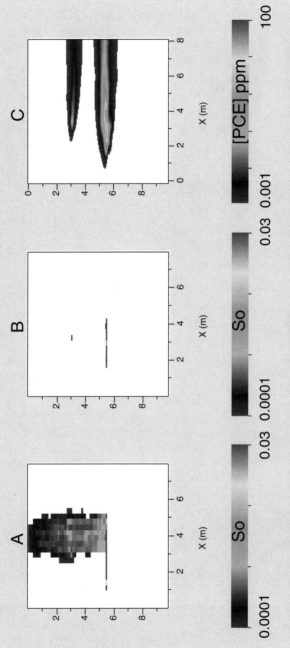

FIGURE 4-2 PCE infiltration and subsequent mass removal behavior in a system with significant pooling behavior: (A) initial PCE saturation, (B) PCE saturation after surfactant flushing, and (C) aqueous PCE concentration after surfactant flushing. SOURCE: Adapted from Lemke and Abriola (2003) and Lemke et al. (2004).

## BOX 4-1 Continued

immediate vicinity of the pooled PCE, reaching 100 mg/L (the aqueous solubility) just above the pools. Thus, if groundwater were sampled at these locations, the results would suggest that little benefit had been gained from mass removal.

If one considers the impact of the source zone mass reduction on mass flux, however, a different picture emerges. Calculation of mass flux at a downstream plane that intersects the entire contaminated plume reveals that mass removal has resulted in a substantial reduction in PCE mass flow rate (approximately 1.5 orders of magnitude in both spill scenarios). In fact, although more mass remains in the second case, the mass flow rate is actually *lower*, since the remaining contaminated zone represents a smaller fraction of the cross-sectional plane.

A flux-averaged downstream concentration can also be computed for these scenarios. The flux-averaged concentration represents the total mass crossing the plane divided by the total volume of groundwater crossing the plane in the same time period. Notice that initial (pre-dissolution) mass-averaged concentrations are quite high (close to the aqueous solubility of PCE), consistent with the local concentrations within the source zone. However, post-mass removal, flux-averaged concentrations have been reduced by more than 1.5 orders of magnitude. Conceptually, one might view these flux-averaged concentrations as more representative of the risk to downstream receptors, since they incorporate the dilution effect, similar to that which would be measured by a fully screened well directly downstream of the source.

Another way to present the results of these types of analyses is to plot the relationship between mass removal and mass flux reduction. Figure 4-3 shows the

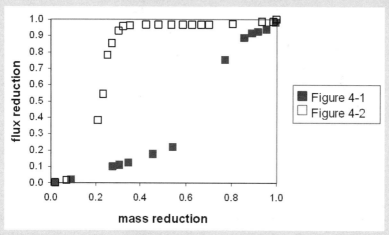

FIGURE 4-3 Potential remediation benefit for the two NAPL distributions shown in Figures 4-1(A) and 4-2(A). SOURCE: Adapted from Lemke and Abriola (2003) and Lemke et al. (2004).

*continued*

## BOX 4-1 Continued

potential benefit of mass removal for the two alternative PCE distributions shown in Figures 4-1(A) and 4-2(A). Here the results for Figure 4-1(A) are indicated by the solid squares and those for Figure 4-2(A) by the open squares. Note the very different shapes of these two curves. In the former, mass flux reduction lags mass removal slightly, while in the latter, initial mass removal results in very substantial benefits in terms of flux reduction. The mass flux reduction "plateau" exhibited by the Figure 4-2 scenario is associated with the persistence of a pool that contains much of the PCE mass.

Although limited simulations are available to extrapolate to other aquifer types, release scenarios, and/or remediation technologies, the illustrations given above suggest that source mass removal may offer substantial benefits, if mass flow rate is the metric of choice. In more heterogeneous formations, one might anticipate that even more of the mass would be distributed in pooled areas, leading to reduced mass flow rates following treatment. For example, Figure 4-4 illustrates predicted PCE distributions in three formations that have identical release rates and average permeability values. The formations differ only in the magnitude of the variance [in $\ln(k)$] of the permeability field. Note that the formation with the highest variance has the most extensive pooling. The results shown above are also expected to be representative of a variety of flushing remedial technologies that remove (or destroy) mass preferentially from high-permeability zones (including pump-and-treat, chemical oxidation, and cosolvent flushing).

As pointed out above, local concentrations in the source zone, subsequent to mass removal, may remain high. Thus, if MCLs in the source zone are used as a metric, little benefit may be realized from treatment. If flux-averaged concentration, however, is employed as a metric, substantial benefits may be achieved from even partial mass removal. The reduction in mass flux can reduce concentrations at downstream receptor wells and may reduce average downstream concentrations to levels where microbial transformation of the chlorinated solvents becomes feasible (Nielsen and Keasling, 1999; Yang and McCarty, 2000; Adamson et al., 2003; Sung et al., 2003).

typically formed when the downward migration of the organic liquid has been impeded by the presence of a capillary barrier or low-permeability layer. DNAPL will tend to spread along such interfaces of contrast until a dynamic equilibrium is reached locally between gravitational, capillary, and pressure forces. Although the pools may be stationary under this dynamic equilibrium, the mass of DNAPL present in these pools cannot be considered immobile. Future disruption to the dynamic balance of forces can induce further migration of the DNAPL to previously uncontaminated areas, enlarging the extent of the source zone. For example, such a disruption could occur as a result of a physical breach in the barrier during

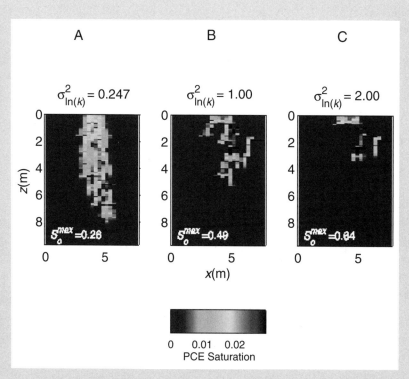

FIGURE 4-4 Potential effect of formation $\ln(k)$ variance on PCE saturation distribution. $k$ = permeability. SOURCE: Reprinted, with permission, from Phelan et al. (2004). © 2004 Elsevier Science.

field characterization efforts or due to aging of the contaminant that leads to alterations in the interfacial properties controlling the capillary forces.

As discussed in Chapter 2, the capillary force that acts to retain DNAPL in a pore is controlled by the pore size, the wettability characteristics of the solid, and the interfacial tension between the water and DNAPL. A DNAPL is truly immobile when this capillary force exceeds the pressure and gravity forces that can act to induce migration. At the larger scale, there is a quantifiable relationship between the average capillary pressure and the DNAPL saturation; in a particular material, capillary forces tend to be smaller at higher saturations. Thus, in a given formation, the mobility of the DNAPL is linked to its saturation, as well as to the porous medium texture. One metric for the mobility of a particular DNAPL, then, is the mass of that DNAPL exceeding a particular local saturation. Such a metric, however, is nearly impossible to quantify in practice, due to the dependence of this saturation threshold on the texture of the medium and the high variability of local source zone saturations. Other metrics that might indicate achievement of this objective include alterations in the DNAPL viscosity or changes in certain soil properties.

Primarily through pool mass removal, source zone remediation activities can result in a reduction in the local saturations of the DNAPL left in place. As noted above, this may result in a reduction of mobility and a reduction in the risk of further migration. It is also possible that chemical oxidation techniques may reduce pool mobility through the formation of reaction "crusts" at the edges of the pool (Li and Schwartz, 2004). This effect has not yet been thoroughly investigated. It should also be cautioned that many source zone remediation technologies are designed to *enhance* DNAPL mobility during the treatment process (often through a reduction in interfacial tension).

*Plume Size Reduction*

Another objective of source zone treatment can be to reduce concentration levels within the downstream contaminant plume and/or to reduce the physical extent of the plume. Reductions in contaminant mass within the treated zone and in mass fluxes from this zone (a reduction in the strength of the source) will theoretically result in a reduction in downstream plume size. Every aquifer has a natural capacity to dilute or attenuate the contaminants. Dilution processes include diffusion and dispersion, while attenuation processes include sorption and chemical/ microbial reactions. Such processes act to limit the rate of migration and growth of a plume. For example, for contaminants that are subject to constant reaction rates (a rather crude but illustrative simplifying assumption) and for a continuous source of a fixed size, there is a maximum size to which the plume will grow. If the source strength is reduced, this maximum size will decrease. Thus, it is possible that a reduction in source strength would eliminate problematic plume discharges to surface waterbodies or would permit natural microbial processes to shrink the plume to a size that fails to reach receptor points of concern.

It is important to recognize the potentially significant time lag that will occur between the reduction of source strength and any recorded changes in concentrations in the plume (which are the primary metrics). Initially, the reduction in flux from the source zone will produce a lower concentration in the dissolved phase plume immediately downgradient of the source, but this effect will migrate only at the rate of dissolved phase contaminant migration. Thus, concentrations from a well in a dissolved phase plume located several hundred meters from the source zone may not be affected for many months or even years. In general, few field data exist to document the benefits of source zone treatment on plume size.

### Changes in Toxicity and/or Mobility of Residuals

In many contamination scenarios, the entrapped NAPL in the source zone exists as a mixture of many compounds. Common examples of NAPL mixtures include coal tars and combined fuels/solvents from spills from degreasing operations. Usually, there are certain compounds within the mixture that are of greater concern, due to their higher toxicity and/or mobility in the subsurface. Thus, another objective of source zone remediation can be to change the composition of the NAPL in situ, resulting in a reduction in overall contaminant toxicity or mobility. Certain technologies, including air sparging, soil heating, water flushing, and enhanced bioremediation, are designed to selectively extract or destroy NAPL components of concern. These technologies take advantage of contaminant component properties, such as solubility, volatility, and biodegradability, to alter the characteristics of the NAPL. Favorable changes can sometimes be achieved without large reductions in total contaminant mass. Furthermore, reductions in the concentrations of target constituents within the NAPL (which is the primary metric) may also reduce both the toxicity and mobility of the downstream contaminant plume.

### Elimination of Barriers to Subsequent Remedial Action

A final physical objective of source zone treatment can be to create a subsurface environment that is conducive to the application of other remediation technologies. For example, in many situations, the high concentration levels or the total mass of a contaminant within a source zone may preclude application of enhanced bioremediation. However, if enough mass is extracted from the zone, the accompanying reduction in concentration levels and mass fluxes may facilitate successful application of bioremediation technologies. Furthermore, some source zone treatment technologies (e.g., surfactant flushing) may leave chemicals in place that alter the biogeochemical environment, making it more conducive to microbial transformation processes. Similarly, concentration reductions may make installation of a reactive barrier a more feasible treatment option. The required thickness of a barrier is a direct function of downstream concentration levels, and reductions in concentrations within the plume will also reduce the risk associated with barrier failure. Under some conditions, where substantial mass removal has been achieved in source zone treatment, monitored natural attenuation may even be a feasible follow-on treatment choice. The metrics associated with this objective are variable, but they commonly include reductions in contaminant concentrations, mass, and mass fluxes.

## Objectives Related to Human Health and Environmental Risk

Risks to human health and/or the environment cannot be directly measured, at least not in any context relevant to the selection and evaluation of remedial technologies for source zones and contaminated groundwater. Accordingly, these objectives inherently involve subsidiary functional objectives and associated metrics, many of which were described in the previous section on physical objectives.

Risk to human health and the environment from contaminants in the subsurface is a function of both the level of exposure sustained and the toxicity of the chemical(s) to which the individual is exposed. Thus, risk reduction can be achieved by reducing or eliminating exposure or by reducing the toxicity of the chemicals present.

### Reducing the Level of Exposure

Reducing an individual's exposure to contaminants is a common functional objective for site cleanup. The level or degree of exposure sustained by a human or ecological receptor to a chemical in the environment is dependent on several factors:

- The spatial extent of contamination (the area affected by the contaminant)
- The concentration of the contaminant present at the point or points of contact
- The frequency and duration a receptor is in contact with the contaminant (e.g., daily, monthly, occasionally)
- Behavioral characteristics of the receptor (e.g., the ingestion of soil by children, other feeding habits, hand washing frequency, degree of skin covering, etc.)
- Fate and transport of chemicals from one environmental medium to another (e.g., migration of vapors from subsurface soil or groundwater into buildings), thereby creating exposure pathways from the contaminant to the receptor

Thus, there are myriad ways to reduce exposure to subsurface contaminants. The ways most commonly encountered during site remediation are (1) to reduce the amount of chemical present at a site (e.g., via any of the previously mentioned physical objectives like mass removal or concentration reduction), (2) to interrupt the exposure pathway (e.g., by constructing containment technologies, or by reducing or eliminating access to the site), or (3) to remove/alter the receptor (e.g., relocation of populations). Thus, the metric of success of the overall objective of exposure reduction may be a physical, measurable property, such as reduction of the concentration of a contaminant at the point of contact with the receptor, or it may be very qualitative in nature, such as an evaluation of the long-term success of institutional controls imposed on a site.

Knowing which of the physical functional objectives is most appropriate for achieving exposure reduction at a site is not a trivial undertaking, as illustrated in Box 4-2. The complex interconnections between mass removal, concentration

## BOX 4-2
## Evaluation of Physical Objectives for Achieving Exposure Reduction: The Role of Environmental Setting

This scenario is intended to highlight the importance of site characterization and site-specific conceptual modeling in the selection of appropriate physical objectives. Consider an aquifer contaminated by a mixture of spent chlorinated solvents (released as a DNAPL). The contaminant has penetrated deep in the saturated zone of an unconfined aquifer, used downstream for potable water. Within the source zone, the contaminant is distributed as a separate phase liquid in pools and ganglia. The aquifer formation is composed of alternating sandy and silty layers of contrasting hydraulic conductivity. Core and drivepoint aqueous phase samples within the contaminated zone reveal pockets of extremely high concentrations of chlorinated solvents (100–1,000 ppm). Suppose that the absolute objective in this scenario is to reduce risk to human health, and that the most important exposure pathway is through water consumption from the supply well.

Two physical functional objectives for remediation are being considered at this site: DNAPL mass removal from the source zone and aqueous concentration reduction within the source zone. Selection of the functional objective is a complex task and, in this scenario, is dependent on the chemical and hydrogeologic setting. Indeed, one, both, or neither of these objectives may be linked to the desired absolute objective of risk reduction.

For example, if the spilled chlorinated solvent had been previously used in dry cleaning, it is likely that it would contain additives that would lower its interfacial tension. Under reduced interfacial tension conditions, the spilled DNAPL would likely penetrate and become entrapped within the finer silty layers in the formation. Under these conditions, a small percentage of mass removal (removal from the higher-permeability zones) may achieve substantial reductions in mass fluxes to the receptor well and may reduce health risks (by reducing concentrations at the well). This mass removal, however, is unlikely to lower maximum local source zone aqueous or solid phase concentrations substantially. Alternatively, since the DNAPL is present primarily in low conductivity zones, reduction of aqueous concentrations within the source zone (particularly within these finer-textured materials) will be extremely difficult and would be a poor indicator of downstream concentrations.

In contrast, if the spilled solvent is reagent-grade (few impurities), it will likely remain pooled within the higher-permeability zones of the formation. Under these conditions, substantial (high percentage) mass removal will likely be necessary to achieve downstream concentration and risk reductions, since most flow through the source zone will be exposed to the DNAPL. In this scenario, however, concentration reductions within the source zone would be a better indicator of risk reduction than would DNAPL mass removal.

Another scenario can be envisioned in which the contaminant source is a solvent that has been used in degreasing operations. In this situation, co-contamination of the solvent with oils and aromatic hydrocarbons is likely. Such co-contaminants can serve as substrates for microbial transformation of the solvents. Microbial transformation may be exerting the primary control on downstream concentrations. In this situation, mass removal or concentration reductions within the source zone may have no discernable influence on receptor well concentrations.

reduction, and mass flux reduction have been previously described. Box 4-2 demonstrates that the appropriateness of these various physical functional objectives is influenced by the hydrogeologic setting and the properties of the contaminant.

### Reducing Chemical Toxicity

A change in toxicity was previously discussed as a physical objective of cleanup, relevant to those cases where entrapped DNAPL in the source zone may exist as a mixture of many compounds. Reducing the concentration of those DNAPL constituents of higher toxicity should result in reduced overall toxicity of the downstream contaminant plume, which directly supports the higher-order functional objective of reducing risk. Because the degree and type of toxicity of a contaminant is an inherent property of the interaction of the contaminant with a particular biological system, to affect a change in toxicity, the DNAPL component must be physically altered. While for inorganic contaminants this can be achieved in some cases by changing a contaminant's chemical form to be less toxic, or by making the contaminant less bioavailable, for DNAPLs the issue is one of changing the proportions of toxic chemicals in a complex mixture. It is worth noting that some transformations (both natural and human-induced) can result in the production of more toxic components, such as when TCE is reductively dechlorinated to produce vinyl chloride.

## Financial Objectives

Cost minimization is typically one of the absolute objectives of any cleanup decision. That is, most stakeholders agree that, *all else being equal*, the lower-cost option should be selected, thereby freeing up funds for other beneficial uses. Typically, therefore, the challenges arise not in stating the objective, but rather in selecting the metrics and estimating the costs of the alternatives.

Cost provides a good example of how a stated objective can be measured by many different metrics, and how use of the same word, that is, "cost," can mask significant differences in stakeholder values. Examples of the different types of metrics routinely used to evaluate cost include annual cost, capital cost, life-cycle cost, cost to the community, cost to the state, project cost, and cost to the federal government. For example, annual cost sometimes may play a large role in decision making; if the annual cost of implementing a given technology is sufficiently large, it may be difficult to fund, and that technology may be abandoned in favor of one with a flatter cost profile. Representatives of local government must be concerned with impacts on the local economy, such as boom–bust cycles and consequent strains on community resources, impacts on property values, and long-term vitality of the community. Stakeholders may also have different perspectives regarding the appropriate discount factor to use in a present value calculation—an issue that may become quite important when the analysis involves long time periods. Some cleanup alternatives may include transferring responsibilities from one organization to another (e.g., from the federal government to state government for long-term monitoring of a site, or from governments to citizen watchdog groups) (NRC, 2000a). In such cases the perspective from which costs are measured may influence the cost metric and resulting cost estimate. Similarly, government cleanup decisions may produce real economic impacts on local and regional economies by affecting local and regional labor markets, property values, community emergency preparedness costs and insurance premiums, and economic development (NRC, 1996). Decisions on whether to accelerate or delay closure of a site may result in boom–bust cycles for the affected communities. In all of these cases, the perspective may influence the choice of metric and ultimately the cost estimate obtained.

Although a variety of different cost metrics are in use, for government decision making the life cycle cost metric is recommended. Life cycle cost analysis represents an attempt to create a comprehensive accounting of the full range of direct and indirect costs and benefits resulting from a course of action over the entire period of time affected by the action. Thus, life cycle cost typically includes all costs associated with an alternative from start-up through long-term stewardship, and it avoids the problems of suboptimization presented by other cost metrics. Even when life cycle cost is defined as the metric to be used, divergent cost estimates may be obtained due to differences in assumptions regarding the scope and boundaries of the analysis and in projections regarding the future of technology, regulations, human and institutional behavior, and other factors, as discussed below.

*Overview of Life Cycle Cost Analysis*

Key to life cycle cost analysis is a full cost inventory that includes all direct and indirect costs ranging from project start-up costs (e.g., design, studies to prove a technology or obtain permits), capital and operating costs, through decommissioning, site closure, and long-term stewardship costs. A life cycle cost analysis requires careful consideration of both the scope of the analysis and the time horizon of analysis, in order to ensure that the full long-term costs and liabilities are factored into decision making. For example, costs that may be borne by other entities (e.g., waste management-related costs, or future surveillance and maintenance costs) should be considered in addition to direct project costs. The time horizon of the analysis should be long enough to include all of the impacts of an alternative. Future and long-term costs such as those needed for continued monitoring, reporting, maintenance, other regulatory compliance-related matters, replacement or corrective maintenance of caps, and other infrastructure should be captured in the analysis. If project time span exceeds the design life of support facilities or other items important to the project's success, the cost of replacing these facilities must be considered. Frequently, a life cycle analysis will reveal that the approach with the lowest initial cost is not the low-cost approach from a life cycle perspective, due, for example, to high long-term operating costs or the need for future replacement of a remedy.

In addition, so-called "hidden costs" should be included in a life cycle cost estimate. Hidden costs are costs not charged to the project actually responsible for incurring them; instead, these costs are charged to indirect or overhead accounts or to other entities. These issues arise frequently in federal facility cleanup decisions because costs associated with a given cleanup may be budgeted for in many separate government accounts. Examples of project-related costs that may not be fully charged to a project include utilities, permitting and regulatory oversight, environmental monitoring, security, long-term surveillance and maintenance, and the full cost of waste disposal. Long-term liability is another form of hidden cost that is sometimes neglected or underestimated; neglecting such a liability may result in a bias in favor of perpetual care alternatives. Finally, economic benefits of an alternative must also be addressed, as, for example, when cleanup leads to beneficial reuse of a building and/or land.

Numerous checklists have been developed to help identify cost elements to aid in producing a full cost accounting (see, for example, NRC, 1997; EPA 1995, 2000; Department of the Army, 2002). Table 4-1 provides an example of an expanded cost inventory that may be appropriate to a federal facility cleanup. A life cycle cost analysis would include all labor, equipment, and material costs associated with the cost elements in the table. In practice, analysts typically use a

TABLE 4-1 Example Cost Elements in a Life Cycle Cost Analysis

| WBS | Element Name |
|---|---|
| **1.0** | **Research, Development, Test, and Evaluation** |
| **1.01** | Design and Engineering |
| **1.02** | Prototype |
| **1.03** | Project Management |
| **1.04** | System Test and Evaluation |
| **1.05** | Training |
| **1.06** | Data |
| **1.07** | Equipment |
| **1.08** | Facilities |
| **1.09** | Other Research, Development, Test, and Evaluation |
| | |
| **2.0** | **Preparation/Mobilization** |
| **2.01** | Planning/Engineering |
| **2.02** | Site Preparation |
| **2.03** | Regulatory Compliance/Permitting |
| **2.04** | Mobilization |

*continued*

TABLE 4-1 Continued

| WBS | Element Name |
|---|---|
| **3.0** | **Capital** |
| 3.01 | Facilities (e.g., buildings and structures such as onsite labs, health and safety offices, monitoring facilities) |
| 3.02 | Equipment (e.g., boiler for steam production, vapor extraction equipment, condenser equipment, pumps, gas–liquid separators, water and gas treatment systems, off-gas treatment equipment, tanks, pumps, blowers, aboveground drainage, containment structures, air or water monitoring equipment) |
| 3.03 | Engineering/Manufacturing/Tooling/Quality Control |
| 3.04 | Project Management |
| 3.05 | System Test and Evaluation |
| 3.06 | Other Construction and Installation (e.g., well installation, barrier wall construction) |
| 3.07 | Training |
| 3.08 | Data |
| 3.09 | Start-up |
| | |
| **4.0** | **Operation and Maintenance** |
| 4.01 | Sampling and Analysis |
| 4.02 | Monitoring/Regulatory Compliance |
| 4.03 | Materials/Chemicals/Consumables |
| 4.04 | Operation |
| 4.05 | Water/Gas Treatment |
| 4.06 | Equipment Repair and Maintenance |
| 4.07 | System Engineering/Project Management/Quality Assurance |
| 4.08 | Safety and Health |
| 4.09 | Training |
| 4.10 | Utilities (electricity, natural gas, water, other utilities) |
| 4.11 | Transportation |
| 4.12 | Waste Management/Disposal |
| | |
| **5.0** | **Site Restoration** |
| 5.01 | Demobilization |
| 5.02 | Capping |
| 5.03 | Decommissioning/Closure |
| 5.04 | Restoration |
| | |
| **6.0** | **Long-Term Management** |
| 6.01 | Institutional Controls |
| 6.02 | Sampling/Monitoring |
| 6.03 | Remedy Failure/Repair/Replacement |
| 6.04 | Natural Resource Damage Liability |
| 6.05 | Other Long-Term Liability |

SOURCE: Adapted from EPA (2000) and Department of the Army (2002).

graded approach to life cycle cost analysis, performing an analysis with a level of detail commensurate with the decision to be made and the level of information available to support the analysis.

### Consideration of Time and Associated Uncertainty

Remedial action projects typically involve construction costs that are expended at the beginning of a project and costs in subsequent years that are required to implement and maintain the remedy after the initial construction period. Present value analysis is a method used to compare alternatives that produce cash flows in different time periods, by transforming the future stream of benefits and costs to a single number, called the present value. The present-value method is based on the concept that a dollar today is worth more than a dollar in the future because, if invested in an alternative use today, the dollar could earn a return.

A present-value analysis of a remedial alternative involves three key steps: (1) define the scope and period of the analysis, (2) estimate the costs and benefits occurring in each year, and (3) select a discount rate to use in calculating the present value of future benefits and costs.[6] The larger the discount rate, the lower is the present value of future cash flows. Discounted values of even large costs incurred far in the future tend to be small. For example, for a 200-year project with a constant annual cost of $500,000 at a 3.2 percent discount rate, 96 percent of the present value cost is incurred in the first 100 years, 79 percent in the first 50 years, and 61 percent in the first 30 years.

Decisions on discount rates can play an important role in remedial action decision making, particularly in determining what remedy to choose to meet the cost objective—for example, contaminant mass removal vs. contaminant isolation or long-term stewardship measures. Such tradeoffs regarding the appropriate discount rate come to the fore when considering sites, like certain DNAPL sites, that would be extremely costly to remediate to a level that would allow unrestricted access, but which would otherwise require indefinite government stewardship. Discount rates may also play a role in decisions related to the time required to close a site, because cleanup costs incurred in the future have a lower present value than the same costs incurred today. In general, differences in discount rates can lead to substantially different conclusions about which of two alternatives is the most cost-effective.

---

[6]For government decision making, the discount rate is the cost of borrowing, that is, the interest rate on Treasury notes and bonds. Office of Management and Budget Circular A-94 provides guidance on discount rates to be used in the analysis of federal projects. For 2003, and for programs of longer than 30-year duration, Appendix C of Circular A-94 reported a real discount rate of 3.2 percent.

To illustrate this, Table 4-2 compares the life cycle cost of five remedial alternatives that have different initial capital costs, annual operating and maintenance (O&M) costs, and project durations. The life cycle cost is calculated using real discount rates of 0 percent, 3.2 percent (typical of government projects), 7 percent, and 12 percent (the latter two are typical of real discount rates used in the private sector to evaluate alternative investment options). Alternative E has the highest cost on an undiscounted basis but the lowest present value for discount rates of 3.2 percent, 7 percent, and 12 percent. This is because much of its cost occurs in the future, and the present value of these future costs is small. Setting aside Alternative E, Alternative D has the lowest present value at a 12 percent discount rate, but Alternative A has the lowest present value at a 3.2 percent discount rate. The undiscounted cost of Alternative B is less than that of Alternative C, but its present value at discount rates of 3.2 percent, 7 percent, and 12 percent is higher than that of Alternative C at these rates due to B's large upfront capital cost. As these examples illustrate, the relative economic benefits of competing alternatives may depend on the choice of discount rate. Low discount rates would tend to make source depletion options appear more attractive, whereas high discount rates would tend to make containment options more attractive. Differences in public sector and private sector discount rates and other financial considerations may lead to different decisions being made at private sector sites and government sites.

Life cycle cost estimation over long timeframes is complex not only due to differences of opinion on the choice of discount rate, but also, and perhaps more important, due to the large uncertainty in projections of future costs given the uncertainties surrounding the site conceptual model, future technology, regulatory policies, societal norms, land use, population density, etc. Major sources of uncertainty in cost estimates relate to the site characterization model and the effectiveness of a selected technology at the specific site. For example, the nature

TABLE 4-2  Effect of Discount Rate on Life Cycle Cost Calculation

| Remedial Alternative | Initial Capital Cost ($1000) | Annual O&M Cost ($1000) | Project Duration (Years) | Life Cycle Cost ($1000) | | | |
|---|---|---|---|---|---|---|---|
| | | | | Real Discount Rate | | | |
| | | | | 0% | 3.2% | 7% | 12% |
| A | $3,650 | $583 | 15 | $12,400 | $10,500 | $8,960 | $7,620 |
| B | $10,800 | $548 | 30 | $27,200 | $21,300 | $17,600 | $15,200 |
| C | $2,850 | $696 | 50 | $37,700 | $20,100 | $12,500 | $8,630 |
| D | $5,500 | $230 | 80 | $23,900 | $12,100 | $8,770 | $7,420 |
| E | $2,000 | $200 | 220 | $46,000 | $8,240 | $4,860 | $3,670 |

and extent of contamination may prove to be greater than anticipated, or the technology may lack sufficient performance history for reliable cost estimation. Furthermore, technology performance may be sensitive to site-specific geologic and contaminant conditions, making cost extrapolations between sites difficult.

There are several instances of wastes that were disposed of decades ago that are now being remediated due to changes in regulatory standards. Cost estimates performed decades ago would not have foreseen the high remediation costs being incurred today. Conversely, development of new technology may result in unforeseen cost reductions. Estimating costs associated with long-term institutional controls adds additional uncertainty because it involves (1) predicting future costs associated with management of the site, (2) predicting potential liabilities (e.g., risk of failure of containment strategies and costs of remedy), (3) making projections regarding the future ability of government or other entities to maintain control, (4) evaluating the ability to maintain both technology and records over long time periods, and (5) predicting the potential costs in the event of institutional control failure at some time in the future.

Cleanup cost estimates are also highly dependent upon decisions regarding the cleanup schedule and the future use of a site. For example, very different cost estimates may result depending on whether the site is to be cleaned up to a level to permit unrestricted residential use or industrial use or whether the site is to be maintained in perpetual government stewardship as, for example, a wildlife preserve. The schedule of cleanup, e.g., the date on which ownership of the site is to be transferred, may also have a substantial effect on the remedial design and on work plans and therefore on cost estimates. Thus, if there is a specific date on which ownership of the site is to be transferred, high annual costs in certain years may be acceptable in order to meet the deadline. Conversely, budgetary pressures and/or a desire to avoid boom–bust cycles and minimize disruption to the local economy may lead to preference for a level funding profile. Either case may have a substantial effect on the remedial design, the work plans, and, consequently, the life cycle cost estimates.

For all these reasons, uncertainty analysis is a critical element of cost estimation, as it is for the characterization of risk (see Box 4-4). Various analytical techniques exist to characterize the uncertainties surrounding cost estimates.[7] Probabilistic techniques such as Monte Carlo analysis (EPA, 1997a) may be used to gain a better understanding of the likely range of costs and their probability of occurrence. When a cost analysis is conducted using probabilistic techniques, parameter distributions that represent the uncertainty inherent in each of the parameters are used as inputs to the cost calculations rather than point estimates. The simulation output is the range of possible costs and the probabilities that they

---

[7]See, for example, NRC (1996), which discusses uncertainty in risk estimation. Much of this discussion applies to cost estimation as well.

will occur, which provides decision makers with a much more complete picture. These tools are used to answer questions such as "What are the chances of this project finishing under budget?" or "What is the probability that the cost of remediation will exceed X amount?" Box 4-3 provides an example of the use of Monte Carlo simulation for the evaluation of remedial alternatives at a hypothetical site.

## Schedule Objectives

Schedule objectives—for example, time to complete cleanup—may vary substantially among stakeholders and often play a significant role in decision making as they relate directly to stakeholders' visions and objectives for the future of a site. For some stakeholders, particularly the military, the objective may be to finish cleanup and transfer the site within a specified timeframe as mandated by base realignment and closure (BRAC) requirements. For other stakeholders, the objective may be to avoid boom–bust cycles and minimize disruption to the local economy, which may lead to a desire to spread remediation out over time and maintain a more level funding profile.

Schedule objectives are complicated by stakeholder values relative to future land uses and future site ownership. For example, one stakeholder may value accelerating cleanup of a site in order to transfer ownership and reuse the site for commercial purposes, while another stakeholder may value maintaining the site in long-term government stewardship as, for example, a wildlife preserve. Schedule objectives may be absolute objectives or functional objectives. For example, accelerating site cleanup is a functional objective when it is used as a means to achieve the absolute objective of reduction of risk to human health and the environment.

Given the numerous and disparate issues related to schedule objectives, it is not surprising that many different metrics have been devised to measure progress. For example, the volume of material removed in a given year (related to a "get on with it" or "get started" objective), the planned year of completion, or the reduction of the amount of contamination present at a site may all be relevant metrics. Indeed, the military uses several temporal milestones as measures of success in the cleanup of DoD facilities. These include the signing of records of decision (RODs) for particular sites, the placement of the remedy (RIP) at sites, the designation of sites as being "response complete," and the closeout of sites.

## Other Objectives

As alluded to under the discussions of financial and schedule objectives above, stakeholders often have a variety of socioeconomic, institutional, and programmatic objectives ranging from maintaining corporate reputation and goodwill to sustained employment or ensuring that cleanup activities are designed

## BOX 4-3
## Example of a Life Cycle Cost Analysis for
## Evaluation of Remedial Alternatives
Based on a hypothetical example developed by
Kyle A. Gorder of Hill Air Force Base

This hypothetical example is based on the economic model being used at Hill AFB. The problem is defined as follows: a plume of dissolved-phase volatile organic compound (VOC) contamination underlies a residential area. The area of groundwater contamination is 50 acres (20 hectares). The groundwater table is generally located about 10 feet (3 m) below ground surface, and the average thickness of contamination is 50 feet (15 m). Two scenarios (out of many possibilities) were evaluated for illustrative purposes:

Scenario 1. Based on site investigations and monitoring indicating that the plume is stable, the management strategy for the site is monitored natural attenuation.

Scenario 2. Additional remedial action is taken at the site that results in reduction in the areal extent of groundwater contamination.

### Scenario 1

Liabilities (Table 4-3) considered for this site include the cost of long-term monitoring, the possibility of finding indoor air contamination in residences and the consequent costs associated with mitigation system operations, maintenance, and monitoring (OMM), the potential for Natural Resource Damage (NRD) claims, and the potential cost of having to obtain remediation easements.

TABLE 4-3 Parameters Used in Analysis of Scenario 1

| Description | Value | Symbol |
|---|---|---|
| Annual Long-term monitoring cost ($) | 50,000 | $C_{LTM}$ |
| Years of long-term monitoring | 30 | $Y_{LTM}$ |
| Average number of homes/acre | 2.25 | |
| Probability of indoor air contamination (%) | 1 to 20 | |
| Number of homes with indoor air contamination | 1 to 23 | N |
| Mitigation system installation and startup cost ($/home) | 10,500 | $C_{MS}$ |
| Annual mitigation system OMM cost ($/home) | 1,500 | $OC_{MS}$ |
| Years of mitigation system OMM | 30 | $Y_{MS}$ |
| Value of groundwater ($/acre-ft) | 919 | |
| NRD liability (value of groundwater x contaminated volume) ($) | 689,250 | $C_{GW}$ |
| NRD settlement probability (%) | 10 | $P_{NRD}$ |
| NRD settlement year | 5 to 30 | $Y_{NRD}$ |
| Value of residential land ($/acre) | 440,000 | |
| Easement liability (land value x area of contamination) ($) | 22,000,000 | $C_L$ |
| Easement settlement probability (%) | 0 to 10 | $P_L$ |
| Easement settlement year | 3 to 30 | $Y_L$ |
| Discount rate (%) | 4 | I |

Values given by ranges in Table 4-3 are variables in the Monte Carlo simulation. Distributions for some of these variables are presented below.

*Number of Homes with Indoor Air Contamination*

Given the area of contamination and the number of homes per acre, indoor air contamination could affect approximately 113 residences. Assuming that between 1 percent and 20 percent of the homes will be affected and that 10 percent is the most likely number, the number of homes affected was assumed to follow the triangular distribution shown in Figure 4-5. The minimum number of homes affected is 1, the maximum is 23, and the most likely is 11 homes.

*Natural Resource Damage*

The NRD settlement year was assumed to follow the uniform distribution shown in Figure 4-6. This distribution indicates that the NRD settlement could occur at any time between year 5 and year 30. All years in this range have equal probability of selection during the Monte Carlo simulation.

*Remediation Easements*

Two variables related to remediation easements (settlement probability and settlement year) were chosen for inclusion in the Monte Carlo simulation. The probability of an easement settlement was assumed to follow the triangular distri-

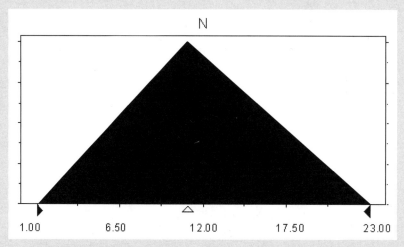

N

| 1.00 | 6.50 | 12.00 | 17.50 | 23.00 |

FIGURE 4-5 Distribution of number of homes potentially requiring indoor air remediation. Vertical axis is relative probability, such that the area under the curve = 1.

*continued*

continued

## BOX 4-3 Continued

bution shown in Figure 4-7. This distribution sets the minimum probability of an easement settlement to zero percent and the maximum probability to 10 percent, with the most likely probability of 5 percent. The easement settlement year was

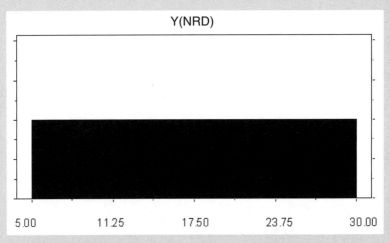

FIGURE 4-6 Distribution of NRD settlement year (min = 5 yrs, max = 30 yrs). Vertical axis is relative probability.

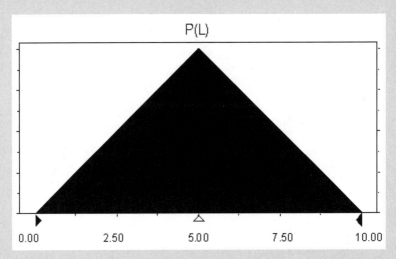

FIGURE 4-7 Distribution of easement settlement probability. Vertical axis is relative probability.

assumed to follow a uniform distribution, based on the assumption that an ease-ment settlement could occur at any time between years 3 and 30. All years in this range have equal probability of selection during the Monte Carlo simulation.

### Scenario 2

Scenario 2 examines the potential liability reduction that could be achieved with the implementation of an aggressive source remediation strategy designed to reduce plume size. For this example, it is assumed that plume reduction from 50 acres (20 hectares) to 12.5 acres (5 hectares) could be achieved over a 30-year timeframe and that this reduction would occur linearly. Note that any plume area reduction curve could be used in the analysis and that this should be based on some understanding of the site conceptual model. Ideally, this curve would be based on detailed analyses of site conditions and/or numerical modeling.

The effect of a smaller plume footprint is incorporated into the analysis in three ways: (1) The average number of years of OMM on indoor air vapor mitigation systems is decreased and follows the distribution shown in Figure 4-8, (2) the volume of contaminated groundwater is reduced to account for the reduction in plume area (volume as of the year of NRD settlement is used), and (3) the land area used to determine the easement liability is reduced, again according to the year of easement settlement.

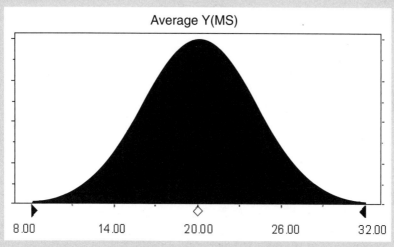

FIGURE 4-8 Average number of years of OMM on indoor air vapor mitigation systems (mean = 20 yrs, standard deviation = 4 yrs, range is from 5 to 35 yrs). Vertical axis is relative probability.

*continued*

## BOX 4-3 Continued

### Calculation of Total Liability

The total liability for each scenario is calculated as follows:

$$\text{Liability} = C_{LTM}Y_{LTM} + N(C_{MS} + OC_{MS}Y_{MS}) + P_{NRD}C_{GW} + P_L C_L$$

The present value of the liability is calculated as:

$$\text{PV Liability} = pa(C_{LTM}, Y_{LTM}, I) + N[C_{MS} + pa(OC_{MS}, Y_{MS}, I)] +$$
$$P_{NRD}[pf(C_{GW}, Y_{NRD}, I)] + P_L[pf(C_L, Y_L, I)]$$

where
$pa(\text{cost, time, } I)$ = the present value of an annual cost over time at discount rate "I"
$pf(\text{cost, time, } I)$ = the present value of a future cost at year "time" and discount rate "I."
All other variables are defined in Table 4-3.

### Results

Table 4-4 presents the results of the liability calculations for the two scenarios. Note that these results do not include the cost (liability) associated with implementing the aggressive source remediation under Scenario 2. The cost ranges presented in the table represent 90 percent confidence intervals estimated from the Monte Carlo analysis.

TABLE 4-4 Estimated Liabilities for the Scenarios

| | Present Value Cost ($1000) | | | |
| | Scenario 1 | | Scenario 2 | |
| Cost Component | Cost Range | Mean | Cost Range | Mean |
| --- | --- | --- | --- | --- |
| Operations, Maintenance, and Monitoring | 865 | 865 | | |
| Indoor Air Contamination Mitigation | 144 to 688 | 407 | 117 to 588 | 342 |
| Natural Resource Damage | 22 to 54 | 36 | 6 to 46 | 23 |
| Remediation Easement | 146 to 1,210 | 597 | 68 to 1,010 | 395 |
| Combined Liability | 1,370 to 2,570 | 1,900 | 312 to 1,420 | 760 |

The results shown in Table 4-4 are interpreted as follows. (Note that mean values are used to simplify this discussion. In practice, the entire distributions resulting from the Monte Carlo simulation for each scenario would be compared.) The mean combined liability present value for Scenario 1 is $1,900,000. The mean combined liability present value for Scenario 2 is $760,000. The difference in these two liabilities ($1,140,000) represents the breakeven point for an investment in aggressive source remediation. An investment in aggressive source remediation with a present value of less than or equal to the breakeven point would be considered a cost-effective investment.

and conducted in a manner that is consistent with community values and long-term vision. Intergenerational equity and long-term land protection are also objectives voiced by some stakeholders. Cost and schedule objectives are sometimes functional objectives used as a means to accomplish some of these larger concerns, as, for example, when communities specify annual funding requests or completion schedules.

Another set of objectives includes those that may be established in order to meet regulatory commitments or avoid the imposition of sanctions or negative consequences. One example of such an objective is the prevention of any offsite contamination that may eventually result from improper handling or disposal of contaminants. Such offsite contamination may result in private or public nuisance lawsuits by owners of neighboring properties, whose use and enjoyment of their land may be adversely affected by the contaminants. It may also result in the imposition of enforcement sanctions by regulatory authorities.

### Communication with Stakeholders Regarding Objectives

The absolute objectives for site remediation can differ significantly, partly because they reflect value judgments made by many different stakeholders. In addition, they depend upon the physical and social environment in which remediation takes place. The first five physical objectives described earlier are singled out in Chapters 5 and 6 for the purposes of evaluating source remediation technologies and developing a protocol for source remediation. These physical objectives can and usually do serve multiple absolute objectives (note that some appear in both the risk- and time-related discussions). Whether these physical objectives can be achieved depends heavily on the technology used and the hydrogeologic setting—a major theme of this report.

In general, the Army, DoD, and other institutions charged with hazardous waste remediation should be cognizant of selecting remedial actions, in particular source remediation, that take into account the absolute and functional objectives held by stakeholders. Decisions as to how to manage historical releases occur within a broad social context, involving multiple parties that may each have diverse and dynamic sets of drivers. A hypothetical example of a set of parties and their primary drivers is presented in Table 4-5, but the actual list of stakeholders is generally much larger. In addition to those stakeholders included in the table, a host of other players may be seeking to satisfy various objectives via remedial decisions. These include persons who are ostensibly agents of one or more of the listed stakeholders (consultants, vendors, researchers, etc.) but who have individual objectives that may profoundly influence their contribution to the decision-making process.

Most of the parties involved in these decisions are working against the background of an organizational policy that sets out criteria for decision making. Decisions are not made on the basis of individual preferences except where an individual is the sole responsible party for a site—typically only true at the very smallest of sites. There may be conflicts between the desires of different stakeholder groups (as Table 4-5 suggests). Indeed, the committee is aware of many cases where programmatic objectives may be in conflict with the technical realities of what can be accomplished at sites. Nonetheless, although decisions depend upon both technical and nontechnical factors, once a decision has been made, the focus must be on the technical goals to determine if remediation is successful.

This report does not revisit the discussion of how to build successful programs for stakeholder involvement at a site, since the difficulties and opportunities involved are discussed in previous NRC reports (1999a,b, 2000a), among other references. Nonetheless, it is essential to the success of source remediation for the RPM to effectively capture the results of stakeholder processes. By noting for each stakeholder whether particular remedial objectives are considered absolute or functional, the RPM will be able to make more informed judgments about the evaluation of source remediation. Thus, if the regulatory authority has a preference for "reducing mass or toxicity," technologies that can be demonstrated to achieve this (and convincing metrics that indicate whether or not they have achieved this) should receive additional consideration. In contrast, if the local community views contaminant migration as "chemical trespass," reduction in contaminant mass may only be relevant insofar as it effectively (and, often,

TABLE 4-5 Hypothetical Example of Stakeholders and Potential Drivers for Determining Objectives

| Stakeholder | Potential Drivers |
| --- | --- |
| Responsible Party | Corporate decision policy and protocol. Protect human health and the environment, manage financial impacts to mission or business, manage reputation |
| Project Manager (typically employees of the Responsible Party) | Corporate decision policy and protocol. Meet schedule and budget commitments, maintain positive relationships with all parties |
| Federal Regulator | Compliance with regulations, decision mechanisms driven by legislation, meet public expectations, meet schedule commitments, manage reputation |
| State Regulator | Compliance with regulations, decision mechanisms driven by legislation meet public expectations, minimize economic liabilities that may pass to state during long-term operations and maintenance, manage reputation |
| Public | Protection of health, preservation of property values, punishment of responsible parties for damages, jobs |

quickly) reduces concentrations at the fence line and beyond. In this case, mass removal technologies may fare poorly in comparison with a comparable investment in plume capture.

By actually drafting a chain of objectives and metrics and presenting it to stakeholders, the RPM's ability to satisfy the range of absolute objectives in the community can be clarified, and the areas where policy-level trade-offs are needed can be separated from those where a more technical evaluation of alternatives is possible.

## EXISTING FRAMEWORKS, THEIR OBJECTIVES, AND ASSOCIATED METRICS

At all hazardous waste sites, cleanup takes place within one or more decision frameworks that provide structure to the activities that occur, from the initial discovery of contamination to the eventual closure of the site. Two broad classes of frameworks are discussed here because they impact cleanup at a high percentage of sites, and because they have defined objectives with associated metrics of success and thus may represent a significant influence on stakeholder formulation of objectives. The major categories of existing frameworks can be broadly classified as regulatory, which defines the legal goals and objectives of the remedial action, and risk assessment, which defines the existing threat to human health and the environment and the level of remediation required to reduce this risk to acceptable levels.

An understanding of these existing frameworks for site remediation is essential to developing an unambiguous set of absolute objectives for site remediation and corresponding functional objectives that define whether those absolute objectives have been obtained. That is, just as different stakeholders may view the same objective in different ways (e.g., as absolute rather than functional, or as serving different absolute objectives), the accustomed cleanup framework(s) of each stakeholder will also shape their perceptions of alternative remedial objectives for a site. This may in turn influence the ability of the stakeholders to achieve a working consensus on remedial objectives.

Knowledge of these conventional/historical frameworks can also aid in the analysis of the status of a potential objective for a stakeholder. An objective in one of the frameworks may be shared by another framework but serve different purposes in each. For example, the objective of mass removal can be found in multiple regulatory frameworks (and thus has been interpreted by some as an absolute objective) as well as in risk assessment frameworks (for the purpose of reducing exposure potential). Indeed, in a risk assessment framework, mass removal is a purely functional goal, and in fact is generally separated from the absolute goal of reducing or eliminating risk by several inferential steps and corresponding functional goals.

The historical frameworks described below tend to inherently weigh certain absolute objectives, and even particular functional objectives, more heavily. This can lead to conflicts between stakeholders or to balancing problems for the RPM, who must apply several frameworks simultaneously. Therefore, it is important for site-specific objectives to be defined with consideration of all relevant frameworks early in the process of remedy evaluation. Early identification of objectives will allow the information that is needed to support the remediation decision to be made available through site investigation, the potential conflicts between the objectives of different frameworks to be addressed, and the potential for establishing mutually agreed upon cleanup objectives to be maximized.

## Regulatory Framework

The cleanup of contaminants at U.S. Army installations takes place within a highly structured, complex regulatory environment. At its core is the Comprehensive Environmental Response, Compensation, and Liability Act (CERCLA), a statute amended in 1986 by the Superfund Amendments and Reauthorization Act (SARA) (P.L. 99-499) to bring all military facilities under the authority of the Superfund Program. SARA established the Defense Environmental Restoration Program (DERP)—managed by the Department of Defense—which includes an Installation Restoration Program (IRP) that conducts environmental cleanups at military bases. In the case of the most contaminated military facilities—that is, those listed on the National Priorities List (NPL)—the cleanups are directly regulated by officials of the U.S. Environmental Protection Agency (EPA) or their state counterparts.

The general procedures and standards to be followed with respect to contaminated facilities have been set forth by the EPA in its National Contingency Plan (NCP) (40 CFR 300 et seq.). These broadly applicable EPA regulations establish the basic framework that responsible parties such as the military follow to investigate, evaluate, and remediate hazardous substance problems at their facilities. Under the NCP, site managers conduct a remedial preliminary assessment (and, where appropriate, a site investigation) to determine whether a particular site should be given priority for long-term remedial response. The results of these evaluations are used to score the site under the EPA's hazard ranking system (HRS) model. If the site scores above the HRS threshold, then the entire facility is placed on the NPL for possible remedial action.

Following that, site managers generally undertake a remedial investigation and feasibility study (RI/FS) to study the nature and extent of the contamination problem at the site and to develop alternative approaches for managing the site problem. In the course of preparing this feasibility study, RPMs along with the appropriate regulatory authority will establish a preliminary remediation objective for the site, and they will prepare a broad list of alternative ways in which the preliminary remediation objective at the site may be attained. This list is screened to eliminate clearly impractical alternatives. The remaining alternatives are studied, compared, and evaluated against a set of nine criteria divided into three categories, described below.

*NCP Threshold Criteria*

The two threshold criteria in the National Contingency Plan are (1) to be protective of human health and the environment and (2) to comply with Applicable or Relevant and Appropriate Requirements (ARARs). In practice, the criterion "protective of human health" has usually if not always been embodied in quantitative risk assessment, specifically as described in the Risk Assessment Guidance for Superfund (RAGS). "Protective of human health" has been interpreted as having a calculated cancer risk between $10^{-6}$ and $10^{-4}$ or a Hazard Index < 1.0. As discussed earlier, meeting an absolute objective of risk reduction is frequently embodied in the more functional objective of preventing human exposure to site-related contaminants during the period that is subject to analysis. A purely administrative remedy (or perhaps a physical barrier) can conceptually meet this functional objective as well as the complete removal of contaminants. "Protection of the environment" is less clearly defined, and although the typical approach employed by EPA and responsible parties has also been risk assessment, the methods used are generally less quantitative and more variable, reflecting the greater complexity of the physical/biological system under consideration.

In contrast to the "protective" criterion, compliance with ARARs is more obviously concerned with absolute objectives, both in philosophy and in practice. This is perhaps most clearly reflected in the ARARs that are directly relevant to this committee's charge. Drinking water Maximum Contaminant Levels (MCLs) and non-zero Maximum Contaminant Level Goals (MCLGs) are considered to be ARARs for groundwater remediation.[8] This designation is independent of whether the particular groundwater is, in fact, currently used as a source of drinking water or is likely to be so used in the future, as long as it is capable of being used as a source of drinking water. Table 4-6 presents MCLs and MCLGs for the chlorinated solvents and drinking water equivalent levels (DWEL) and lifetime health advisory levels for the explosives of concern in this report. These values are commonly set as the objectives of source remediation.

Requiring groundwater to meet specific concentration targets *independent* of its uses is clearly not an attempt to protect human health, at least as far as toxic risk is concerned, since less stringent rules are set for actual public water sup-

---

[8]If MCLs or non-zero MCLGs are exceeded, action generally is warranted (EPA OSWER Directive # 9355.0-30, April 22, 1991). Where the cumulative carcinogenic site risk to an individual based on reasonable maximum exposure for both current and future land use is < $10^{-4}$, and the noncarcinogenic hazard quotient is < 1.0, action generally is not warranted unless there are adverse environmental impacts.

TABLE 4-6 Toxicological and Regulatory Benchmarks for Source Zone Chemicals

| Chemical | Oral RfD (mg/kg/d) | MRL (mg/kg/d) | Carcinogen Class | $10^{-6}$ Conc. (µg/L) | MCLG (µg/L) | MCL (µg/L) |
|---|---|---|---|---|---|---|
| **Chlorinated Solvents** | | | | | | |
| Tetrachloroethene (perchloroethylene, PCE) | 0.01 | 0.02 Acute | 2A / — | — | 0 | 5 |
| Trichloroethene (TCE) | — | 0.2 Acute | 2A / — | — | 0 | 5 |
| cis-1,2-Dichloroethene | — | — | — / D | — | 70 | 70 |
| 1,1-Dichloroethene (1,1-DCE) | 0.05 | — | — / C | — | — | — |
| 1,1,1-Trichloroethane (TCA) | — | — | 3 / D | — | 200 | 200 |
| 1,2-Dichloroethane (DCA) | — | 0.2 Int. | 2B / B2 | 0.4 | 0 | 5 |
| Tetrachloromethane (carbon tetrachloride) | 0.0007 | 0.007 Int. | 2B / B2 | 0.3 | 0 | 5 |
| Trichloromethane (chloroform) | 0.01 | — | 2B / B2 | — | 0 | 80[a] |
| Dichloromethane (methylene chloride) | 0.06 | — | 2B / B2 | 5.0 | 0 | 5 |
| **Other Hydrocarbons** | | | | | | |
| Naphthalene | 0.02 | — | 2B / C | — | — | — |
| Benzo(a)pyrene | — | — | 2A / B2 | 0.005 | 0 | 0.2 |
| Aroclor 1254 (PCB mixture) | 0.00002 | (0.00002) | — / (B2) | (0.1) | (0) | (0.5) |
| Aroclor 1260 (PCB mixture) | — | (0.00002) | — / (B2) | (0.1) | (0) | (0.5) |
| **Explosives** | | | | | | |
| TNT | 0.0005 | — | — / C | 1.0 | 20[b] | 2[b] |
| 2,4-DNT | 0.002 | — | [c] | 0.05 | 100[b] | |
| HMX | — | — | — / — | 4.0 | 2,000[b] | 400[b] |
| RDX | — | — | — / — | 0.3 | 100[b] | 2[b] |

NOTES: Dashes mean there is no information on a particular source. Parentheses are used where the table discusses a specific chemical, but where the given value was specified for a broader chemical class.

Oral RfD = Oral Reference Dose: an EPA estimate of a dose at which chronic exposure is not expected to cause adverse effects.

MRL = Minimum Risk Level: an ATSDR estimate of a dose at which chronic exposure is not expected to cause adverse effects. Int. and Acute refer to intermittent and acute exposure, which are sometimes specified rather than chronic exposure.

Carcinogen Class: A ranking of the weight of evidence that the substance is carcinogenic in humans. The first is the IARC (1 = known, 2A = probable, 2B = possible) classification, while the second is EPA's (A = known, B1 = probable – human evidence, B2 = probable – animal studies, C = possible, D = not classifiable as to carcinogenicity) (EPA, 1986). EPA's system has undergone several changes over the years, and no consistent classification by EPA appears to be available.

$10^{-6}$ Conc. = Concentration of the chemical in water that would result in an excess cancer risk value of one-in-a-million from ingesting 2 liters of water a day.

MCLG = Maximum Contaminant Level Goal

MCL = Maximum Contaminant Level

[a]This is the MCL for total trihalomethanes, of which chloroform is one.

[b]Last two columns for the explosives refer to DWEL (μg/L) and Lifetime Health Advisory (μg/L), respectively.

[c]None recommended, potential human carcinogen (Group B2)

SOURCE: EPA (2002).

plies. That is, under the Safe Drinking Water Act, a public water supply (i.e., a utility with more than 15 service connections, or serving more than 25 customers) is *not* prohibited from exceeding MCLs in the water it supplies. Rather, it is simply required to *notify* the state and its customers when they are exceeded (40 CFR 141.31, 141.32).

Although MCLs have the advantage of being quantifiable (i.e., the concentrations of contaminants in water can readily be accurately measured), MCLs have major limitations with respect to both selection of remediation technologies and evaluation of benefits of source zone remediation efforts. As noted in EPA (2003b), the use of MCLs may inhibit source zone remediation attempts. This is because in the most prevalent hydrogeologic settings, minute amounts of DNAPL, sorbed mass, and dissolved mass in stagnant zones will remain after source zone remediation, such that desorption and reverse diffusion from the source zone into the dissolved phase plume region may maintain concentrations in some locations above MCLs for some time. Thus, even if removal of DNAPL is quite complete, attainment of MCLs throughout the source zone can almost never be expected immediately after source zone remediation. As a result, where attainment of MCLs is the absolute objective, the general conclusion might be that no technology is capable of meeting the objective and thus source zone remediation may not be attempted.

Furthermore, depending on where they are measured, MCLs can constitute a confusing absolute objective. Consider a case in which 99.9 percent of the DNAPL is removed, and some wells in the source zone show concentrations below MCLs. Nonetheless, a sample taken from a well far downgradient of the remaining DNAPL may continue to be above MCLs for a considerable period, given the time required for the effects of the remedy to be felt away from the source. In fact, downgradient concentrations may initially be virtually unchanged even though both the mass flux from the site and the time required for the site to return to precontamination conditions have been dramatically reduced. The position of the monitoring well with respect to any remaining contamination is also a factor. In this case, a well that is directly downgradient of the remaining mass may not show a large change in concentration compared to a well located directly downgradient of a portion of the source that was effectively treated. Furthermore, contaminant concentrations in groundwater from downgradient wells are highly dependent upon specific well design (screen interval) and location, making it difficult to interpret the significance of measured concentrations. Thus, the relationship between monitoring well concentrations and source remediation is complex, which should be kept in mind when selecting MCLs as metrics.

*NCP Balancing Criteria*

The five balancing criteria in the NCP, designed to guide the selection of the most appropriate among several remedies that could meet the threshold criteria, are (3) long term effectiveness and permanence, (4) reduction of mobility, toxicity, or volume, (5) short term effectiveness, (6) implementability, and (7) cost. These criteria, in the aggregate, address multiple types of objectives. Cost is likely an absolute objective and is clearly independent of either of the two "threshold" criteria. In contrast, effectiveness primarily addresses a judgment of functionality, presumably relative to the threshold criteria. Implementability appears to address both functional and procedural objectives.

## NCP Modifying Criteria

The modifying criteria help to clarify cases where more than one alternative is judged suitable according to the threshold and balancing criteria. These are (8) state acceptance and (9) community acceptance. That EPA has placed them in the least valued of its categories clearly indicates that these objectives are not, from an agency point of view, absolute objectives.

Applying the above-described criteria, EPA attempts to select the remedial alternative that utilizes "permanent solutions and treatment" to the "maximum extent practicable" and that is "cost-effective," in the sense that the costs it entails are in proportion to the treatment effectiveness. In making this decision, EPA has considerable discretion and flexibility. In consultation with state officials, EPA next issues, for public comment, a proposed plan that sets forth the agency's recommended remedial alternative. After public review and comment, EPA will make a final remedy selection, which it will document in a formal ROD.

Once the ROD has been published, EPA (or one or more responsible parties, sometimes including the Department of Defense) goes about designing, constructing, and implementing the selected remedy. If at the close of that cleanup phase contamination left in place is at levels above those allowing for unrestricted land use, then long-term monitoring and sometimes institutional controls are required. Such monitoring and controls must be in place until the site no longer poses an unacceptable risk to the environment or human health. At some sites, in fact, they may be required indefinitely. The effectiveness of long-term monitoring and institutional controls over long periods is discussed elsewhere (EPA, 1998, 1999; NRC, 1999b, 2000a,b, 2003).

## The RCRA Corrective Action Program

Although the Superfund regulatory system described above is the source of most of the regulatory requirements that affect the cleanup of military facilities, Superfund is supplemented in that respect by other governmental requirements. In 1984, Congress enacted a comprehensive set of amendments to the Resource Conservation and Recovery Act (RCRA). Among other things, those amendments made it clear that owners and operators of treatment, storage, and disposal facilities (TSDFs) (including military facilities of this type) must investigate and, as necessary, clean up past as well as present releases of hazardous wastes from their properties (hence, RCRA Corrective Action Program).

This program is administered by EPA and by those states that have been authorized by EPA to administer their own state hazardous waste programs. Although RCRA does not expressly require source treatment, in the early 1990s, EPA proposed corrective action regulations that adopted essentially the CERCLA remedy selection factors for RCRA sites (these regulations have not yet been finalized). EPA's One Cleanup Program initiative also reinforces its view that one set of rules should apply to all similar sites regardless of the statutory program. Under RCRA, state-administered hazardous waste programs are required to be "equivalent" to the federal hazardous waste program (Section 6926(b)). Nonetheless, such programs contain numerous details, and they include considerable variations in their corrective action components as well as in their other features. In view of this substantive diversity, this report does not attempt to summarize the corrective action requirements imposed by particular states.

Two "environmental indicators," developed by EPA, suggest that the absolute objectives of RCRA are to eliminate the most immediate public health and environmental risks. The two indicators, "Current Human Exposures under Control" and "Migration of Contaminated Groundwater under Control," measure whether people are currently being exposed to unacceptable levels of environmental contamination, and whether existing groundwater contamination is growing and/or affecting nearby surface waterbodies. The functional objectives that are frequently called for in site-specific agreements between owners and operators of TSDFs and regulatory authorities are typically defined in terms of concentrations of particular contaminants as measured at the boundaries of given units of real property.

### Human Health and Environmental Risk Assessment Framework

CERCLA and the NCP define a regulatory process for characterizing the level of hazard presented by site contaminants and identifying the degree of cleanup required. This regulatory process has historically been focused on the metric of risk (NRC, 1999b). During the site investigation process, information is collected to identify the sources of contamination, the extent of contamination, and the environmental characteristics and conditions contributing to exposure and potential risk. This information is used to conduct human health and ecological risk assessments during the RI/FS, following the guidance provided by the EPA known as Risk Assessment Guidance for Superfund (RAGS) (EPA, 1989, 1991a). Other risk-based methodologies, including the ASTM Risk-Based Corrective Action approach (ASTM, 1998), which is similar to the RAGS process, may also be used.

Risk assessment applied to environmental cleanup of hazardous waste sites is the process of determining the level of risk posed by chemicals at the site to (primarily local) human and ecological receptors. The risk assessment process integrates information on the physical conditions at the site, the nature and extent of contamination, the toxicological and physical–chemical characteristics of the contaminants, the current and future land use conditions, and the dose–response relationship between projected exposure levels and potential toxic effects (see Table 4-6 for toxicological data on DNAPL constituents and chemical explosives). The end result of human health and ecological risk assessment is a numerical value of potential additional risk to the hypothetical receptor from the contaminant source. The calculated risk values are compared to an acceptable target risk level or to a range of acceptable risk defined by the NCP or by state regulations. If the risk estimate is greater than the acceptable target risk level, target cleanup level objectives are identified for the site using the assumptions developed in the risk assessment related to potential levels of exposure.

The overall purpose of risk assessment is to address the absolute objective of protecting human health and the environment. The risk assessment will determine if site-specific risk is above acceptable limits and the extent to which site risk needs to be reduced to meet the absolute objective. The risk assessment will also provide information that will support development of functional objectives, such as identifying which chemicals and exposure pathways contribute most to elevated risk. It will also help define metrics of success related to remediation. For example, the risk assessment may determine that levels of chlorinated solvents in groundwater would result in unacceptable risk if the groundwater is used as a source of potable water. If the ability to reduce the groundwater contaminant concentrations were limited by a lack of available technologies, the metric of risk reduction might be met by supplying an alternate source of potable water to residents.

RAGS and the other risk-based methods commonly used to evaluate site risk and establish cleanup levels provide a standardized, systematic approach for estimating site risk. The standardized approach allows for relatively easy implementation of the methods at a large number of sites and allows sites to be prioritized for cleanup action. These methodologies and their strengths and weaknesses for different applications have been described and evaluated in other NRC reports (e.g., NRC, 1983, 1999b). Since all contaminated military facilities conduct their site investigations and cleanups under either RCRA or CERCLA (NRC, 2003), it is likely that at identified Army sites where source remediation is an option, a risk assessment has already been conducted or will be conducted in the future.

*Distinctions between Human Health and Ecological Risk Assessment*

The methods typically used in human health risk assessment are highly prescribed by RAGS and similar risk-based methodologies. RAGS requires that risk estimates for humans be protective of individuals and be based on the maximum exposure that is reasonably likely to occur. This risk estimate tends to be conservative, that is, it is more likely to overestimate than underestimate true risk. Site-specific information concerning exposure can be used in calculating the risk estimate when the information is well documented. More typically, however, default exposure assumptions identified by the standardized risk-based method-ologies and toxicity criteria developed by EPA based on laboratory animal testing data are used. Because of the uncertainly inherent in using animal data to predict toxicity in humans, the toxicity criteria recommended for use by EPA have incor-porated modifying factors that result in far lower allowed chemical intake in humans. For practical reasons, the use of these standardized assumptions for exposure and toxicity in evaluating human health risk is encouraged by the regula-tors. The outcomes of this approach are relatively limited flexibility in accounting for site-specific conditions and risk estimates that represent higher-than-average exposure conditions. Box 4-4 further discusses the role of uncertainty in risk assessment calculations, and how uncertainty can be more quantitatively assessed in lieu of using the default assumptions discussed above.

The ecological risk assessment process is far less prescribed in the published risk-based methodologies than the human health risk evaluation process for several reasons. In the ecological risk assessment process, ecological risk does not exist unless receptors and habitat are currently present at a site or are likely to be present in the future. Highly developed industrial sites are less likely to sustain ecological receptors and habitat. Unlike human health risk assessment, where risk to only one species is evaluated, ecological risk assessment must consider all ecological receptors present or potentially present in all environmental media potentially impacted. This evaluation requires a site-specific survey to determine what types of receptors (plants, animals, invertebrates, etc.) are present in each medium (soil, surface water, sediment, etc.). Finally, ecological risk assessment evaluates risk to the population of each species present, being concerned with risk to individual members of the population only when the receptor is classified as a threatened or endangered species by state or federal regulations. The available risk-based methodologies present a general framework for ecological risk evalu-ation (EPA, 1991b, 1992c, 1994, 1997b), but the type of evaluation conducted for a specific site is typically negotiated with the regulatory agency having responsi-bility for the site.

### Exposure Pathways at Army Facilities

Explosives and DNAPL contamination at Army facilities can represent very long-term sources of contamination for soil, groundwater, and surface water. If the explosives or DNAPL contamination is present in relatively shallow soil (4–6 meters below ground surface), direct contact with contaminated soil (inges-tion, dermal contact) could occur through or as a result of excavation activities that might bring contaminated subsurface soil to the surface. Army facility occu-pants and offsite occupants may indirectly contact contaminants in shallow soil

---

## BOX 4-4
## Evaluation of Variability and Uncertainty in Risk Assessment

The inherent variability in exposure variables or population response and the lack of knowledge about specific parameters used in estimating risk can both affect the outcome of a risk assessment and the degree of confidence associated with the results. Evaluation of these sources of uncertainty is necessary to allow risk assessment results to be viewed in the appropriate context. "Variability" refers to the true heterogeneity or diversity that occurs within a population or a sample. Examples of factors that have associated variability include contaminant concentration in a medium (air, water, soil, etc.), differences in exposure frequencies or duration, or, in the case of ecological risk assessments, inter- and intraspecies variability in dose–response relationships. EPA risk assessment guidance (EPA, 1989) states that risk management decisions at Superfund sites will generally be based on an individual that has a *reasonable maximum exposure* (RME). The intent of the RME is to estimate a conservative exposure case (i.e., well above the average case) that is still within the range of possible exposures based on both quantitative information and professional judgment. In addition, EPA recommends conducting a *central tendency exposure* estimate (CTE), which is a measure of the mean or median exposure. The difference between the CTE and RME gives an initial impression of the degree of variability in exposure and risk between individuals in an exposed population (EPA, 2001a).

If a risk assessment has been conducted using a point estimate approach, a range of point estimates can be developed to represent variability in exposures. To calculate RME risk estimates using this approach, EPA has developed recommended default exposure values to use as inputs to the risk equations (EPA, 1992a, 1996a, 1997b, 2001b). A CTE risk estimate is calculated using central estimates for each of the exposure variables, which are available from EPA guidance and other sources. For both RME and CTE risk estimates, site-specific data are used if they are available. The point estimate approach to risk assessment does not determine where the CTE or RME risk estimates lie within the risk distribution, and the likelihood that an estimated risk will be sustained cannot be determined. This leads to uncertainty as to what level of remedial action is justified or necessary.

If a risk assessment has been conducted using probabilistic techniques, parameter distributions are used as inputs to the risk equations rather than single values. These distributions characterize the interindividual variability inherent in each of the exposure assumptions, and they are used with mathematical processes such as Monte Carlo simulation to estimate risk. The simulation output is a distribution of risks that would occur in the population, which provides a better understanding of where the CTE and RME risks occur in the distribution. A technique known as one-dimensional Monte Carlo analysis can be used to estimate the probability of occurrence associated with a particular risk level of concern (e.g., cancer risk of $10^{-6}$) (EPA, 2001a).

Uncertainty is also inherent in every human health and ecological risk assessment because one's knowledge of actual exposure conditions and receptor

*Continued*

## BOX 4-4 Continued

response to chemical exposure is imprecise. The degree of uncertainty depends to a large extent on the amount and adequacy of the available facility-specific data. Typically, the most significant areas of uncertainty associated with receptor exposure include exposure pathway identification, exposure assumptions, assumptions of steady-state conditions, environmental chemical characterization, and modeling procedures. The toxicity values used in risk assessment must also be viewed in light of uncertainties and gaps in toxicological data. Information concerning the effect of a chemical on humans is often limited. Toxicity data are often based on data derived from high-dose studies using a specially bred homogeneous animal population. These data are extrapolated for use in predicting risk to a heterogeneous human population that is more likely to experience a low-level, long-term exposure (EPA, 2001a).

Ideally, the uncertainty associated with each parameter used in the risk assessment would be carried through the evaluation process in order to characterize the uncertainty associated with the final risk estimates. However, since actual exposure conditions cannot be fully described, a variety of modeling strategies are available to evaluate uncertainty. If a risk assessment has been conducted using a point estimate approach, parameter uncertainty is usually addressed in a qualitative manner for most variables (EPA, 2001a). For example, the uncertainty section of a point estimate risk assessment document might note that soil sampling conducted may not be representative of overall contaminant concentrations and, as a result, the risk estimate may over- or underestimate actual risk. Uncertainty in the environmental concentration term is addressed quantitatively to a limited extent in a point estimate approach by using the 95 percent upper confidence limit (UCL) for the arithmetic mean concentration in the risk estimate, which accounts for uncertainty associated with environmental sampling and site characterization (EPA, 1992b, 1997c, 2001a). The 95 percent UCL is combined in the same risk calculation with various central tendency and high-end point estimates for other exposure factors.

If a risk estimate is conducted using probabilistic methods, the uncertainty associated with the best estimate of the exposure or risk distribution can be quantitatively estimated using a two-dimensional Monte Carlo analysis. This analysis can provide a quantitative measure of the confidence in the fraction of the population with a risk exceeding a particular level. Additionally, the output from this analysis can provide a quantitative measure of the confidence in the risk estimate for a particular fraction of the population (EPA, 2001a).

Compared to a point estimate risk assessment, a probabilistic risk assessment based on the same state of knowledge can provide a more complete characterization of variability in risk and a quantitative evaluation of uncertainty. In deciding whether a probabilistic assessment of risk should be performed, the key question is whether this type of analysis (vs. a point estimate assessment) is likely to provide information that will help in risk assessment decision making. To assist site managers in deciding what type of risk assessment is best suited to their site, decision-making tools such as a tiered approach developed by EPA based on "scientific management decision points" are available to help identify the complexity of analysis that may be needed (EPA, 2001a).

through inhalation of DNAPL contaminants that have volatilized and migrated to the ground surface.

If groundwater contaminated with explosives or DNAPLs is used as a potable water source at an Army facility, exposure to facility occupants can occur through direct contact with the water during ingestion and via dermal contact. Indirect contact through inhalation of DNAPL volatile organic compounds that become airborne during water use or through migration to the ground surface or into occupied structures is also a possibility. The same type of exposure can occur offsite if contaminant migration has occurred or could occur in the future, and if groundwater is used by offsite residents as a source of potable water.

Ecological receptors are most likely to contact contaminants from explosives or DNAPL after the contaminants have migrated through groundwater and have discharged to surface water. In these cases, the threat may be somewhat less given the dilution of the contaminant that is likely to occur once it is discharged to the surface water and given the rapid volatilization of many DNAPL contaminants to air. Ecological receptors are not likely to contact explosive or DNAPL contaminants in soil below the top meter unless excavation activities bring contaminated soil to the surface.

The physical extent of the contamination and the timeframe required for its reduction to levels that represent an acceptable risk affect several elements of exposure assessment and subsequent risk characterization:

• The higher the concentration of the contaminant in the environmental medium, the higher the potential intensity of the exposure.
• The more widespread the source of contamination, the larger the potential population of receptors that may contact the contaminant and/or the higher the potential exposure frequency.
• The longer it takes to remove contamination from the environment, the longer the potential exposure duration.

In many cases, these factors require that the overall objective of protecting human health and the environment be met through a combination of treatment and long-term site management actions.

*Time-Scale Considerations for Risk Assessment*

The risk-based methods typically used at contaminated sites evaluate carcinogenic and noncancer risk to a hypothetical individual over the course of the person's lifetime. These methods do not factor the lifetime of a source of contamination into risk estimates. They do not typically evaluate the size of the population potentially at risk, nor do they consider risk beyond the lifetime of an individual (i.e., they do not consider cumulative risk to the entire population exposed for the lifetime of the source of contamination). These shortcomings are serious, given that source zone cleanup may take decades to complete for technical and financial reasons, and some level of contamination is likely to be left in place[9] for an extended period of time.

Some types of chemical sources that represent particularly long-term problems in groundwater (e.g., chlorinated solvents) are known or are presumed to be highly toxic to humans. Only very low concentrations of these contaminants would be allowed in the groundwater if it were a source of drinking water. The timeframe required to achieve these low-level concentrations, either through natural site recovery or various remedial alternatives, may be so long as to be inconsistent with the timeframes implicit in risk-based methodologies. This can severely limit the ability of the risk assessor to differentiate what may be significant public health impacts, if they occur some time in the future.

The RAGS model, for example, is a static examination of risk for a fixed population, assuming constant conditions for 30 years (40 years for a family farm) under Reasonable Maximum Exposure conditions. More conservative variants of this model may address full lifetime exposure. There are also more realistic models that address changes in contaminant concentration over time, as well as residential mobility, aging, and other demographic factors influencing exposure (e.g., Price et al., 1996; Wilson et al., 2001), but these also fail to address timeframes of contamination that may span centuries.

Accordingly, existing risk metrics may be unable to demonstrate benefits from source remediation efforts, if the primary effect of those efforts is to reduce the time over which the source contributes to elevated contaminant concentrations in groundwater. If, hypothetically, a remedy were to have the effect on contaminant concentrations after an interval of several decades, it would not be detectable with current risk metrics. For the population that will reside in the area in the future, however, the risks from use of the groundwater have been substantially reduced. That is, in the absence of a remedy that will be effective within 30 years, existing analytical frameworks obscure important distinctions between remedies that are effective in 100 vs. 500 years.

Techniques are available for longer-term types of risk evaluation; these techniques and associated models have been used for many years to evaluate risk associated with Department of Energy legacy waste sites where very long-lived radionuclides will be present in the environment for thousands of years (Yu et al., 1993; EPA, 1996b). Unfortunately, with chemical contaminants, estimation of population risk over the lifetime of the contaminant source is not typically conducted because there is no regulatory requirement to conduct such an evaluation, nor is there a currently prescribed regulatory context for considering the results of

---

[9]"Contamination left in place," as used in this report, is consistent with the interagency definition as hazardous substances, pollutants, or contaminants remaining at the site above levels that allow for unlimited use and unrestricted exposure (Air Force/Army/Navy/EPA, 1999).

such an evaluation. [It should be noted, however, that even though tools for evaluating long-lived contaminants are available, they are considered imperfect in predicting long-term risk because they rely on unverifiable assumptions about the future behavior of people and institutions (NRC, 2000a).] Existing risk assessment frameworks are badly in need of explicit reconsideration to better reflect the physical realities at sites if the best attainable remedies for these sites are to be selected.

## CONCLUSIONS AND RECOMMENDATIONS

As mentioned in the opening of this chapter, clear definitions of absolute and functional objectives and metrics for success are not evident in most of the reports (both Army and non-Army) reviewed by the committee. This has made it difficult to determine the "success" of projects under any consistent definition. Reports of early (pre-2000) projects seldom contained sufficient rationale for how and why certain technologies were selected. More recent projects discuss objectives such as concentration reduction in the dissolved phase plume or reduction of source mass, but there is seldom evidence to suggest that the technology selected would meet those specific objectives. Indeed, within the Army several source remediation technologies have been piloted and then selected for scaling up in the absence of having specific cleanup objectives prior to the pilot projects. As an illustration of this, in situ chemical oxidation might have been attempted for a small portion of the source zone during a pilot study and found to achieve a certain percentage of mass removal. The committee observed that this would subsequently lead to full-scale implementation of the technology (1) without considering whether mass removal would meet the objectives of full-scale cleanup (which may be, for example, protection of human health) or (2) in the absence of any objectives for full-scale cleanup. Thus, in many cases observed by the committee, the decision to proceed with larger-scale remediation was not based on a demonstrated ability to achieve cleanup objectives. Rather, if the pilot test showed significant concentration reductions or mass removal, it was simply assumed that a larger-scale project would bring more widespread reductions.

The following recommendations regarding objectives for source remediation are made.

**Remedial objectives should be laid out *before* deciding to attempt source remediation and selecting a particular technology.** The committee observed that remedies are often implemented in the absence of clearly stated objectives, which are necessary to ensure that all stakeholders understand the basis of subsequent remediation decisions. Failure to state objectives in advance virtually guarantees stakeholder dissatisfaction and can lead to expensive and fruitless "mission creep" as alternative technologies are applied. This step is as important as accurately characterizing source zones at the site.

**A clear distinction between functional and absolute objectives is needed to evaluate options.** If a given objective is merely a means by which an absolute objective is to be obtained (i.e., it is a functional objective), this should be made clear to all stakeholders. This is particularly important when there are alternative methods under consideration to achieve the absolute objectives, and when it is known or is likely that different stakeholders have a different willingness to substitute objectives for one another.

**Each objective should result in a metric; that is, a quantity that can be measured at the particular site in order to evaluate achievement of the objective.** Objectives that lack metrics should be further specified in terms of subsidiary functional objectives that do have metrics. Furthermore, although decisions depend upon both technical and nontechnical factors, once a decision has been made, the focus should be on the technical metric to determine if remediation is successful.

**Objectives should strive to encompass the long time frames characteristic of many site cleanups that involve DNAPLs.** In some existing frameworks, timeframes are very short (rarely longer than 30 years) relative to the persistence of DNAPL (up to centuries), such that alternative actions with significant differences in terms of the speed with which a site can be remedied cannot be distinguished. Within life cycle cost analysis, the chosen timeframe and discount rate can significantly affect cost estimations for different remedies. Decision tools with a more realistic temporal outlook have been developed in other areas of environmental science (e.g., storage and disposal of radioactive materials). Their application to DNAPL problems needs to be considered by the Army and by the site restoration community as a whole.

## REFERENCES

Abriola, L. M., K. Rathfelder, M. Maiza, and S. Yadav. 1992. VALOR Code Version 1.0: A PC code for simulating immiscible contaminant transport in subsurface systems. TR-101018. Palo Alto, CA: Electric Power Research Institute.

Abriola L. M., C. D. Drummond, E. J. Hahn, K. F. Hayes, T. C. G. Kibbey, L. D. Lemke, K. D. Pennell, E. A. Petrovskis, C. A. Ramsburg, and K. M. Rathfelder. 2005. Pilot-scale demonstration of surfactant-enhanced PCE solubilization at the Bachman Road site: (1) site characterization and test design. Environ. Sci. Technol. (In press).

Abriola, L. M., C. D. Drummond, L. M. Lemke, K. M. Rathfelder, K. D. Pennell, E. Petrovskis, and G. Daniels. 2002. Surfactant enhanced aquifer remediation: application of mathematical models in the design and evaluation of a pilot-scale test. Pp. 303–310 *In*: Groundwater Quality: Natural and Enhanced Restoration of Groundwater Pollution. S. F. Thornton and S. E. Oswald (eds.). IAHS Publication No. 275. Wallingford, Oxfordshire, UK: International Association of Hydrological Sciences.

Abriola, L. M., C. A. Ramsburg, K. D. Pennell, F. E. Löffler, M. Gamache, and E. A. Petrovskis. 2003. Post-treatment monitoring and biological activity at the Bachman road surfactant-enhanced aquifer remediation site. ACS preprints of extended abstracts 43:921–927.

Adamson, D. T., J. M. McDade, and J. B. Hughes. 2003. Inoculation of a DNAPL source zone to initiate reductive dechlorination of PCE. Environ. Sci. Technol. 37:2525–2533.

Air Force/Army/Navy/EPA. 1999. The Environmental Site Closeout Process Guide. Washington, DC: DoD and EPA.

American Society for Testing and Materials. 1998. Standard Provisional Guide for Risk-Based Corrective Action (PS 104-98). Annual Book of ASTM Standards. West Conshohocken, PA: ASTM.

Barnthouse, L., J. Fava, K. Humphreys, R. Hunt, L. Laibson, S. Noesen, J. Owens, J. Todd, B. Vigon, K. Weitz, and J. Young. 1997. Life-Cycle Impact Assessment: The State-of-the-Art. Pensacola, FL: SETAC Press.

Department of the Army. 2002. Cost Analysis Manual. U.S. Army Cost and Economic Analysis Center.

Environmental Protection Agency (EPA). 1986. Guidelines for Carcinogen Risk Assessment. Federal Register 51:33991.

EPA. 1989. Risk Assessment Guidance for Superfund (RAGS): Volume I. Human Health Evaluation Manual (HHEM) (Part A, Baseline Risk Assessment). Interim Final. EPA/540/1-89/002. Washington, DC: Office of Emergency and Remedial Response.

EPA. 1991a. Risk Assessment Guidance for Superfund (RAGS): Volume I-Human Health Evaluation Manual Supplemental Guidance: Standard Default Exposure Factors. Interim Final. OSWER Directive No. 9285.6-03. Washington, DC: EPA Office of Solid Waste and Emergency Response.

EPA. 1991b. Ecological Assessment of Superfund Sites: An Overview. Publication No. 9345.0-051. Washington, DC: US EPA Office of Solid Waste and Emergency Response.

EPA. 1992a. Final Guidelines for Exposure Assessment. EPA/600/Z-92/001. Washington, DC: EPA.

EPA. 1992b. Supplemental Guidance to RAGS: Calculating the Concentration Term. OSWER Directive No. 9285.7-081. Washington, DC: EPA Office of Solid Waste and Emergency Response.

EPA. 1992c. Developing a Work Scope for Ecological Assessments. Publication No. 9345.0-051. Washington, DC: EPA Office of Solid Waste and Emergency Response.

EPA. 1994. Field Studies for Ecological Risk Assessment. Publication No. 9345.0-051. Washington, DC: EPA Office of Solid Waste and Emergency Response.

EPA. 1995. Federal facility pollution prevention project analysis: a primer for applying life cycle and total cost assessment concepts. EPA 300-B-95-008. Washington, DC: EPA Office of Enforcement and Compliance Assurance.

EPA. 1996a. Final Soil Screening Guidance: User's Guide. EPA 540/R-96/018. Washington, DC: EPA Office of Solid Waste and Emergency Response.

EPA. 1996b. Fact Sheet: Environmental Pathway Models-Ground-Water Modeling in Support of Remedial Decision Making at Sites Contaminated with Radioactive Material. EPA/540/F-94-024. Washington, DC: EPA Office of Radiation and Indoor Air.

EPA. 1997a. Guiding Principles for Monte Carlo Analysis. EPA/630/R-97/001. Washington, DC: EPA.

EPA. 1997b. Ecological Risk Assessment Guidance for Superfund: Process for Designing and Conducting Ecological Risk Assessments. Interim Final. EPA/540/R-97/006, OSWER Directive No. 9285.7-25. Edison, NJ: EPA Environmental Response Team.

EPA. 1997c. Lognormal Distribution in Environmental Applications. EPA/600/R-97/006. Washington, DC: EPA Office of Research and Development and Office of Solid Waste and Emergency Response.

EPA. 1998. Institutional Controls: A Reference Manual (Working Group Draft).

EPA. 1999. Department of Defense (DoD) Range Rule. Letter from Timothy Fields, Jr., Acting Assistant Administrator, U.S. Environmental Protection Agency, to Ms. Sherri Goodman, Deputy Under Secretary of Defense, Department of Defense, April 22, 1999.

EPA. 2000. A Guide to Developing and Documenting Cost Estimates During the Feasibility Study. EPA 540-R-00-002. Washington, DC: U.S. Army Corps of Engineers and EPA Office of Emergency and Remedial Response.

EPA. 2001a. Risk Assessment Guidance for Superfund: Volume III – Part A, Process for Conducting Probabilistic Risk Assessment. EPA 540-R-02-002, OSWER 9285.7-45. Washington, DC: EPA Office of Emergency and Remedial Response.

EPA. 2001b. The Role of Screening-Level Risk Assessments and Refining Contaminants of Concern Baseline Risk Assessments. 12th Intermittent Bulletin, ECO Update Series. EPA 540/F-01/014. Washington, DC: EPA Office of Solid Waste and Emergency Response.

EPA. 2002. 2002 Edition of the Drinking Water Standards and Health Advisories. EPA 822-R-02-038. Washington, DC: EPA Office of Water.

EPA. 2003a. Abstracts of Remediation Case Studies, Volume 7. Washington, DC: EPA Federal Remediation Technologies Roundtable.

EPA. 2003b. The DNAPL Remediation Challenge: Is There a Case for Source Depletion? EPA 600/R-03/143. Washington, DC: EPA Office of Research and Development.

Gregory, M. 1993. Health Effects of Hazardous Waste: An Environmentalist Perspective. Presented to ATSDR Hazardous Waste Conference in 1993. http://www.atsdr.cdc.gov/cx2b.html.

Lemke, L. D. 2003. Influence of alternative spatial variability models on solute transport, DNAPL entrapment, and DNAPL recovery in a homogeneous, nonuniform sand aquifer. Ph.D. dissertation, University of Michigan, Civil and Environmental Engineering.

Lemke, L. D., and L. M. Abriola. 2003. Predicting DNAPL entrapment and recovery: the influence of hydraulic property correlation. Stochastic Environmental Research and Risk Assessment. 17:408–418, doi 10.1007/s00477-003-0162-4.

Lemke, L. D., L. M. Abriola, and J. R. Lang. 2004. DNAPL source zone remediation: influence of hydraulic property correlation on predicted source zone architecture, DNAPL recovery, and contaminant mass flux. Water Resources Research 40, W01511, doi:10.1029/2003WR001980.

Li, X. D., and F. W. Schwartz. 2004. DNAPL remediation with in situ chemical oxidation using potassium permanganate: part I—mineralogy of Mn oxide and its dissolution in organic acids. Journal of Contaminant Hydrology 68:39–53.

Londergan, J. T., H. W. Meinardus, P. E. Mariner, R. E. Jackson, C. L. Brown, V. Dwarakanath, G. A. Pope, J. S. Ginn, and S. Taffinder. 2001. DNAPL removal from a heterogeneous alluvial aquifer by surfactant-enhanced aquifer remediation. Ground Water Monitoring and Remediation 21:57–67.

National Research Council (NRC). 1983. Risk Assessment in the Federal Government: Managing the Process. Washington, DC: National Academy Press.

NRC. 1996. Understanding Risk. Washington, DC: National Academy Press.

NRC. 1997. Innovations in Ground Water and Soil Cleanup. Washington, DC: National Academy Press.

NRC. 1999a. Groundwater and Soil Cleanup: Improving Management of Persistent Contaminants. Washington, DC: National Academy Press.

NRC. 1999b. Environmental Cleanup at Navy Facilities: Risk-Based Methods. Washington, DC: National Academy Press.

NRC. 2000a. Long-Term Institutional Management of U.S. Department of Energy Legacy Waste Sites. Washington, DC: National Academy Press.

NRC. 2000b. Natural Attenuation for Groundwater Remediation. Washington, DC: National Academy Press.

NRC. 2003. Environmental Cleanup at Navy Facilities: Adaptive Site Management. Washington, DC: National Academies Press.

Nielsen, R. B., and J. D. Keasling. 1999. Reductive dechlorination of chlorinated ethene DNAPLS by a culture enriched from contaminated groundwater. Biotechnology and Bioengineering 62:160–165.

Phelan, T. J., S. A. Bradford, D. M. O'Carroll, L. D. Lemke, and L. M. Abriola. 2004. Influence of textural and wettability variations on predictions of DNAPL persistence and plume development in saturated porous media. Advances in Water Resources 27(4):411–427.

Price, P. S., C. L. Curry, P. E. Goodrum, M. N. Gray, J. I. McCrodden, N. W. Harrington, H. Carlson-Lynch, and R. E. Keenan. 1996. Monte Carlo modeling of time-dependent exposures using a Microexposure event approach. Risk Anal. 16(3):339–348.

Rao, P. S. C., and J. W. Jawitz. 2003. Comment on "Steady state mass transfer from single-component dense nonaqueous phase liquids in uniform flow fields" by T. C. Sale and D. B. McWhorter. Water Resources Research 39: 1068, doi: 10.1029/2001WR000599.

Rao, P. S., J. W. Jawitz, G. C. Enfield, R. W. Falta, M. D. Annable, and L. A. Wood. 2002. Technology integration for contaminated site remediation: clean-up goals and performance criteria. Pp. 571–578 *In:* Groundwater Quality: Natural and Enhanced Restoration of Groundwater Pollution. S. F. Thornton and S. E. Oswald (eds.). IAHS Publication No. 275. Wallingford, Oxfordshire, UK: International Association of Hydrological Sciences.

Rathfelder, K., and L. M. Abriola. 1998. On the influence of capillarity in the modeling of organic redistribution in two-phase systems. Advances in Water Resources 21(2):159–170.

Rathfelder, K. M., L. M. Abriola, T. P. Taylor, and K. D. Pennell. 2001. Surfactant enhanced recovery of tetrachloroethylene from a porous medium containing low permeability lenses. II. Numerical simulation. Journal of Contaminant Hydrology 48:351–374.

Sung, Y., K. M. Ritalahti, R. A. Sanford, J. W. Urbance, S. J. Flynn, J. M. Tiedje, and F. E. Löffler. 2003. Characterization of two tetrachloroethene-reducing, acetate-oxidizing anaerobic bacteria and their description as *Desulfuromonas michignensis* sp. nov. Applied and Environmental Microbiology 69:2964–2974.

Taylor, T. P., K. D. Pennell, L. M. Abriola, and J. H. Dane. 2001. Surfactant enhanced recovery of tetrachloroethylene from a porous medium containing low permeability lenses. I. Experimental studies. Journal of Contaminant Hydrology 48:325–350

Udo de Haes, H., R. Heijungs, P. Hofstetter, G. Finnveden, O. Jolliet, P. Nichols, M. Hauschild, J. Potting, P. White, and E. Lindeijer. 1996. Towards a Methodology for Life Cycle Impact Assessment. Brussels: SETAC-Europe.

Wilson, N., P. Price, and D. J. Paustenbach. 2001. An event-by-event probabilistic methodology for assessing the health risks of persistent chemicals in fish: a case study at the Palos Verdes Shelf. J. Toxicol. Environ. Health 62:595–642.

Yang, Y., and P. L. McCarty. 2000. Biologically enhanced dissolution of tetrachloroethene DNAPL. Environ. Sci. Technol. 34:2979–2984.

Yu, C., A. J. Zielen, J.-J. Cheng, Y. C. Yuan, L. G. Jones, D. J. LePoire, Y. Y. Wang, C. O. Lourenro, E. K. Gnanapragasam, E. Faillace, A. Wallo III, W. A. Williams, and H. Peterson. 1993. Manual for Implementing Residual Radioactive Material Guidelines Using RESRAD Version 5.0. ANL/EAD/LD-2. Argonne, IL: Argonne National Laboratory.

# 5

# Source Remediation Technology Options

Many aggressive source remediation technologies have become increasingly popular in the last five years, which partly underlies the Army's request for this study. This chapter presents those technologies that have surfaced as leading candidates for source zone remediation, including a description of each technology, a discussion of the technology's strengths and weaknesses, and special considerations for the technology. The following sections are not necessarily equivalent because information on each technology is complete to varying degrees. For example, numerous case studies are available for surfactant flooding, chemical oxidation, and steam flushing, while almost none exist for chemical reduction. The uneven treatment of the innovative technologies in this chapter is thus largely a reflection of the amount of data available.

Because source zone remediation is the focus of this discussion, technologies that target remediation of the dissolved plume are not discussed. Thus, for example, permeable reactive barriers, which primarily treat the plume, are not included. In addition, excavation, containment, and monitored natural attenuation are only briefly touched upon. While they may well be used in combination with a source zone remedial activity, these remedies are not considered to constitute in situ source zone remediation.

In addition to describing the state of the art for each individual technology, the chapter provides a qualitative comparison of the technologies, first by assessing the types of contaminants for which each technology is suitable, and then by qualitatively evaluating each technology's relative potential for mass removal, concentration reduction, mass flux reduction, source migration, and changes in toxicity—physical objectives discussed extensively in Chapter 4. It should be noted that effectiveness data that would be pertinent to these objectives and others discussed in Chapter 4 are infrequently gathered. Most pilot- and field-scale studies of source remediation measure effectiveness in terms of mass removal and occasionally concentration reduction (although these latter data can be very difficult to interpret). Mass flux and source migration measurements have rarely been documented. Indeed, virtually no data exist on the life cycle costs associated with the technologies. Furthermore, most reports of case studies are not published in the peer reviewed literature. These facts should be kept in mind throughout this chapter, especially when interpreting summary tables. The qualitative comparison is conducted for each of the hydrogeologic settings described in Chapter 2. Because these settings are generalizations, whether a certain technology will work for a given site depends on a complex integration of a wide range of site and contaminant properties.

The two contaminant types of concern in this report—dense nonaqueous phase liquids (DNAPLs) and chemical explosives—have varying characteristics and have been handled differently with respect to source remediation. This chapter covers DNAPLs in greater detail than explosives because most of the research to date has focused on DNAPL contamination. However, when a certain technology has been used or is applicable to chemical explosives, it is mentioned. The discussion of DNAPL treatment focuses on contamination of the saturated zone, as this medium presents the greatest difficulties in terms of site cleanup. Thus, technologies that target the unsaturated zone (e.g., soil vacuum extraction, bioventing, biosparging, etc.) are not discussed here.

Table 5-1 provides an overview of the technologies discussed in this chapter. Although excavation, containment, and pump-and-treat are considered conventional approaches for addressing DNAPL contamination, they are discussed here

TABLE 5-1 Source Remediation Technologies in This Chapter

| Technology | Approach | Page # |
|---|---|---|
| Excavation | Extraction | 180 |
| Containment | Isolation | 182 |
| Pump-and-Treat | Extraction/Isolation | 185 |
| Multiphase Extraction | Extraction | 187 |
| Surfactant/Cosolvent | Extraction | 194 |
| Chemical Oxidation | Transformation | 206 |
| Chemical Reduction | Transformation | 218 |
| Steam Flushing | Extraction/Transformation | 224 |
| Conductive Heating | Extraction/Transformation | 236 |
| Electrical Resistance Heating | Extraction/Transformation | 242 |
| Air Sparging | Extraction | 250 |
| Enhanced Bioremediation | Transformation | 256 |
| Explosives Technologies | Extraction/Transformation | 288 |

to provide a baseline for the more innovative technologies that follow. Multiphase extraction is an approach for removing as much of the mobile DNAPL as possible. The remaining technologies target residual or trapped DNAPL, and their approach is categorized as either extraction, transformation, or both. An extraction technology seeks to improve the rate at which the DNAPL can be recovered from the subsurface, while transformation technologies seek to alter the form of the DNAPL in situ. Many technologies do both. The final section in this chapter discusses technologies for treating explosive contaminants.

## CONVENTIONAL TECHNOLOGIES

The conventional technologies that play a significant role in managing source areas at hazardous waste sites include excavation, containment, and pump-and-treat. To a certain extent, excavation (if completely successful) and containment represent the extreme ends of what is possible with source remediation, in that one technology completely removes the source, while the other removes no mass whatsoever. Pump-and-treat and all of the innovative technologies discussed subsequently fall between these two extremes in their intent.

### Source Area Excavation

Excavation is commonplace for source remediation at hazardous waste sites, and is thus mentioned briefly for completeness. Excavation is carried out by heavy construction equipment that can dig out the source materials and place them into shipping containers. The containers are then shipped to an appropriate site for treatment or disposal, which may include designated onsite locations. Backfilling the excavation is required and necessitates having available clean backfill material and carefully and safely placing the backfill so that cross-contamination is avoided. All of these activities require extensive physical access to the source area.

For excavation to succeed it is essential to know the areal extent, depth, and general distribution of source materials, which suggests an intensive source characterization effort prior to the commencement of excavation. Indeed, if pre-excavation investigations are flawed, then some portion of the source zone may be unintentionally left in place. These same characterization tools are also used later to verify that all of the source material has been removed and to classify materials encountered during the excavation as contaminated or uncontaminated. In addition to information on the size and shape of the source zone, basic geotechnical information is also important to predicting the success of excavation. For example, one should determine whether bedrock is present, as it is hard to excavate. Excavation below the water table is difficult due to the influx of groundwater, which is contaminated by contact with the source material and must be treated. Saturated sandy soils tend to liquefy during excavation (the jargon for this phenomenon is running sands) and can dramatically raise the complexity of excavation—in some cases sheet piling or dewatering systems must be employed around the source materials to reduce the water flow and to stabilize the side walls and bottoms of excavations. Finally, the ability to completely excavate a source is highly dependent on having adequate physical access to the source zone. If physical access is restricted by nearby foundations or buildings, complete removal may not be possible without serious damage to surrounding structures. In any case where an excavation is planned near a foundation or a building, it is particularly important to have a high quality-investigation before excavation begins. It is also difficult to excavate on steep slopes with a thin layer of contaminated soil because the construction equipment tends to slip in dangerous ways.

Certain hydrogeologic settings are more amenable to excavation as a remedial action. Shallow source zones in hydrogeologic settings I, II, and III can readily be excavated with standard equipment. Some Type IV sedimentary bedrock—for example, soft sandstone or shale—can be excavated. However, excavation of source zones in bedrock that falls into hydrogeologic settings IV and V is generally difficult, especially if the source zone is in igneous or metamorphic rock. Overall, experience has shown that excavation works best and is most cost-competitive at sites where confining layers are shallow, soil permeabilities are low, the volume of source materials is under 5,000 m$^3$, and the contaminants do not require complex treatment or disposal. Many other references, including NRC (2003), discuss innovative and adaptive ways of excavating sites to ensure more complete capture of the entire source zone.

As suggested in Chapter 2, excavation is the principal remediation measure for near-surface explosives source areas. When there is risk of detonation, telerobotic remote excavation equipment can be used to increase the standoff distance between the field teams and the source areas. For very high explosives concentrations, removed soil must be blended with clean soil as a pretreatment, followed by incineration or composting, the latter of which has become the principal technology for treating soils highly contaminated with explosives.

The primary advantage of excavation is that source materials are taken out of the groundwater system very quickly. Migration of contaminants out of the source area is cut off as soon as excavation is finished. Excavation may be inexpensive compared to in situ treatments, and is often preferred by potentially responsible parties (PRPs) and stakeholders because of its perceived simplicity. There are also many disadvantages to excavation, especially the need for an area that can receive the excavated material, the dangers of working with heavy excavation equipment, worker exposure to potential volatile organic compound (VOC) releases, and the inability to predict source area volumes. Indeed, experience with excavation is that projects often remove up to twice the volume of source material predicted before the excavation began because of faulty initial source characterization. Deep excavations may require benching, which greatly increases the volume of soil excavated. Furthermore, when water tables are lowered for excavation, it is likely that DNAPL will be remobilized and will flow into the excavation, creating a worker exposure hazard. Finally, cost can be a disadvantage if the excavated volume is large or if the source materials removed are subject to land disposal restrictions that lead to high ex situ treatment costs (e.g., incineration of Resource Conservation and Recovery Act/Toxic Substances Control Act wastes).

A properly planned and executed excavation carried out in an appropriate hydrogeologic setting should completely remove all mass in a source zone. In these cases, mass flux reduction, concentration reduction, and reduction of source migration potential will also be complete. Excavation produces no change in contaminant toxicity because the contaminants are transported offsite for treatment or disposal, so that is shown as "not applicable" in the comparison table presented at the end of the chapter.

## Containment

Containment, both physical and hydraulic, is a common remedy applied to contaminant source areas. This section discusses physical containment of a source zone, while the following section on pump-and-treat technology encompasses hydraulic containment. The goal of a containment remedy is to reduce risks by greatly minimizing contaminant migration via containment of the source so that there can be no direct route of exposure to the source. Physical containment is accomplished by creating impermeable barriers on all sides of the source zone with standard heavy construction methods and equipment. Thus, a typical containment remedy consists of a very low-permeability vertical barrier surrounding the source on all sides, a clay aquitard below the source, and a low-permeability cap on top. Vertical barriers can be constructed using bentonite slurries, slurries combined with polymer sheets, sheet pilings with sealed joints, pressurized injection methods, or cryogenic systems that freeze the soil. Constructed-bottom barriers can be emplaced by several drilling methods, but such barriers are uncommon.

Top barriers are used to minimize infiltration of rain water and subsequent dissolution of contaminants. Most top barriers are multilayer systems that include polymer sheeting and drainage layers. Typical operating practice for a containment system is to keep groundwater levels inside the container low relative to the adjacent aquifer by operating a small pump-and-treat system that withdraws groundwater from inside the system. This creates an inward groundwater gradient that helps ensure that contaminants will not migrate outward. Top barriers are very helpful in maintaining an inward gradient and in lowering pumping and treating costs. More recently vegetative/evaporative caps are becoming popular for controlling infiltration.

*Applicability of the Technology*

**Contaminants.** Containment systems are broadly applicable to organic contaminants. They can be used to contain any contaminants that are not expected to react with or leach through the components of the containment system. Source materials with extreme pH values are the most likely to create problems.

**Hydrogeology.** Two types of characterization related to hydrogeology are essential to containment: the areal extent and the depth of source areas to be contained, which must be known so that all source materials are indeed inside the containment system. There is no need to understand the internal structure of the source materials or the mass or concentration of contaminants present. Knowing the depth and thickness of the underlying aquitard is critical to making the vertical barriers deep enough to key into the aquitard. The aquitard topography must also be known so that any depth variations can be taken into account during barrier construction. Subsurface obstructions should be carefully mapped in advance so that barrier construction is not interrupted and so that they do not cause worker safety concerns.

Groundwater modeling is necessary during the designing of a containment system because the flow of groundwater will be changed by the new barrier. Adjacent sites could be affected as water diverts around the barrier, and some groundwater mounding will happen upgradient of the barrier. If modeling predicts that mounding will be substantial, then groundwater overflowing the top of the barrier and flooding of low areas or basements up gradient would be significant concerns, and a diversion/drainage method might have to be implemented.

Containment systems typically work well in unconsolidated soil (hydrogeologic settings I, II, and III) due to the relative ease of construction. Environmental conditions that can limit the applicability of containment include the presence of boulders or cobbles in soil, which can make installing vertical barriers difficult and costly. Containment is difficult in Type IV and V bedrock environments, and often relies on grout curtains. Grout curtains are difficult to install and do not yet provide the same level of assurance as vertical barriers constructed by trenching in unconsolidated soils. Verification of construction quality is also more difficult in Type IV and V settings. At sites where no natural bottom exists, there is little experience in constructing bottom barriers. Finally, it should be noted that a containment system creates a permanent subsurface wall that eliminates that part of the aquifer as a potential water source.

Barriers and other structural enhancements used for containment can be constructed to depths of approximately 30 meters, using such equipment as augers, draglines, clamshells, and special excavators with extended booms. As with all technologies discussed in this chapter, the cost of containment rises as the depth of treated subsurface increases.

**Health, Safety, and Environmental Considerations.** The main safety concerns of containment are those associated with operation of the heavy equipment necessary for construction. Once a containment system is in place, it is paramount that it remain effective in order to avoid potential health or other problems in the surrounding areas. This requires continuous and rigorous inspections during the construction of the remedy and subsequent long-term monitoring. Even though there is a loss of the use of the area for any intrusive activities, alternative land uses such as parks and golf courses built on top of containment systems are becoming more common, and newer construction technologies such as jetting are reducing land-use restrictions around contained systems. If the source materials have the potential to generate gasses, a system to control gas migration should be built to avoid future exposures.

*Potential for Meeting Goals*

Compared to most of the technologies discussed in this chapter, containment is simple and robust. When constructed well, a containment system almost completely eliminates contaminant transport to other environmental compartments and thus prevents both direct and indirect exposures. In Type I, II, and III environments, containment systems provide a very high degree of mass flux reduction and a very high reduction of source migration potential. Nonetheless, monitoring of containment systems is essential for assuring no migration of the contaminants. Containment systems do not reduce source zone mass, concentration, or toxicity unless they are deliberately used with treatment technologies. (In most cases only limited treatment will be provided by the pump-and-treat systems installed to control groundwater infiltration.) It is possible to combine containment systems with in situ treatment, since most in situ technologies that can clean up a free-range source can operate inside a contained zone—for example, the Delaware Sand & Gravel cometabolic bioventing system. In some cases, containment may allow for the use of treatments that would constitute too great a risk (e.g., with respect to migration of either contaminants or reagents) in an uncontrolled aquifer, though there would need to be substantial drivers to cause installation of two remedies in the same source zone.

*Cost Drivers*

The cost drivers for containment all relate to the types and quantities of construction necessary. They are the depth to aquitard, the total length of vertical barrier necessary, the type of barrier wall construction selected, the type of cap selected, and the need (if any) to construct a bottom. Monitoring systems are necessary, but they are not complex or costly. Containment systems are typically inexpensive compared to treatment, especially for large source areas.

*Technology-Specific Prediction Tools and Models*

Containment systems are very predictable because they are basically standard construction projects. Their long-term performance is currently predicted by models. The same techniques have been used in the construction industry for water control for some time, and have a good track record. Bench studies are typically used to define the best components of slurry mixtures.

*Research and Demonstration Needs*

Given its status as a conventional technology, the research needs of containment are minimal. However, better monitoring techniques would be helpful, and better ways to confirm the integrity of vertical barriers and the bottom containment would raise confidence. More information on the longevity of barrier materials in contact with contaminants would be helpful in the design of better barrier materials.

## Hydraulic Containment

Hydraulic containment is one of the most widely used methods to limit the movement of contaminants from DNAPL source zones. Through the use of extraction wells, contaminated groundwater emanating from source zones can be captured and treated ex situ, a technique commonly referred to as pump-and-treat (NRC, 1994, 1999). To reduce ex situ treatment costs, injection wells can be used in conjunction with extraction wells to hydraulically isolate contaminant source zones. It is generally accepted that in most cases, hydraulic containment will not be very effective for source remediation due to the limited solubilities of most contaminants of concern and due to limitations in mass transfer to the aqueous phase (NRC, 1994, 1999; EPA, 1996; Illangesekare and Reible, 2001). Therefore, the current discussion is focused only on hydraulic containment of source zones, rather than their remediation. (Some small measure of source depletion may result from the water flow through the source zone that may accompany hydraulic containment.)

*Case Studies*

Because pump-and-treat is the most widely used technology applied at contaminated sites, detailed case studies are numerous and are best summarized elsewhere (e.g., NRC, 1994, and EPA, 1998a). At most sites where pump-and-treat has been used, decreases in contaminant concentrations in extracted water were observed during pumping, but cleanup targets were not met. However, at almost all sites hydraulic containment was achieved, demonstrating that the technology can be effective in simply halting the spread of contaminants from source zones to groundwater.

*Applicability of the Technology*

**Contaminants**. Hydraulic containment is not limited to a particular contaminant type. However, any concomitant mass removal from the source zone that might occur during pump-and-treat operations will be greater for contaminants with higher solubilities.

**Hydrogeology**. The effective design of hydraulic containment systems requires a thorough understanding of site hydrogeology in order to choose the optimal locations of and pumping rates for wells. Incomplete hydrogeologic characterization can lead to systems that do not achieve complete containment or that pump excessive amounts of groundwater, leading to increased pump-and-treat costs. Thus, the more complex the hydrogeologic setting, the more challenging will be the design of an optimal hydraulic containment system. There are no depth limitations associated with hydraulic containment other than those associated with well drilling, although costs are expected to increase as well depth increases.

In systems with high hydraulic conductivities (such as gravel or coarse sand), hydraulic containment may be difficult to achieve because high pumping rates may be required from closely spaced wells. In low-permeability formations (such as clays, silts) it may also be difficult to obtain effective hydraulic containment due to the high gradients required to achieve significant capture zone size. In highly heterogeneous systems, effective hydraulic containment is limited by the lack of hydraulic connectivity resulting from the presence of lower-permeability zones. This may be particularly problematic for fractured systems and karst, for which the connectivity can be difficult to determine.

**Health, Safety, and Environmental Considerations**. The primary health, safety, and environmental considerations for hydraulic containment relate to the treatment and disposal of contaminants removed from the subsurface. Precautions must be taken to ensure that exposure to extracted contaminated groundwater does not occur, particularly due to off-gassing of contaminant vapors. Typical ex situ treatment technologies involve activated carbon, catalytic oxidation, and biological treatment. When contaminants are transferred to another medium, as with activated carbon treatment, the additional steps that are involved in the ultimate disposal of the contaminants may present health and safety risks.

*Potential for Meeting Goals*

Given its widespread use, the effectiveness of hydraulic containment for meeting various objectives for different types of sites is widely known. Regardless of hydrogeologic setting, hydraulic containment will not achieve significant mass removal due to the low aqueous phase solubilities of most contaminants of concern relative to the amounts of mass typically present in the organic phase

(Illangasekare and Reible, 2001) and sorbéd to the soil. These solubilities are primarily controlled by the slow mass transfer from the organic and soil phases to the aqueous phase. For some highly soluble contaminants such as DCA (solubility of 8,600 mg/L), hydraulic containment that maximizes water flow through the source zone may produce significant mass removal in homogeneous permeable settings. In heterogeneous media settings, removal of contaminants from the permeable zones may also contribute to a reduction in contaminant flux, while local average concentrations may not be significantly reduced due to the channeling of water through the high-permeability regions.

Due to the low mass removal expected with hydraulic containment, reduction of source migration potential is not significant, although maintenance of upward gradients has been proposed as a means of preventing downward migration of DNAPL in fractured rock (Chown et al., 1997).

*Cost Drivers*

The costs of hydraulic containment are associated with the operation and maintenance of a pumping system and with treatment of extracted water.

*Technology-Specific Prediction Tools and Models*

In the majority of cases, the design of the hydraulic containment system and the associated modeling are focused on simulating water flow, not on contaminant transport and removal. There are a large number of tools and models that can be used to design hydraulic containment systems. These range from simple analytical solutions for homogeneous steady-state systems (EPA, 1996) to complex numerical models that can incorporate heterogeneities and transient boundary conditions.

## EXTRACTION TECHNOLOGIES

Two technologies commonly used for source remediation work primarily by physically extracting the contaminants from the subsurface. Multiphase extraction employs a vacuum or pump to extract NAPL, vapor, and aqueous phase contaminants, which may then be disposed of or treated. Surfactant and cosolvent flushing are somewhat akin to pump and treat (discussed earlier) in that a liquid is introduced into the subsurface into which the contaminant partitions, and then the mixture is extracted out of the subsurface and subsequently treated.

### Multiphase Extraction

Multiphase extraction involves the extraction of water, gas, and possibly NAPL from the subsurface through the application of a vacuum to wells. In one variant referred to as two-phase extraction (EPA, 1997), a high vacuum of 18–26 inches (46–66 cm) of Hg is applied to the extraction well through a suction pipe (slurp tube) to extract a mixture of soil vapor, groundwater, and possibly NAPL (see Figure 5-1). Turbulent multiphase fluid flow in the suction pipe may enhance transfer of VOC vapors to the gas phase. The second variant, referred to as dual-phase extraction, uses a submersible, or pneumatic, pump to extract the liquids from the well, while a vacuum (3–26 inches or 8–66 cm of Hg) extraction blower is used to remove soil vapor (see Figure 5-2). Because application of a vacuum in multiphase extraction induces atmospheric air infiltration that can stimulate aerobic biodegradation, it is sometimes also referred to as bioslurping.

Multiphase extraction wells are usually installed with at least a portion of their screened sections in the vadose zone. Thus, the vacuum applied creates vapor flow through the vadose zone to the multiphase extraction wells, thereby removing volatile organic vapors in the soil gas. The extraction of water lowers the water table, and therefore exposes a greater portion of the subsurface to vapor stripping. The extraction of groundwater also removes dissolved contaminants from the subsurface. The application of a high vacuum to the extraction well enhances groundwater flow to the well by increasing pressure gradients around the well, without substantial drawdown of the water table. If LNAPL is present at the site, this can reduce the smearing of LNAPLs in the soil around the well that can occur when there is significant water table lowering. NAPL present in the zone of influence of the multiphase extraction well may also be captured, particularly in the case of LNAPL pools sitting on the water table.

Design of a multiphase extraction system requires determining the zones of influence of wells for given vacuum levels, determining gas and liquid extraction rates, and determining optimal well spacing. Preliminary design can be done with hydraulic models for gas and water flow, but pilot tests are advisable. The required aboveground equipment includes pumps, gas–liquid separators, and gas and liquid treatment trains. A variety of proprietary designs for multiphase extraction have been developed (EPA, 1999), which typically involve variations in the details of fluid extraction from the wells.

*Overview of Case Studies*

Multiphase extraction has been applied to a variety of sites contaminated with either halogenated or nonhalogenated VOCs. The only documented examples of using this technology specifically for NAPL recovery have been sites where the contaminant was an LNAPL (and these cases are not described further). There are a limited number of chlorinated solvent case studies available (as described below), and there appears to have been little monitoring for contaminant concentration rebound after treatment.

A one-year multiphase extraction treatability study was conducted at the Defense Supply Center in Richmond, Virginia, during 1997–1998. The contaminants were

189

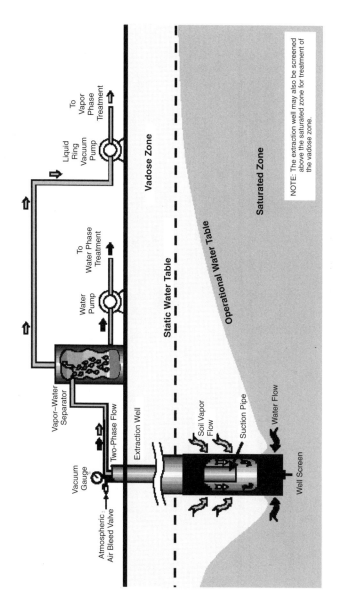

FIGURE 5-1 Two-phase extraction system. SOURCE: EPA (1997).

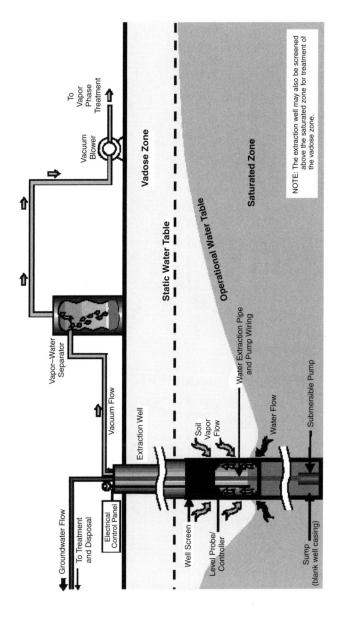

FIGURE 5-2 Dual-phase extraction system. SOURCE: EPA (1997).

primarily PCE (3.3 mg/L in groundwater) and TCE (0.9 mg/L in groundwater), extending from the ground surface to a depth of 25 feet (7.6 m) below the ground surface (bgs). Soil layers included silty clay, fine-grained sand, coarse-grained sand, and interlayered gravel. The water table was 10–15 feet (3–4.6 m) bgs. Twelve dual-phase extraction wells and six air injection wells were installed at the site on a rectangular grid to depths of 22–28 ft (6.7–8.5 m) bgs. Water was removed from wells using submersible recovery pumps. Vacuum applied was approximately 11 kPa. The zone of influence of the well as indicated by water table drawdown was 600–800 ft (183–244 m) downgradient and 1,800–2,500 ft (549–762 m) upgradient, with an average water extraction rate of 37 gpm. A total of 117 pounds (53 kg) of VOC were removed in the vapor phase, and 28 pounds (13 kg) were removed in extracted water. At the end of the study, parts of the treatment area were below the remedial goals of 5 mg/L for PCE and TCE, while other areas on the outer edge of the treatment zone remained above this level. The treatment was deemed to be successful, and continued operation to meet the remedial goals was recommended.

Multiphase extraction was implemented at a manufacturing facility in Santa Clara, California, in 1996 to remediate TCE (46 mg/kg in soil, 37 mg/L in groundwater) in silts and clays to depths of 20 ft (6 m) bgs. Pneumatic fracturing was conducted to increase air flow rates. Twenty dual-phase single-pump extraction wells were installed at the site. Groundwater extraction rates were 35 gpm, which was much higher than expected due to the presence of high-permeability lenses in the treatment area. Total VOC removal was 382 lb (173 kg) in extracted groundwater and 785 lb (356 kg) in extracted vapor. Extraction rates declined significantly after about two months of operation, which continued for two years. After a six-month shutdown, little rebound in concentrations in extracted water and vapor was observed, indicating effective removal of contaminants from the treatment zone. VOC concentrations in one well declined from 4 mg/L to 0.7 mg/L by the end of treatment.

Multiphase extraction at Tinkham's Garage Superfund Site, in Londonderry, New Hampshire, was implemented to treat 9,000 cubic yards (6,881 m$^3$) of soil contaminated with PCE, TCE, and benzene, toluene, ethylbenzene, and xylene (BTEX) (maximum total VOC contamination of 652 ppm in soil, 42 ppm in groundwater). The site geology consisted of inorganic and organic silty clay and sand overlying weathered metamorphic bedrock at 14 ft (4.3 m) bgs. The extraction system consisted of 25 shallow wells screened in the overburden, and eight wells screened in the upper bedrock and overburden. A dual-pump configuration was used. Vapor extraction flow rates averaged 500 standard ft$^3$/min (scfm) (14 m$^3$/min) at a vacuum of 68 inches (173 cm) water column (WC), while water extraction rates averaged 2.5 gpm. At the end of the ten-month treatment period, all soil borings had been remediated below the targets of 1 ppm total VOCs. Total VOC concentrations in groundwater averaged 82 ppb at the end of the treatment period. A total of 48 lb (22 kg) VOC were removed in the vapor phase, and 5 lb

(2.3 kg) were removed in the aqueous phase. The treatment was judged to have been successful in meeting the remedial goal of lowering soil total VOC concentrations to below 1 ppm, and a long-term migration control remedy involving pump-and-treat is now in operation.

The three case studies cited indicate that multiphase extraction can achieve some removal of VOCs from shallow source zones. In particular, multiphase extraction was judged to be more effective than individual application of either soil vapor extraction or pump-and-treat. In all cases, contamination remained at the end of the treatment period, and continued treatment or control remedies were recommended.

*Applicability of the Technology*

**Contaminants.** Multiphase extraction is most effective for volatile organic compounds (i.e., organic compounds with a vapor pressure > 1 mm Hg at 20°C), as contaminant vapor stripping is one of the primary removal mechanisms. Highly viscous NAPLs, such as creosotes and coal tars, are not effectively recovered during multiphase extraction. In the case of semivolatile organic compounds (SVOCs), multiphase extraction may enhance aerobic biodegradation through increased supply of oxygen to the contaminated zone, but other methods for enhancing bioremediation may be more effective.

**Hydrogeology.** Multiphase extraction is most appropriate for soils of moderate permeability ($10^{-5}$ to $10^{-3}$ cm/sec) (EPA, 1996). If the permeability is too low, there are difficulties in dewatering the soils due to high air entry pressures (Baker et al., 1999), and the flow rates and zones of influence will be too low. If the permeability is too high, then water withdrawal and corresponding water treatment costs will be high. This was shown during pilot tests at three Comprehensive Environmental Response, Compensation, and Liability Act (CERCLA) sites that involved single-pump extraction wells with slurp (suction) tubes for the removal of both gas and water (Baker et al., 1999). At high-permeability sites, the single-pump wells with suction tubes flooded with water, reducing their effectiveness.

The effectiveness of the technology is also reduced in highly heterogeneous soils due to channeling. The two-phase extraction configuration of multiphase extraction, whereby both gases and liquids are removed by the vacuum pump, is limited to depths of approximately 50 ft (15 m) (EPA, 1997). In theory, the multiphase extraction configurations that employ separate pumps for liquid recovery can operate at any depth. However, if the contaminated zone is too deep, the influence of the vacuum on recovery may be limited, and the system will essentially be a pump-and-treat system. Thus, the technology is only applicable to source zones near the water table.

Placement of wells for optimal recovery of DNAPL is a major challenge to the effective implementation of multiphase extraction. The wells must be located close to, or in the zone of, mobile DNAPL, and site stratigraphy must be characterized to determine appropriate pumping rates and capture zones.

**Health, Safety, and Environmental Considerations.** The major concern with the use of multiphase extraction is the proper treatment of extracted gases and liquids. Vapors extracted may be treated by activated carbon, catalytic oxidation, or other technologies for gas phase treatment. The water phase may be treated by air stripping, activated carbon, or biological treatment. Vapor–liquid separation facilities are required to separate the gas and liquid streams, and if NAPL recovery is also expected, then NAPL–water separation will also be required.

*Potential for Meeting Goals*

Although the goals of multiphase extraction were met in the three examples reviewed above, these goals were usually partial mass removal to reduce source migration potential and provide some reduction in aqueous and vapor concentrations. In all cases, it was acknowledged that subsequent treatment by, for example, pump-and-treat and natural attenuation would be required. In one case, pneumatic fracturing was used to increase air flow rates and VOC removal, while difficulties in predicting groundwater extraction rates were encountered in another case. With respect to reductions in the potential for DNAPL movement, such reductions can be very difficult to ascertain, especially where the source is not well characterized. Usually one can only infer that as DNAPL flow rates to the well decrease with time, and as DNAPL thicknesses in monitoring wells decrease, that DNAPL mobility in the capture zone is being reduced.

The effectiveness of this technology depends on the well spacing, flow rates, channeling due to soil heterogeneities, and mass transfer limitations. As with most flushing technologies, the risk of failure increases with increasing site heterogeneity. Water table lowering has the potential to spread NAPL contamination downward, although this is more of a concern with LNAPLs than with DNAPLs (which would have likely existed below the water table prior to the application of multiphase extraction). Iron precipitation due to increased subsurface aeration has also been reported as a problem (Rice and Weston, 2000). Finally, difficulties in balancing applied vacuum and liquid extraction rates may occur. Emulsification of liquid–gas mixtures can create problems for aboveground treatment.

Although there is little reported experience with multiphase extraction in fractured media, this technology would not be expected to be very effective in achieving any of the objectives listed in the comparison table at the end of the chapter (Table 5-7), due to severe flow channeling along high-permeability fractures.

*Cost Drivers*

The capital costs for multiphase extraction are associated with well installation, pump equipment, gas–liquid separators, and water and gas treatment systems. Costs are also associated with energy for pump operation and for operation of the treatment systems.

*Technology-Specific Prediction Tools and Models*

The multiphase extraction process involves multiphase flow and transport of water, gas, and possibly organic phases, with interphase mass transfer. Biodegradation may also occur. The flow rates generated may be quite high, and nonequilibrium interphase mass transfer effects may be significant. Numerical models exist for simulation of multiphase flow and transport with either equilibrium (Sleep et al., 2000) or kinetic mass transfer and biodegradation (McClure and Sleep, 1996). However, these processes are highly nonlinear and are difficult to model, particularly under conditions of high flow rates. Multiphase models also require many soil parameters that can be very difficult and costly to measure for the different soils present at a site, such as parameters for capillary pressure–saturation constitutive relationships. It is not surprising that these multiphase models are primarily in the research domain, rather than in a format amenable to usage by nonspecialists. Simplified models for gas or water flow may be used to predict zones of influence of extraction wells, although these models may be of limited accuracy, as they ignore the multiphase nature of the flow system.

*Research and Demonstration Needs*

There is limited understanding of the applicability of multiphase extraction specifically for DNAPL removal from the subsurface. At many field sites where DNAPL is suspected to be present, DNAPL is never found in wells, so it is unlikely that DNAPL would be recovered from a multiphase extraction well. In general, the impacts of DNAPL distribution, soil permeabilities, heterogeneities, and rate-limited interphase mass transfer on the effectiveness of multiphase extraction are not well understood.

## Surfactant/Cosolvent Flushing Systems

Surfactants (commonly known as soaps or detergents) and alcohols (cosolvents) are amphiphilic molecules, having both water-like and oil-like parts. Because they are amphiphilic, surfactants and alcohols are surface-active molecules, and thus accumulate at interfaces of multiphase systems, with the water-like part of the molecule in the polar water phase and the oil-like part of the molecule in the nonpolar oil or less polar air phase. In this way, both parts of the molecule are in a preferred phase, and the free energy of the system is minimized (Rosen, 1989; Myers, 1999).

Although they are in some ways similar, there is a unique characteristic that differentiates surfactants from alcohols. When the aqueous surfactant concentration exceeds a certain level—called the critical micelle concentration or CMC—surfactant molecules self-aggregate into clusters known as micelles, which contain 50 or more surfactant molecules (Rosen, 1989; Myers, 1999; Holmberg et al., 2003). Micelle formation is unique to surfactant molecules; alcohols do not form such aggregates. Surfactant micelles, with a polar exterior and nonpolar interior, can increase the aqueous concentration of low-solubility organic compounds by providing a hydrophobic sink into which the organic compound partitions. Thus, by adding surfactant at concentrations above the CMC, the micelle concentration increases, as does the contaminant's apparent solubility. It is therefore desirable to be well above the CMC (e.g., 10 or even 100 times the CMC, or more) in order to maximize the contaminant solubility and thus extraction efficiency.

Alcohols can also increase the solubility of organic compounds, albeit in a somewhat different manner. As opposed to forming aggregates with nonpolar interiors, water-miscible alcohols render the aqueous phase less polar, thereby increasing the aqueous concentration of sparingly soluble organic compounds. This can be understood by realizing that a sparingly soluble organic compound will dissolve to a much higher degree in ethanol than in water. Thus, as more and more ethanol is added to the water, the solution takes on more of the properties of ethanol and the contaminant solubility increases above the water-only case—a process referred to as cosolvency (Rao et al., 1985). The solubility enhancement is less dramatic for alcohols than for surfactants, such that much higher alcohol concentrations are required to achieve high contaminant solubility (nominally an order of magnitude or more alcohol is required than surfactant). At the same time, alcohol costs per unit mass are often much lower than surfactant costs, helping to equalize the economics of these two approaches.

Using a single surfactant or alcohol to achieve higher solubility of organic contaminants is referred to as enhanced solubilization. While this is a fairly straightforward approach, it is not necessarily the most efficient. Using a mixture of surfactants, alcohols, and/or other cosolvents, it is possible to further enhance solubility while also further reducing the water–NAPL interfacial tension (Martel and Gelinas, 1996; Jawitz et al., 1998; Dwarakanath et al., 1999; Falta et al., 1999; Sabatini et al., 1999; Knox et al., 1999; Dwarakanath and Pope, 2000; Jayanti et al., 2002). The former is certainly desirable, but the latter may be undesirable, especially for DNAPLs that, if released, may settle or penetrate into deeper regions not previously contaminated. Approaches in which the interfacial tension is intentionally lowered so as to displace the trapped NAPL are referred to as mobilization approaches. Mobilization is particularly effective in the remediation of LNAPLs because vertical migration will tend to be upward. Laboratory research has investigated using alcohols that partition into the DNAPL and convert it into an LNAPL prior to surfactant flooding, thereby mitigating the vertical migration concerns (Ramsburg and Pennell, 2002); however, this concept has yet to be demonstrated at the field scale. The more efficient process of mobilization has been successfully demonstrated at sites where there was a sufficient flow barrier below the source zone to prevent the DNAPL from migrating downward (Hirasaki et al., 1997; Delshad et al., 2000; Holzmer et al., 2000; Londergan et al., 2001; Meinardus et al., 2002). However, a much greater degree of site charac- terization is required in such cases to satisfy both technical and regulatory requirements.

Although the main mechanisms underlying surfactant/cosolvent flooding are typically thought of as being either solubilization or mobilization, an alternate approach exists between these two extremes—supersolubilization. Here, the solubility enhancement is maximized while still maintaining a sufficiently high interfacial tension so as to mitigate the potential for mobilization and vertical migration (Jawitz et al., 1998; Sabatini et al., 2000). Site-specific conditions help dictate the best approach for a given site.

When designing a surfactant/alcohol system, one should consider the com- patibility of the additive with the subsurface environment, including the porous medium, the groundwater, and the NAPL itself. Failure to consider these interac- tions may result in excessive loss of the additive (e.g., surfactant sorption or precipitation, phase separation, or even partitioning into the NAPL), rendering the system highly inefficient. When designing these systems, one must also con- sider factors such as the viscosity and density of the flushing solutions, both prior to and after contacting the NAPL, the temperature and salinity/hardness impacts on the system, the biodegradability of the additive and its metabolites, and the potential impacts of the additive, both in the zone of flushing and in other regions that may be impacted (Fountain et al., 1996; Jawitz et al., 1998; Falta et al., 1999; Holzmer et al., 2000; Sabatini et al., 2000). The additive must also be introduced in such a way that it efficiently contacts the trapped NAPL. In highly heteroge- neous systems (e.g., Type III hydrogeologic settings), special design features (e.g., use of polymers, foam, or unique hydraulic schemes, such as vertical circu- lation wells) may be required. Finally, when conducting multiple pore-volume flushes, economic considerations may mandate decontaminating the surfactant/ alcohol system aboveground prior to reinjection (Sabatini et al., 1998). Whatever approach is used, the site is flooded with water following remediation to flush out surfactant/cosolvent and associated contaminants. There have been two manuals published on best practices and design of surfactant/cosolvent systems (AATDF, 1997; NFESC, 2002).

*Overview of Case Studies*

According to a recent U.S. Environmental Protection Agency (EPA) survey, there have been at least 46 field demonstrations of surfactant/cosolvent flooding, with roughly three-fourths of these studies being surfactant-based approaches (www.cluin.org). Of these sites, roughly one-third were LNAPLs, one-third were DNAPLs, one-sixth were mixtures of LNAPLs and DNAPLs, and the remainder were non-liquid organic contaminants. Roughly two-thirds of the sites were federally funded, with the remainder being largely state funded. The majority of the sites (roughly half) were 25–50 ft (7.6–15 m) in depth and less than 3,000 cubic feet (85 m³) in volume.

Table 5-2 summarizes results from 12 different surfactant and cosolvent projects. These studies cover a range of locations (including Utah, California, North Carolina, Hawaii, and Canada) and thus a range of hydrogeologic conditions and contaminant matrices. The swept pore volume ranged from five to a few hundred cubic meters. The mass removed, as estimated by pre- and post-core and partitioning interwell tracer tests, was in the mid 70 percent to the high 90 percent range. The high removal efficiencies experienced in field studies conducted in the 1990s are in contrast to early field studies conducted in the 1980s, where little of the surfactant or contaminant was recovered (e.g., Nash, 1987). The poor performance of these early studies can be attributed in part to poor surfactant selection—for example, failure to consider surfactant behavior under field conditions. These early studies are not included in Table 5-2 because insufficient characterization does not allow comparison with the thoroughly characterized tests listed in the table. Thus, the successes listed in Table 5-2 should not mislead the reader into thinking that this technology is easy to design and implement. It is only because of thorough site characterization, experienced design, and careful implementation that the studies in Table 5-2 were successful, unlike previous efforts. Nonetheless, the relatively high efficiency of the systems reported in Table 5-2 is very encouraging, especially given that the studies were conducted by a range of investigators, addressing a variety of contaminant matrices in a range of hydrogeologic conditions. Case studies of surfactant and cosolvent flooding are presented in Boxes 5-1 and 5-2, respectively, as well as in Box 3-4.

*Applicability of the Technology*

**Contaminants.** Both surfactant and cosolvent flushing have been successfully applied to a wide range of contaminants. The NAPLs at the sites listed in Table 5-2 range from PCE and TCE to mixtures of chlorinated solvents, and in some cases include mixtures of widely varying contaminants (DNAPL and LNAPL mixtures). While insights can be garnered from previous studies, such as those cited in Table 5-2, to maximize performance the surfactant or cosolvent system must be designed for the site-specific contaminant of interest. Weathering and alteration of the NAPL will impact this optimization; thus, design of the surfactant or alcohol system should be made using actual NAPL and geological material from the site in laboratory batch and column studies (e.g., Sabatini et al., 2000; Dwarakanath and Pope, 2000; Rao et al., 2001).

TABLE 5-2 Summary of Well-Designed Field Tests of Surfactant and Cosolvent Flooding

| Year | Location/Additive | Geology | NAPL | Swept Pore Volume (m³) | Reduction in NAPL Mass (%) | Post-NAPL Saturation (%) | Reference |
|---|---|---|---|---|---|---|---|
| 1991 | Borden, Ontario 14 PV, 2% Surf. | Sand | PCE | 9.1 | 77 | 0.2 | Fountain et al., 1996 |
| 1994 | L'Assomption, Quebec 0.9 PV, Surf./Alcohol/Solvent | Sandy Gravel | DNAPL | 6.1 | 86 | 0.45 | Martel et al., 1998 |
| 1995 | Hill AFB, UT, OU1 9 PV, 82% Alcohol | Sandy Gravel | LNAPL[a] | 4.5 | 85 | 0.9 | Rao et al., 1997 |
| 1996 | Hill AFB, UT, OU1 9.5 PV, 3% Surf. / 2.5% Alcohol | Sandy Gravel | LNAPL[a] | 4.5 | 78 | 0.8 | Jawitz et al., 1998 |
| 1996 | Hill AFB, UT, OU1 6.5 PV, 4.3% Surf. | Sandy Gravel | LNAPL[a] | 4.5 | 86 | 0.4 | Knox et al., 1999 |
| 1996 | Hill AFB, UT, OU2 2.4 PV, 8% Surf. | Sand | DNAPL | 57 | 99 | 0.03 | Brown et al., 1999 |
| 1996 | Hill AFB, UT, OU1 4 PV, 95% Alcohol | Sandy Gravel | LNAPL[a] | 4.5 | 80 | 0.4 | Falta et al., 1999 |
| 1997 | Hill AFB, UT, OU2 4% Surf. & Foam | Sand | DNAPL | 31 | 90 | 0.03 | Szafranski et al., 1998 |
| 1999 | Camp Lejeune, NC 5 PV, 4% Surf. | Silt | PCE | 18 | 72 | 0.5 | Holzmer etal., 2000 |
| 1999 | Alameda Point, CA 6 PV, 7% Surf. | Sand | DNAPL | 32 | 98 | 0.03 | Hasegawa et al., 2000 |
| 1999 | Pearl Harbor, HI 10 PV, 8% Surf. | Volcanic Tuff | Fuel Oil | 7.5 | 86 | 0.35 | Dwarakanath et al., 2000 |
| 2000 | Hill AFB, UT, OU2 2.4 PV, 4% Surf. | Sand | DNAPL | 188 | 94 | 0.07 | Meinardus et al., 2002 |

NOTE: PV = pore volume; Surf = surfactant

[a]LNAPL means an LNAPL with sufficient DNAPL components present, such that in the absence of the LNAPL, the waste would be a DNAPL

## BOX 5-1
## Surfactants Case Study

In 1996, a surfactant field test was conducted at Hill Air Force Base Operational Unit 2 to remediate DNAPL contamination. The DNAPL consisted primarily of trichloroethylene, 1,1,1-trichloroethane, and tetrachloroethylene. Sheet piling was installed to isolate the treatment zone, which was 6.1 by 5.4 m in cross section, with a treatment zone thickness of 6.2 m. The subsurface geology includes an alluvial sand aquifer that is confined on its sides and below by thick clay deposits that form a capillary barrier to DNAPL migration. The hydraulic conductivity of this alluvium is in the range of $10^{-2}$ to $10^{-3}$ cm/s. Based on extensive field characterization, laboratory testing, and simulation efforts using UTCHEM, the remedial system was designed and implemented. The design called for a NaCl pre-flood (0.7 pore volumes) followed by 2.4 pore volumes of the surfactant flood and finishing with post-water flooding. Treatment performance was assessed by pre- and post-partitioning interwell tracer tests. The surfactant system consisted of 7.55% sodium dihexyl sulfosuccinate, 4.47% isopropanol, and 7,000 mg/L of NaCl. The surfactant removed approximately 99 percent of the DNAPL from the swept zone, leaving a residual DNAPL saturation of about 0.0003. The concentration of dissolved contaminants was reduced from 1,100 mg/L to 8 mg/L in the central monitoring well (Londergan et al., 2001; Brown et al., 1999). Overall, the model simulations were able to predict the trends observed in the field results, although the actual concentrations varied somewhat, as shown in Figure 5-3. Nonetheless, use of the model to design the field implementation resulted in excellent system performance.

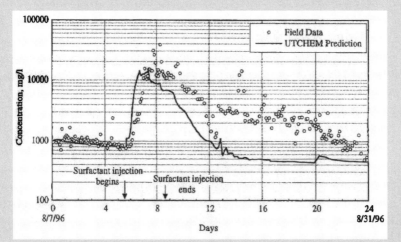

FIGURE 5-3 Contaminant concentration produced at extraction well SB-1 during Phase II test. Comparison of UTCHEM predictions with field data. SOURCE: Reprinted, with permission, from Brown et al. (1999). © 1999 American Chemical Society.

---

**BOX 5-2**
**Cosolvents Case Study**

In 1997, a cosolvent field test was conducted at Hill Air Force Base Operational Unit 1 to remediate LNAPL contamination. The original LNAPL spill resulted from disposal of petroleum hydrocarbons (e.g., jet fuel) and spent solvents (e.g., chlorinated hydrocarbons) into chemical disposal pits. The aged LNAPL is now a complex mixture of aromatic and aliphatic hydrocarbons and chlorinated solvents.

Sheet piling was installed to isolate the treatment zone, which was 3 by 5 m in cross section, with a treatment zone thickness of 2 m. The subsurface geology is a sandy gravel with a lower clay layer. The hydraulic conductivity of this material is as high as $10^{-1}$ cm/s. The cosolvent flood consisted of injecting four pore volumes (one pore volume equals 7000 L) of the cosolvent mixture, which was 80% tert-butanol and 15% n-hexanol. Treatment performance was assessed by pre- and post-coring and partitioning interwell tracer tests. The cosolvent removed more than 90 percent of the more soluble compounds (tricholorethane, toluene, ethylbenzene, xylenes, trimethylbenzene, naphthalene) and upwards of 80 percent of the less soluble constituents (decane and undecane), with an overall NAPL removal of 80 percent. The unextracted NAPL mass was highly insoluble, resulting in extremely low contaminant concentrations after the remedial effort (Falta et al., 1999). These results are similar to those of a previous study conducted in a different portion of the same formation. The previous study used a 70% ethanol, 12% pentanol, and 18% water flushing solution, and an 85 percent removal of the NAPL mass was achieved (Rao et al., 1997).

---

Failure to design for the site-specific contaminant can cause poor performance (low solubility enhancements), or even result in system failure (significant loss of the additive(s) into the trapped oil phase, or formation of a gel phase with the oil). More hydrophobic oils (e.g., coal tar or creosote) may require a mixture of surfactants, alcohols, or other solvents, or even a combination of surfactants/ alcohols with increased temperature (Dwarakanath et al., 2000), to maximize system performance.

Design and implementation of this technology requires careful site characterization to assess the potential impacts of vertical migration of the contaminant. The better the resolution of contaminant distribution, the more effectively the surfactants/cosolvents can be targeted to the contaminant and the more economical will be the design.

**Hydrogeology.** Site hydrogeology can pose at least two challenges to implementing surfactant and cosolvent flushing: low flow and flow bypassing. In tight formations such as fine silt or clay (Type II hydrogeologic settings), flushing any solution, even water, will be challenging. At the same time, surfactants and cosolvents have been successfully applied in silty formations, although the time scale is obviously extended. Flow bypassing occurs when the heterogeneous nature of the geology causes preferential flow paths (e.g., due to layering as in a Type III hydrogeologic setting). The flow bypassing can be further amplified during the remedial effort as regions that are cleaned up first will attract even more flushing solution through them, thereby bypassing the remaining contaminant. Flow bypassing can be addressed by increasing the viscosity of the flushing solution (e.g., with polymers) or by intermittently injecting air to create foam in the preferential pathways, thereby temporarily blocking off these pathways and forcing the flow through the less-preferred regions (Hirasaki et al., 1997; Dickson et al., 2002; Meinardus et al., 2002; Jackson et al., 2003). Since alcohols act as antifoams, their use should be avoided when designing a foam-augmented surfactant system. If they are not properly designed, mobilization systems can significantly increase viscosity, which may make it difficult to flush contaminants through the porous media; at the same time, when they are properly designed and implemented, this concern can be mitigated. Like multiphase extraction, there are no depth limitations associated with surfactant flooding other than those associated with drilling wells.

**Health, Safety, and Environmental Considerations.** Although at low concentrations both surfactants and alcohols are relatively innocuous, at higher concentrations they can pose a risk to health, safety, and the environment. For example, very high surfactant concentrations, typical of the form in which the surfactant might be delivered, can be harsh to skin. Alcohols can pose a flammability risk that must be accounted for. Accidental releases of surfactant to surface waters can result in fish kills. As described above, mobilization (excess lowering of the interfacial tension) can result in vertical migration of the DNAPL into previously uncontaminated zones, which is obviously an environmental concern. In contrast, solubilization (with minimal reduction in interfacial tension) helps mitigate this concern. All these risks can be avoided when the system is properly designed and implemented, as evidenced by the successful field-scale studies summarized above.

*Potential for Meeting Goals*

As summarized in Table 5-2, properly designed surfactant and cosolvent systems have achieved greater than 85 percent to 90 percent mass removal in the relatively homogeneous hydrogeological settings reported in Table 5-2, with a number of cases exceeding 97 percent removal. Concentration and mass flux reductions have generally not been documented, although mass flux reductions are expected in more heterogeneous systems even though the mass removal is lower. This is because a portion of the remaining mass in heterogeneous systems is expected to exist in diffusion-limited stagnant regions. These concepts have been demonstrated in modeling efforts that are described in Box 5-3. Field efforts to validate these models have only recently been attempted. Because surfactant/cosolvent technologies are extraction or mass removal technologies, they do not transform the contaminants into less toxic forms. Finally, it should be noted that while surfactant/cosolvent technologies have been widely evaluated in porous media systems, much less is known about the performance of these systems in fractured media systems.

Assuming that proper design considerations have been addressed (e.g., making sure that the surfactant does not phase separate or that the alcohol does not override the contamination due to density considerations), both surfactant and cosolvent systems are fairly robust. Even when the goal is to achieve a mobilization system, which is more sensitive to implementation conditions than solubilization, good performance can still be realized even when optimal conditions are not achieved.

One must consider how surfactants and alcohols impact other aspects of the overall remediation strategy (e.g., impacts on aboveground treatment processes). While the increased contaminant solubility produced by the surfactant/alcohol is highly desirable in removing contaminants from the subsurface, this same phenomenon will decrease the stripping efficiency in commonly used air stripping processes. In addition, the presence of certain surfactants will cause significant foaming in the air stripper. These issues have been successfully addressed where the surfactant system is properly designed (or modified, for an existing system) and operated (Brown et al., 1999; Hasegawa et al., 2000). For example, one can modify the air stripper design equations to account for the surfactant reduction in system performance, as corroborated in field-scale studies (Sabatini et al., 1998). In addition, modification of the air stripper, use of antifoams in the air stripper, or use of hollow-fiber-membrane air stripping can mitigate foam formation in the air stripper (Sabatini et al., 1998; O'Haver et al., 2004).

It is important to consider how the presence of these additives might impact follow-on activities at the site (e.g., natural or enhanced bioremediation). The answer is strongly dependent on the surfactant/alcohol concentration present. Whereas the high concentrations present during the remedial activity are likely to inhibit microbes, the lower concentrations present after post-remediation water flushing may not inhibit microbes and may even stimulate them. In fact, several recent field activities have successfully used lower surfactant or alcohol concentrations as a carbon source to stimulate post-remedial bioactivity (Rao et al., 2001; Abriola et al., 2003).

As evidenced in Table 5-2, the percentage of mass removed is a common metric for evaluating the success of surfactant/cosolvent technologies, although other metrics such as concentration reduction or mass flux reduction may be more appropriate. Indeed, the relationship between mass removal and mass flux reduction, first mentioned in Chapter 4, has been best explored for surfactant flushing technologies (see Box 5-3). Two additional metrics that should be

## BOX 5-3
## Depletion Profiles for Surfactant Flushing

Depletion profiles, first introduced in Box 4-1, are receiving increasing attention as a means for assessing and designing source zone remedial systems. Depletion profiles seek to demonstrate the relationship between mass flux (mass leaving a source zone per unit area per time) and mass removed at a site. Mass flux is selected as the parameter of interest because it has a significant impact on the risk experienced by a downgradient receptor. Higher fluxes have a greater likelihood of overpowering any natural attenuation processes, and thus have a greater likelihood of causing an undesirable contaminant exposure. Removal of contaminant mass from the source zone may reduce the mass flux, and thus the risk, emanating from the source zone. However, until recently, there has been little information on the relationship between mass removed and mass flux reductions.

Figure 5-4 shows several possibilities for the relationship between mass removal and mass flux reduction during surfactant flushing, which have been determined in

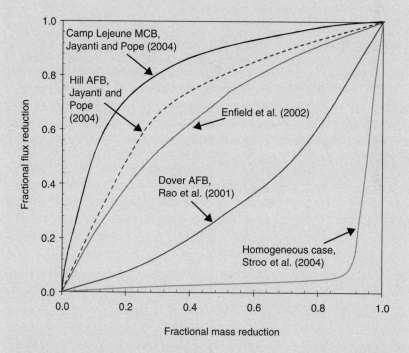

FIGURE 5-4 Depletion profiles from modeling of different cases of surfactant flushing.

*continued*

---

### BOX 5-3 Continued

different modeling studies. As can be seen, the relationship is very site-specific and is clearly highly dependent on the nature of the contaminant distribution and the level and type of heterogeneities at a given site. The uppermost curve would apply to a site that is highly heterogeneous. In this case, mass is initially removed from the more permeable and mobile regions of the source zone, leaving the remaining mass behind in diffusion-limited regions that make little contribution to the overall mass flux. Thus, in such systems a significant reduction in mass flux can be experienced even though the mass removal is not so dramatic. At the other extreme is the case where a site is more homogeneous and all of the contaminant is equally accessible by the chosen technology. In this situation, almost all of the mass must be removed before a noticeable change in mass flux levels is observed. These two depletion curves define the extremes, with the reality for a given site likely to fall someplace in between. Part of the challenge of using this approach is defining the particular curve that applies to a given site. With the addition of field data to corroborate these modeling results, it may be possible to establish a general relationship between heterogeneity of a site and the approximate depletion curve, or range of curves, that could apply to a particular site. Ongoing research is evaluating this approach for several sites that have been remediated with surfactant/cosolvent technologies (Jayanti and Pope, 2004). Depletion profiles have yet to be developed for other source remediation technologies.

Another challenge associated with using the depletion curves is that one must know exactly how much mass existed prior to the remedial activity, and how much mass has been removed, or will be removed, during a given remedial activity, which is extremely difficult.

Again, with the development of new characterization techniques, and as additional data and experience with the remedial activities are gained, our ability to determine the mass flux reduction will improve. The availability of such tools and data will make it much easier to take advantage of the depletion profiles as a way to assess the extent to which a given remedial activity will achieve risk reduction at a given site.

---

considered during surfactant flushing are (1) verifying that uncontrolled vertical migration has not occurred and (2) verifying that the surfactants/cosolvents do not negatively impact another water resource.

### Cost Drivers

The costs of surfactant and cosolvent systems have steadily declined as these technologies have progressed, with costs being competitive with the long-term pump-and-treat systems (although economic discounting can favor these longer-term projects). While not necessarily as efficient or cost-effective, the solubilization approach is somewhat less complex to design and implement than is mobilization. Surfactant costs can be a significant component of the total cost, especially if surfactant concentrations of 4 to 8 wt% are used. However, as surfactant concentrations are lowered toward 1 wt% or lower, and as surfactant recovery and reuse are implemented (which has been demonstrated—e.g., Sabatini et al., 1998; Hasegawa et al., 2000), costs become more economical.

*Technology-Specific Prediction Tools and Models*

Experience has shown that laboratory treatability studies and modeling efforts can successfully be used in designing field-scale surfactant/cosolvent systems. This section, which provides a brief overview of several of these simulators, is in no way intended to be exhaustive.

Abriola et al. (1993) discuss the development of a simulator to describe surfactant-enhanced solubilization of NAPLs. The model incorporates transport equations for organic and surfactant constituents, as well as a mass balance for the organic phase. The rate-limited surfactant-enhanced solubilization process is described by a linear driving force expression. The surfactant sorption is described by a Langmuir isotherm. The model is implemented in a Galerkin finite element simulator, where the trapped oil is idealized as a collection of spherical globules. This code was later extended to consider geologic heterogeneities (e.g., low-permeability lenses), as described in Rathfelder et al. (2001).

Delshad et al. (1996) describe a three-dimensional, multicomponent, multi-phase compositional finite-difference simulator for evaluating surfactant-enhanced aquifer remediation. An important feature of this simulator is the ability to describe the many types of micellar/microemulsion phases that are possible with mixtures of surfactant, water, and NAPL, and to capture the dependence of these phases on system properties such as temperature and salinity/hardness. Additional surfactant properties that are incorporated into this simulator include adsorption, interfacial tension, capillary pressure, capillary number, and microemulsion viscosity. In addition to its widespread application in subsurface remediation, this simulator was first developed for and has been widely used to model surfactant-enhanced oil recovery. Brown et al. (1999) and Londergan et al. (2001) describe the use of this simulator for the case study presented in Box 5-1. Delshad et al. (2000) used this same simulator to design and interpret the surfactant-enhanced aquifer remediation (SEAR) to remove PCE DNAPL at the Camp Lejeune site. The largest and most significant use of this simulator to date has been its use to design the full-scale SEAR applications to the DNAPL source zone at Hill Air Force Base (AFB) (Meinardus et al., 2002).

Mason and Kueper (1996) developed a one-dimensional numerical model that simulates surfactant-enhanced solubilization of pooled DNAPLs. Two non-equilibrium expressions were used for capturing mass transfer processes. The nonwetting phase saturation distribution is calculated as a function of the hydraulic gradient, allowing determination of the local velocity. The simulator was applied in an upward flow fashion in an attempt to overcome the potential for downward pool migration in response to a lowering of the interfacial tension. Model predictions agreed well with experimental results.

Thus, several simulators exist for predicting the efficiency of surfactant/cosolvent flushing technologies. These tools have been validated against and have been used to predict both laboratory and field data. In the hands of a skilled user, these simulators can be used to design and assess the field implementation of surfactant/cosolvent technologies.

*Research and Demonstration Needs*

One of the great challenges facing surfactant/alcohol systems is achieving good sweep efficiency—that is, making sure the injected solution flows uniformly through the media. Effective sweep efficiency becomes more difficult as the hydrogeology and the contaminant distribution become more heterogeneous. A number of methods already mentioned have been proposed for addressing this challenge (e.g., foams, polymers, vertical circulation wells) and have received limited research at the demonstration level. Additional studies are necessary to further demonstrate the viability and increased efficiency of these methods.

Additional research is also necessary to optimize the implementation of surfactant/cosolvent technologies in karst and fractured bedrock formations, to evaluate the combination of these technologies with other source zone and/or plume remedial technologies, and to evaluate the long-term impact of the mass removal on such activities as post-flushing water flooding and natural attenuation. Many of these research needs are germane to most of the source remediation technologies.

## CHEMICAL TRANSFORMATION TECHNOLOGIES

Two technologies that attempt to transform subsurface contaminants in situ include chemical oxidation and chemical reduction. In both cases, chemicals introduced into the subsurface react with the compounds of concern, leading to their transformation or degradation into less toxic breakdown products.

### In Situ Chemical Oxidation

In situ remediation of groundwater contamination by chemical oxidation (ISCO) involves the addition of strong oxidants such as peroxide, ozone, permanganate, or persulfate to the subsurface (GWRTAC, 1999; ITRC, 2001). These compounds can oxidize a wide variety of dissolved contaminants including halogenated and nonhalogenated aliphatic and aromatic compounds to comparatively less harmful compounds, thereby promoting mass transfer from sorbed or NAPL phases to the aqueous phase and consequently shrinking the source mass.

Hydrogen peroxide, in the presence of added or naturally occurring ferrous iron, produces Fenton's reagent (Glaze and Kang, 1988; Ravikumar and Gurol, 1994; Gates and Siegrist, 1995; Watts et al., 1999; Tarr et al., 2000). The ferrous iron catalyzes the breakdown of hydrogen peroxide into a hydroxide ion and a hydroxyl radical in what known as the Fenton's reaction:

$$H_2O_2 + Fe^{2+} \rightarrow Fe^{3+} + OH^- + OH$$

The hydroxyl radicals are very reactive toward organic compounds, with final breakdown products being carbon dioxide, water, and, in the case of chlorinated solvents, hydrochloric acid. For example, the reaction of Fenton's reagent with TCE is:

$$C_2HCl_3 + 3H_2O_2 \rightarrow 2CO_2 + 2H_2O + 3HCl$$

Typically, hydrogen peroxide is added to the subsurface as an aqueous solution (10–50 wt % $H_2O_2$) with ferrous sulfate. The greatest reactivity occurs in the pH range of 2–4, so pH amendment is often included in application of Fenton's reagent to in situ remediation.

Ozone ($O_3$) gas is another typical oxidant, and it is added to the subsurface through sparge wells. Ozone is very reactive and can oxidize contaminants directly or via formation of reactive hydroxyl radicals (Liang et al., 1999, 2001). For example, the reaction of ozone with TCE is:

$$C_2HCl_3 + O_3 + H_2O \rightarrow 2CO_2 + 3HCl$$

Like hydrogen peroxide, ozone is most effective under acidic conditions. Ozone is the most complex of the common oxidants, requiring the use of onsite ozone generation and operation of sparge wells, some variants of which involve specialized equipment.

Permanganate ($MnO_4^-$) is most commonly used as an aqueous solution of potassium or sodium permanganate. The permanganate ion can oxidize a variety of organic compounds; for example, the reaction of potassium permanganate with TCE is:

$$2KMnO_4 + C_2HCl_3 \rightarrow 2CO_2 + 2MnO_2 + 2KCl + HCl$$

The $MnO_2$ produced by this reaction precipitates in the soil. The reaction rates of permanganate with organic compounds are slower than rates of reaction of ozone and Fenton's reagent (see kinetic studies of Yan and Schwartz, 1999; Hood et al.,

1999; and Huang et al., 1999, 2002). Permanganate is an effective oxidant over a pH range of 4–8 (Yan and Schwartz, 1999).

Persulfate ($S_2O_8^{2-}$) has been proposed as an oxidant suitable for remediation of chlorinated solvents (Liang et al., 2003), but research and field implementations of persulfate are quite limited compared to hydrogen peroxide, ozone, and permanganate. At ambient temperatures, oxidation of chlorinated organics such as TCE by persulfate is not expected to be significant (Liang et al., 2003). However, at temperatures above 40°C, persulfate ions may be transformed to highly reactive sulfate free radicals:

$$S_2O_8^{2-} + Heat \rightarrow 2SO_4^{-}\bullet$$

The sulfate free radicals can oxidize chlorinated organics such as TCE, producing carbon dioxide, water, chloride ions, and sulfate ions. Liang et al. (2003) found that half-lives for TCE decreased from 385 hours at 20°C to 1.44 hours at 40°C and to 0.15 hours at 60°C.

*Overview of Case Studies*

There have been many field applications of chemical oxidation technologies in recent years (see Table 5-3 for select cases and EPA, 1998b) that allow some generalizations to be made about the technology. For the studies summarized in Table 5-3, the contaminants treated include chlorinated ethenes, BTEX, polycyclic aromatic hydrocarbons (PAHs), and methyltertbutylether (MTBE), and DNAPL presence is inferred from high dissolved contaminant concentrations. Sites include highly permeable settings, as well as some silt/clay soils and fractured rock. At most sites, reductions in contaminant concentrations were achieved, but complete cleanup was not reported at any site. In many of the case studies in Table 5-3, performance is based on reductions in dissolved concentrations of contaminants measured within or near the treatment area shortly after completion of the treatment. In most cases initial and final contaminant masses were not determined.

The greatest difficulties in the application of chemical oxidation were encountered in heterogeneous soils and low-permeability soils. In the Kansas City example from Table 5-3, soil mixing was used to overcome limitations associated with the low permeability of the clay soils. Siegrist et al. (1999) investigated the effectiveness of permanganate oxidation in low-permeability silty clay soil by emplacing permanganate in hydraulic fractures. After ten months, they demonstrated that the reactive zone had extended only about 40 cm into the matrix from the fracture. As TCE loadings were increased, removal efficiencies declined. There have been very few applications of in situ chemical oxidation in fractured rock, although application of permanganate to remediation of TCE in fractured rock at Edwards Air Force Base (Morgan et al., 2002) resulted in reductions in TCE and DCE concentrations to below detection in the treatment zone.

TABLE 5-3 Field Applications of Chemical Oxidation

| Location | Oxidant | Media and Contaminants | Application Method | Results |
|---|---|---|---|---|
| Anniston Army Depot | $H_2O_2$ | TCE (1,760 mg/kg) in shallow clay backfill (1997) and in soil, epikarst, and bedrock (1999). | The shallow project (1997) involved 109,000 gal (413,000 L) of 50% $H_2O_2$ injected through 255 injectors [8- to 26-ft (2.4- to 8-m) depth] over 120 days for remediation below lagoons (EPA, 1998b). The deeper full-scale project (1999) involved 20 days of injection, followed by another 7 days of injection two months later (Abston, 2002). | The 1997 full-scale project was considered to be successful, with soil concentrations of TCE reduced to below detection (EPA, 1998b). 1999 full-scale implementation was not successful, as 11 of 42 wells [31–81 ft (9.4–24.7 m)] remained above target of 28,000 ppb TCE (Abston, 2002). |
| Swift Cleaners Jacksonville, FL | $H_2O_2$ | PCE, TCE, DNAPL reported. Dissolved concentrations of 4,400 to 10,000 µg/L for PCE, and 24 to 382 µg/L for TCE. | 400 to 600 gal (1,514 to 2,271 L) of 14%–15% hydrogen peroxide (plus catalyst) injected in two separate areas (12 wells in one area, 13 in second area, 2 injections per well). Radius of influence was 7.5 ft (2.3 m). In third injection, a total of 600 gal (2,271 L) of 15% hydrogen peroxide was injected in 11 wells. Treatment area was 4,500 ft² (418 m²). Treatment depth was 35 to 45 ft (10.6 to 13.7 m). | PCE concentrations were reduced to below 200 µg/L in the first quarter. In the second quarter, PCE concentrations rose to 1,050 µg/L. Contaminant rebound continued in third quarter. |
| Former News Publisher, Framingham, MA | $H_2O_2$ | 1,1,1-TCA, 1,1-DCE, and VC in groundwater, fine-grained silty sand. | Solution of $H_2O_2$, iron catalyst, and acid injected through two 4-in-diameter PVC wells. | In 3 weeks TCA dropped from 40,600 to 440 µg/L and DCE from 4,800 to 2,300 µg/L. VC dropped to below 85 µg/L. |

*continued*

TABLE 5-3  Continued

| Location | Oxidant | Media and Contaminants | Application Method | Results |
|---|---|---|---|---|
| Active Industrial Facility, Clifton, NJ | $H_2O_2$ | TCA in groundwater, fill. | $H_2O_2$, iron catalyst, and acid applied through 12 4-in-diameter PVC wells into the fractured bedrock. | TCA dropped in the most contaminated well from 101 mg/L to 2 mg/L. Average total VOC dropped from 44 mg/L to 15 mg/L. |
| Westinghouse Savannah River Site, Aiken, SC | $H_2O_2$ | 600 lb (272 kg) DNAPL (TCE, PCE) in sand, clay. | $H_2O_2$, ferrous sulfate added over a 6-day period over 27-ft (8.2-m) radius. One batch (500–1000 gal or 1,893–3,785 L) injected per day. | 94% of DNAPL destroyed in treatment zone. Average final [PCE] = 0.65 mg/L, [TCE] = 0.07 mg/L. |
| Cape Canaveral Air Force Station, Launch Complex 34 | $KMnO_4$ | 6,122 kg of TCE in test plot area, with 5,039 kg as DNAPL. Test plot size was 75 ft by 50 ft by 45 ft (23 m by 15 m by 13.7 m). Sandy soils, heterogeneous. | 842,985 gal (3.2 million L) of permanganate solution (1.4–2%) injected into test plot in 3 phases over 8 months. After first injection, monitoring showed local heterogeneities limited oxidant distribution in some areas. Second and third injection phases focused on areas where monitoring showed insufficient oxidant delivery. | Mass of TCE and DNAPL reduced by 77% and 76%, respectively. The best distribution of the oxidant was in upper sandy soils. Distribution of oxidant more difficult in finer-grained soils. Local geologic heterogeneities and native organic matter content may have limited oxidant distribution in some regions. |
| U.S. Army Cold Regions Research and Engineering Lab, Hanover, NH | $KMnO_4$ | TCE (170 mg/kg at one site, 60,000 mg/kg at another site) in sand, silt. | 1.5% $KMnO_4$ solution (15 g/L) injected via two direct-push wells. Site 1: 200 gal (757 L) in 53 days; Site 2: 358 gal (1,355 L) in 21 days. | Chloride concentrations increased from 20 to 6,420 mg/L. More oxidant required for complete cleanup. |

| Site | Oxidant | Contaminants | Application | Results |
|---|---|---|---|---|
| Canadian Force Base Border Ontario | $KMnO_4$ | TCE (1,200 mg/kg), PCE (6,700 mg/kg), DNAPL in sand. | Six injection and five oxidant recovery wells used to flush DNAPL source zone with a solution of 8 g/L $KMnO_4$ for 500 days. | 99% reduction in peak concentrations for both PCE and TCE. Mass flux of dissolved contaminants reduced by 4–5 orders of magnitude. |
| Kansas City Plant | $KMnO_4$ | TCE (81 mg/kg), 1,2-DCE (15 mg/kg), VC, TPH (7,000 mg/kg), PCBs (10 mg/kg) in clay. | Soil mixing using 8- to 10-ft (3-m) diameter blades and a 4–5% $KMnO_4$ solution over 15 soil columns, treated to 25- to 47-ft (7.6- to 14.3-m) depth. | Results showed 83% TCE removal from unsaturated soil, 69% removal from saturated soil. |
| Portsmouth Gaseous Diffusion Plant, Piketon, OH | $KMnO_4$ | TCE (maximum 300 mg/kg in soil, 800 mg/L in groundwater) in sand and gravel. | Parallel horizontal wells, 90 ft (27 m) apart, 200-ft (61-m) screened sections. Water from up-gradient well amended with 1.5–2.5% $KMnO_4$, reinjected into downgradient well. | TCE reduced to below detection in treated areas; heterogeneities impacted treatment coverage. |
| Edwards Air Force Base | $KMnO_4$ | TCE in fractured rock. | 7,500 gallons (28,391 L) of 1.8% $KMnO_4$ injected into bedrock over 4 days from 8 wells. A 40:1 ratio of $KMnO_4$ to VOC was used. | TCE, DCE fell to below detection in treatment zone. Acetone, elevated metals were detected. |
| Former Service Station, Commerce City, CO | $O_3$ | TPH (90–2,380 mg/kg), BTEX (7,800–36,550 µg/kg) in sand/gravel. | 50-ft (15-m)-deep C-Sparge wells (14–20 psi sparge pressure). | TPH reduced from free product to 37 mg/L in one well. TPH and BTEX below detection in all other monitoring wells. |

continued

212

TABLE 5-3 Continued

| Location | Oxidant | Media and Contaminants | Application Method | Results |
|---|---|---|---|---|
| Dry Cleaning Facilities, Hutchinson, KS | $O_3$ | PCE (30–600 µg/L in groundwater) in sand, silt, clay. | C-Sparge well at 35-ft (11-m) depth | PCE was reduced from 34 to 3 µg/L. Air-only injections gave a 71% reduction, in-well stripping an 87% reduction, and air sparging/SVE a 66% reduction. Many operational problems with C-Sparge wells. |
| Former Industrial Facility, Sonoma, CA | $O_3$ | PAHs (1,800 mg/kg), PCE (3,300 mg/kg) in sand, clay. | 4 multilevel ozone injection wells in vadose zone, SVE wells were used outside the treatment areas to control fugitive ozone emissions. | PAHs reduced 67–99.5%, PCE reduced 39–98%. 90% ozone utilization achieved. |
| Park, Utrecht, Netherlands | $O_3$ | Halogenated VOCs (HVOCs) at 1,450–14,500 µg/L; BTEX at 62–95 µg/L in fine sand. | C-Sparge well. | In a 10-day field test, HVOC was reduced from 14,500 to below 1,000 µg/L. Mean BTEX levels fell from 54 to 17 µg/L. |

SOURCE: EPA (1998b, 2003).

Two case studies are presented in Boxes 5-4 and 5-5. The first case study, at NAS Pensacola, involved the use of Fenton's reagent to remediate TCE in a fairly homogenous soil. In this case, rebound of TCE after the first round of treatment was observed. At some locations in the treatment zone, TCE concentrations were still above maximum contaminant levels (MCLs), but the treatment was deemed to have met remediation objectives. The second study, at the Portsmouth Gaseous Diffusion Plant, involved potassium permanganate addition to remove TCE. This study demonstrates the difficulties encountered in using chemical oxidation in heterogeneous soils, as some areas of the treatment zone were not effectively remediated by the permanganate.

*Applicability of the Technology*

**Contaminants.** Peroxide and ozone are suitable for oxidation of BTEX, PAHs, phenols, and alkenes, while permanganate is suitable for BTEX, PAHs, and alkenes. All are suitable for treatment of NAPLs. Some classes of contaminants such as alkanes and polychlorinated biphenyls (PCBs) are resistant to chemical oxidation. Highly reactive chemicals such as explosives are also not

---

**BOX 5-4**
**In Situ Oxidation of TCE DNAPL with**
**Fenton's Reagent at NAS Pensacola**

The site was the Wastewater Treatment Plant Sludge Drying Bed at NAS Pensacola, Florida. The source area was estimated to be 50 ft by 50 ft (15 m by 15 m) in fairly homogeneous sands. TCE concentrations were 3,600 $\mu$g/L, and an estimated 5,000 pounds (2,268 kg) of chlorinated hydrocarbons existed in the source area. In the first phase of remediation, 14 injections wells (10- to 40-ft or 3- to 12-m depth) were used to inject 4,000 gallons (15,141 L) of $H_2O_2$ and 4,000 gallons (15,141 L) of 100 ppm ferrous sulfate over one week. TCE concentrations were reduced from 3,000 to 130 ppb in one well and from 1,700 ppb to below detection limits in another well. One month later, TCE concentrations at the non-detect well had rebounded to pretreatment levels in several locations. A second week of Fenton's reagent injection [6,000 gallons (22,712 L) of $H_2O_2$] at the 35- to 40-ft (10.6- to 12-m) depth was conducted, bringing maximum TCE concentrations down to 90 ppb. Thirty (30) days later, maximum TCE concentrations had rebounded to 180 $\mu$g/L, and then to 198 $\mu$g/L eight months after treatment. It was concluded that this met the remediation objectives and that natural attenuation would be sufficient to control this level of TCE. The cost of the remediation was $250,000.

SOURCE: enviro.nfesc.navy.mil/erb/erb_a/support/wrk_grp/raoltm/case_studies/rao_pensacola.pdf and NAVFAC (1999).

---

**BOX 5-5**
**In Situ Oxidation of TCE DNAPL with Permanganate**

A well-documented application of permanganate to treat TCE present as DNA-PL and as a dissolved plume (54 mg/kg in soil, as high as 820 mg/L in groundwater) occurred at the Portsmouth Gaseous Diffusion Plant in 1997 (DOE, 1999). A 2% permanganate solution was injected into the subsurface for one month through two parallel horizontal wells [200-ft (61-m) screened sections] installed in the center of the plume located in a 5-ft (1.5-m)-thick silty, gravel aquifer. The site stratigraphy consisted of a 25- to 30-ft (7.6- to 9.1-m)-thick silt and clay layer, overlying a 2- to 10-ft (0.6- to 3-m)-thick layer of sand and gravel above bedrock. The sand and gravel aquifer was the target for treatment [a volume of 90 ft by 220 ft by 6 ft (27 m by 67 m by 1.8 m) deep]. It was later found that vertical heterogeneities in the aquifer led to channeling that reduced the effectiveness of treatment. Possible plugging of the midsection of one of the well screens was also suspected to have caused additional delivery problems. An additional vertical well was used to inject additional permanganate. A total of 206,000 gal (780,000 L) of oxidant solution (12,700 kg of $KMnO_4$) was injected into the treatment region. Good treatment was achieved (< 5 ppb TCE) in treated areas, while little change in TCE concentration was observed in areas not swept by oxidant. The average TCE groundwater concentrations in the treatment area were 176 mg/L before treatment, 110 mg/L at completion of treatment, and 41 mg/L two weeks after recirculation. TCE concentrations increased to 65 mg/L and 103 mg/L at 8 and 12 weeks after recirculation, respectively. The gradual increase in TCE concentrations was attributed to dissolved TCE flowing into the area or diffusing out from finer-grained, less-permeable regions.

---

candidates for oxidation technologies due to the potential for causing explosions and fires and for creating hazardous byproducts.

Source zones with high saturations of NAPL may not be good candidates for in situ chemical oxidation, as they will have a very large oxidant demand. The reaction of oxidizing compounds with the NAPL may lead to the generation of excessive amounts of heat in the case of Fenton's reagent or ozone, and to the generation of excessive $MnO_2$ in the case of permanganate. In the case of chlorinated solvents, high levels of acidity may also be generated. Generation of large amounts of carbon dioxide, chemical precipitation, and other geochemical and physical changes may lead to reduced mass transfer from the NAPL to the water phase, limiting the effectiveness of chemical oxidation (Schroth et al., 2001; Mackinnon and Thomson, 2002; Lee et al., 2003).

**Hydrogeology.** Hydrogeologic considerations are perhaps the most important factor for the design of in situ chemical oxidation treatment systems. Peroxide and permanganate are delivered as aqueous solutions through horizontal or vertical wells or vertical injection probes (which control the depth to which in situ chemical oxidation can occur). Rates of injection are therefore limited by soil permeability. In low-permeability soils such as clays, soil mixing may be necessary, as was done at Kansas City (Table 5-3). Hydraulic fracturing has also been used to allow emplacement of oxidants in low-permeability soils (Siegrist et al., 1999).

Treatment effectiveness is also highly influenced by heterogeneities in soils due to layering or fracturing. As peroxide reacts very rapidly, effectiveness of peroxide treatment is particularly sensitive to flow channeling. Permanganate is more stable and reacts more slowly, allowing time for diffusion into low-permeability zones. Ozone is injected as a gas through sparge wells in the vadose or saturated zones. In the saturated zone, channeling of sparged ozone due to viscous instabilities and soil heterogeneities may significantly reduce effectiveness of treatment.

Natural organic matter, reduced minerals, carbonate, and other free radical scavengers in the subsurface consume oxidant, thereby reducing the amount available to degrade the target compounds. Thus, the background oxidant demand must be considered when determining dosage requirements for oxidants. Background oxidant demand should be determined from laboratory tests with soil from the site and the same oxidant dosages as planned for the field.

**Health, Safety, and Environmental Considerations.** Peroxide, permanganate, persulfate, and ozone are all hazardous chemicals that must be handled properly. Application of ozone or Fenton's reagent can generate excessive amounts of heat and a significant amount of gas (Nyer and Vance, 1999). In particular, ozone, being a gas, requires special precautions. The oxidation of soil organic matter and of contaminants generates acidity, and can therefore reduce the pH of the groundwater if sufficient buffering capacity is not present naturally or is not added. There is also the potential for mobilization of redox-sensitive and exchangeable sorbed metal ions. This was observed at Pueblo, Colorado, where application of Fenton's reagent for remediation of TNT, 1,3,5-trinitrobenzene (TNB), and RDX resulted in increases in concentrations of Cr, Se, Mn, and Hg (May, 2003).

Many of these by-products of oxidation reactions may have detrimental effects on the environment. If natural attenuation is desired as a polishing step after the source remediation phase, oxidants may not be the best choice of technology, as they may destroy indigenous microbial populations, particularly redox-sensitive anaerobic microbial communities associated with chlorinated solvent biodegradation. Kastner et al. (2000) found that application of Fenton's reagent reduced microbial populations in groundwater and soil, particularly methanotrophs. There has been very little additional study of the impact of oxidation technologies on subsurface indigenous microbial activity.

Reductions in permeability as a result of in situ chemical oxidation may be caused by the formation of colloidal materials. Permanganate reaction with organics leads to the precipitation of manganese dioxide, which can reduce soil permeability, remain as a long-term source of manganese in the soil, and cause problems in some sensitive environments.

*Potential for Meeting Goals*

The likely effectiveness of oxidation technologies with respect to various objectives for different hydrogeologic conditions is summarized in Table 5-7. Oxidation technologies have the potential for achieving significant mass destruction of organics in the subsurface. However, as demonstrated by the field applications listed in Table 5-3, and by a variety of laboratory-scale studies (Schnarr et al., 1998; Gates-Anderson et al., 2001; MacKinnon and Thomson, 2002; Lee et al., 2003), complete removal of contaminants is not likely to be achieved with oxidation technologies even under optimal conditions.

The installation and operation of in situ chemical oxidation technologies is relatively straightforward, once subsurface conditions are defined, injection well locations are determined, and oxidant requirements are estimated. Of the various technologies employed, ozone sparging is the most difficult to operate due to the reactivity of ozone and the difficulties in operating sparging wells. The use of bench- and pilot-scale tests is recommended to evaluate the potential effectiveness of oxidants for the soils and contaminants to be treated.

Assessment of the effectiveness of chemical oxidation should include monitoring of groundwater geochemistry (pH, redox, dissolved metals), oxidant concentrations, reaction products such as chloride, and temperature. In addition, post-oxidation monitoring should be conducted to evaluate possible rebound of contaminant concentrations, release of metals, dissipation of oxidants, and rebound of microbial populations.

Chemical oxidation technologies most often fail because of ineffective delivery of oxidants caused by subsurface heterogeneities or by poor delineation of contaminant distribution in the subsurface. In heterogeneous soils, the transfer of oxidants into low-permeability zones where contaminants reside may be problematic, resulting in very low efficiency of contaminant destruction. There has been very little study of the use of chemical oxidants in fractured clay and rock, but these technologies are not expected to be very effective in these environments due to the diffusion-limited mass transfer rates of oxidants into the clay and rock matrices, particularly in the case of the unstable ozone and peroxide oxidants (Struse et al., 2002). In the case of permanganate, the use of oxidation emplacement technologies may offer some promise (Siegrist et al., 1999), but treatment times are expected to be lengthy, with significant difficulties in emplacing oxidants with any certainty. Alteration of subsurface permeability due to formation of gases or colloidal materials, or from manganese dioxide precipitation in the case of permanganate, may further reduce the efficiency of in situ chemical oxidation.

*Cost Drivers*

The major costs of oxidation technologies are associated with injection well installation, chemical (oxidant) costs, and post-treatment sampling and monitoring. The costs will therefore be highly influenced by the well depths, the size of the treatment zone, the background oxidant demand, the amount of contaminant to be oxidized, and the effectiveness of delivery of oxidant to the contaminant. Costs and the likelihood of failure are likely to increase with increasing heterogeneity of the subsurface. Treatment costs are also likely to be higher if the subsurface is poorly characterized and the contaminant distribution is poorly delineated.

*Technology-Specific Prediction Tools and Models*

Design of in situ chemical oxidation systems requires selection of injection well spacing and injection rates, as well as prediction of rates of removal of target contaminants. For peroxide and permanganate, injection systems can be designed with conventional groundwater models, as these oxidants are injected in aqueous solution. For injection of ozone into the vadose zone, vapor flow models can be used. However, for ozone sparging into the saturated zone, reliable models for accurate prediction of movement of sparged gases do not exist. Predicting the rates of contaminant oxidation requires modeling the distribution of contaminants, the movement of injected oxidants, and the contact and kinetic reactions between oxidants and contaminants. A few conference proceedings of the modeling of these processes have been published (Hood and Thomson, 2000; Reitsma and Dai, 2000; Zhang and Schwartz, 2000). In the case of dissolved contaminants, the modeling of the processes is straightforward mathematically, although accounting for the impact of small-scale soil heterogeneities on subsurface transport of oxidants and target contaminants can be challenging. In the case of oxidation of NAPLs, complex models may be required to account for the effective kinetics changing as the amount of NAPL, and thus the contact area, is reduced. Changes in soil permeability due to oxidant reactions, such as the formation of manganese dioxide from permanganate, also present a significant modeling challenge that has not been addressed.

*Research and Demonstration Needs*

Continued research on the effectiveness of oxidation technologies is required. In particular, continued research is needed on the interactions of the oxidants with subsurface media (soil, rock), and on the impact of oxidants on soil permeability and on mass transfer from NAPL phases (e.g., impact of $MnO_2$ precipitation on soil permeability and on NAPL dissolution and reaction). The ultimate removal levels possible for the various oxidants in various hydrogeologic settings (particularly fractured media) have not been well demonstrated. In addition, little research has been conducted on the impact of oxidants on metal release and on microbial activity and related intrinsic bioremediation after oxidant flushing.

## Chemical Reduction

Source treatment by chemical reduction consists of mixing granular iron (also known as zero valent iron or ZVI) and clay into a source zone to react with and treat chlorinated solvents, typically at a 95:5 weight ratio of clay to ZVI. The treatment mechanism is the well-documented reductive dehalogenation process used in permeable reactive barrier applications. The purpose of mixing of clay into the source zones is to create a stagnant hydrologic environment to inhibit transfer of contaminants from the source zone to groundwater while the reaction with ZVI occurs inside the source zone. To date, there are a very small number of sites where this technology has been used and none have been documented in peer-reviewed literature. DuPont completed one project using high-pressure jetting as the slurry delivery method and another project using an auger-based soil mixing process. This case study of chemical reduction is presented in Box 5-6.

### *Applicability of the Technology*

**Contaminants.** ZVI has been demonstrated in column and field studies to dechlorinate a wide variety of chlorinated and fluorinated compounds (EPA, 1998c). Reaction with ZVI degrades carbon tetrachloride to chloroform and then to dichloromethane, with some of the original completely degrading to unknown non-toxic products. In long exposures at Martinsville VA, dichloromethane appears to degrade to chloromethane and then methane. PCE and TCE are dechlorinated by ZVI to dichloroethene and vinyl chloride and then to a mixture of ethene and acetylene. Although ZVI is known to react with highly chlorinated ethane compounds (e.g., hexachloroethane and 1,1,1-trichloroethane), ZVI/clay treatment is unlikely to be effective for treatment of source zones containing dichloroethane.

The essential site characterization parameter is the extent of the source area, both vertically and horizontally. Because chemical reduction can be used both above and below the water table, the source zone needs to be defined in both environments. A detailed knowledge of the DNAPL or adsorbed solvent distribution is not necessary because the whole source zone will be homogenized during treatment. However, if a potentially mobile pool of DNAPL is present, the disruption of any confining layers that may occur during treatment should be considered. A rough estimate of the contaminant mass is useful for selecting the amount of ZVI, though engineers may choose to be conservative and inject an excess of ZVI.

**Hydrogeology.** Chemical reduction is practical in any hydrogeologic environment where soil mixing is economically feasible. It has not been practiced in

**BOX 5-6**
**In Situ Treatment of a Vadose-Zone TCE Source Area**
**Using a Jetted Slurry of ZVI and Clay**

Combined chemical reduction/containment technology uses granular iron to degrade chlorinated solvents via reduction and dechlorination reactions, and clays to reduce the permeability of the soils. This combination both treats the source area and reduces groundwater flow through the source area. A field trial of chemical reduction/containment was carried out at DuPont's Martinsville, Virginia, site. Several patents on this technology were granted to the DuPont company. DuPont donated all rights to this technology to Colorado State University in August 2003.

The test was carried out at a former acid neutralization area known as Unit I, which received various laboratory wastes including spent nitric and formic acids, phenol, and carbon tetrachloride (CT). The laboratory waste pits were used between 1958 and 1974. They were closed in 1974 by backfilling with soil. The pits had concrete walls and a concrete cover with two surface openings, one used to discard spent acid and one used to discard solvents. The pits were approximately 12 ft (3.7 m) deep and had open bottoms that were lined with limestone cobbles.

A detailed site assessment showed that the pit area was a continuing source of CT in groundwater. The source area was then carefully delineated. The surface footprint of the source area was approximately 70 ft (21 m) by 100 ft (30 m), and the unsaturated contaminated soil depth extended to approximately 30 ft (9 m) below grade. The source area volume was estimated to be approximately 88,000 cubic yards (67,281 $m^3$). Soil concentrations of CT as high as 30,000 ppm were observed. Much lower concentrations of PCE, TCE, and dichloromethane were found. Based on the site assessment data, DuPont estimated that the source area contained about 22,000 kg of CT. The highest CT concentrations were generally near the contact between alluvial soils and saprolites, located approximately 15–20 ft (4.6–6 m) below grade.

The ultimate goal for remediation in this area was to improve the quality of down gradient groundwater. Groundwater is not used as a potable source, but eventually discharges to the Smith River.

The decision was made to remediate the pit rather than contain it. A number of remedial technologies were considered. Laboratory studies were carried out on chemical reduction, and field pilot evaluations were conducted for soil vapor extraction (SVE) and chemical reduction with containment. The decision was made to proceed with chemical reduction with containment because the laboratory results were promising and because there was a perceived need for a field trial of that technology.

The laboratory studies were conducted with up to 30,000 ppm CT. They showed that the iron reacted rapidly with the carbon tetrachloride and degraded it to about 1,500 ppm of dichloromethane. The dichloromethane appeared to persist in the laboratory studies. A surprise in the laboratory studies was the appearance of up to 1,500 ppm of tetrachloroethene, and trace amounts of hexachlorobutadiene. Because dichloromethane biodegrades very rapidly in soil (NRC, 2000), it was predicted to be a transient compound in field implementation. Tetrachloroethene is well known to react with iron, so it was also predicted to degrade in the field.

*continued*

### BOX 5-6 Continued

Several methods of mixing the iron and clay into the source area soil were evaluated. Deep soil mixing using large augers was selected after competitive bidding. Based on site assessment information and the laboratory studies, three treatment zones were designed. The most contaminated zone would be treated with 6 pounds of iron per cubic foot (96 kg/m$^3$) of soil, a less contaminated zone would be treated with 4 pounds of iron per cubic foot (64 kg/m$^3$), and the third zone would receive 2 pounds of iron per cubic foot (32 kg/m$^3$).

The source area treatment was conducted in October 2002. Before source treatment began, the pit concrete walls and cover were excavated and removed, and buried utilities were located, abandoned, and removed. The reagent injection and soil mixing was done using a Link-Belt crane equipped with a Casagrande mixing unit with an 8-ft (2.4-m)-diameter auger. An 8- to 10-person crew was needed to support the operations. Supporting equipment included an excavator, a batch mixing plant, and a forklift. Seventy six (76) 8-ft (2.4-m)-diameter columns of soil were mixed and treated to a depth of 35 ft (11 m). Productivity increased over time, increasing by the end of the project to four columns per day mixed and treated. A significant excess of granular iron was added during the mixing. The treatment reactions are expected to continue until no solvents remain within the mixed material.

Photo Courtesy of David Ellis, DuPont.

Quality control was maintained by sampling each mixed column at several depths using a push tube. The concentrations of clay and iron were measured at each depth. If not enough iron was found, more iron was added and the column was re-mixed. The only operating problem encountered was that the augers were unable to penetrate and mix a thin boulder layer that was encountered in a small area at the south side of the source area. The boulders had to be excavated before treatment could continue. Some additional soil volume was created during the project, so a small mound was left in place at the end of the project.

The amounts of additives injected were 225 tons of granular iron, 340 tons of kaolinite, and 250,000 gallons (946,353 L) of water. Cement was added to the top five feet (1.5 m) of each column to improve the soil's bearing properties. The remedy required 10 weeks to construct and cost roughly $700,000. This cost includes site preparation, utility location and removal, mobilization, start-up, materials, oversight, quality control, air monitoring, demobilization, paving, and report preparation.

A series of soil cores were collected from the treated area one year after treatment to monitor the progress of the treatment. Forty-four (44) pretreatment cores had been analyzed; 18 posttreatment core samples were collected and analyzed. The following table summarizes the observed average concentrations and estimated contaminant masses observed before and after treatment. As with all DNAPL sites, these estimates are based on the best information available. The posttreatment estimates are thought to be more rigorous than the pretreatment estimates because the site was homogenized by the mixing equipment.

| | Pretreatment | | Posttreatment | |
|---|---|---|---|---|
| | Concentration (ppm) | Mass (kg) | Concentration (ppm) | Mass (kg) |
| Carbon Tetrachloride | 1,250 | 22,000 | 0.7 | 2.5 |
| Chloroform | 11 | 184 | 1.1 | 18 |
| Dichloromethane | 2 | 33 | 29 | 502 |
| Chloromethane | ND | ND | 1.7 | 3.8 |
| Tetrachloroethene | 0.4 | 6 | 5.4 | 87 |
| Trichloroethene | 0.4 | 6 | 0.15 | 2.3 |

any bedrock environments to date, as it is anticipated that bedrock environments would be very difficult to mix adequately and economically. The bearing properties of the soil should be known, both before and after treatment. If the bearing capacity will be too low after treatment, a small amount of cement can be mixed into the soil during treatment in order to restore the bearing properties. There are no physical limits on the area or volume that could be treated with chemical reduction, although there may be limits imposed by economics and the relative costs of competing technologies like containment. The ease of soil mixing decreases with depth, such that soil mixing is rarely used at depths greater than 35 meters.

It should be noted that the mechanical process of soil mixing will first increase local permeability before clay is added to reduce permeability. Thus, there is the potential for mobilizing DNAPL by removing capillary barriers during the mixing process. The mixing process may also induce local pressure gradients, potentially increasing DNAPL mobility. With virtually no data available on the mechanics of the process under saturated conditions, the resulting risks are largely unknown and must be evaluated on a site-specific basis.

**Health, Safety, and Environmental Considerations.** The primary health and safety considerations are the dangers of working with the heavy construction equipment used to conduct the mixing essential to this treatment. Prior to mixing, all underground utilities should be identified and either deactivated, or preferably removed. Worker exposure to potential VOC releases during the mixing process must be monitored and controlled. Once mixing is completed, the potential for VOC release is very low. Neither the ZVI nor the clay used in this process is believed to present a hazard to worker health.

*Potential for Meeting Goals*

Chemical reduction is believed to have a high potential for meeting a variety of remediation goals when it is used on appropriate sites. Both the chemistry of the contaminant degradation reactions that this technology depends upon and soil mixing are well-documented and established. In unconsolidated media of Types I, II, and III, the potential for this technology is high for achieving mass removal, concentration reduction, mass flux reduction, reduction of source migration potential, and a substantial reduction in toxicity. However, this technology is being rated based on a very small number of field studies, and without peer-reviewed documentation.

The technology should only be used at sites where there is reason to believe that soil mixing can be successful. Soil mixing works best when there are no large objects in the subsurface (e.g., large cobbles or boulders) and when there are no surface structures. Mixing of bedrock (Types IV and V) is very difficult, such that this technology would not be appropriate in these hydrogeologic settings.

Finally, the simplicity of chemical reduction makes it a very robust technology. However, it must be noted that it cannot be easily combined with technologies other than containment or excavation because of the loss of permeability that is a consequence of chemical reduction.

### Cost Drivers

The volume and depth of the source zone are the primary drivers of treatment cost for chemical reduction. The depth of the necessary mixing is key—shallow mixing can be done more rapidly than deep mixing, so the productivity of the mixing equipment in cubic meters mixed per hour will be considerably higher at shallow sites. The amount of ZVI that will need to be added during mixing contributes to costs.

### Technology-Specific Prediction Tools and Models

At this time there are no modeling tools specifically adapted to predict the success or failure of chemical reduction using ZVI and clay, though some parts of the treatment can be modeled. The residence time of water within the treated zone can easily be predicted assuming homogeneous mixing is accomplished. Simple kinetic models of ZVI may roughly predict the rates of treatment. Laboratory treatability studies appear to be good tools for designing treatment mixtures and for predicting the success of this treatment on a site-specific basis.

### Research and Demonstration Needs

The impacts of soil type and composition on chemical reduction are not known. Whether contaminant mixtures present a problem has not been tested, especially for cases where hydrocarbons are a co-contaminant. Catalysis of the ZVI degradation reactions has been examined for permeable reactive barriers, but little is known about the ability or impact of catalysis on a ZVI/clay treatment. A detailed knowledge of the kinetics of reaction and of the impacts of mass transfer kinetics would be useful in predicting the performance of chemical reduction remedies. Finally, the potential for mobilizing DNAPL by mechanical disruption of confining layers or by hydraulic displacement resulting from pressure gradients induced during mixing needs to be the subject of future studies.

## SOIL HEATING TECHNOLOGIES

The three most widely applied soil heating methods used for source remediation are steam flushing or flooding, thermal conduction heating, and electrical resistance heating. All of these technologies are intended to increase the partitioning of organic chemicals into the vapor or gas phase where they can be extracted under vacuum, which is a form of enhanced soil vapor extraction (SVE). In addition, there is evidence that some organic contaminants can be destroyed in situ at sufficiently high temperatures.

The three heating methods draw on different physical processes to transport energy in the soil, and as a result each is particularly appropriate for certain site conditions. Steam flooding uses a hot fluid to carry heat into the subsurface. Steam follows high-permeability pathways through the subsurface, however, so it preferentially heats those paths and leaves the less conductive soil relatively cool. Thermal conduction creates the highest temperatures and is relatively insensitive to soil properties. Electrical resistance heating passes an electrical current through the soil, heating formations where the electrical current flow is the greatest. Electrical current flows through clays and silts more readily than through sand, so electrical resistance heating preferentially warms the clay-bearing horizons that are not swept or are poorly swept during steam injection.

All thermal methods rely on contaminant flow and transport in the gas phase. While control of the gas phase above the water table usually is not a problem, gas flows below the water table may be strongly dominated by buoyancy forces. In cases where high temperatures are required for DNAPL evaporation (above 100°C), the inflow of groundwater into the thermal treatment zone is potentially a problem. This could be a limiting factor in high-permeability aquifers unless barrier walls or other means are used to prevent such inflow.

## Steam Flushing

Steam injection was first used for tertiary petroleum recovery in 1933 (White and Moss, 1983), and it is still widely used today, particularly for recovery of heavy oil from tar sands. Steam injection assists in recovery of viscous oils, primarily by reducing oil viscosity and allowing more effective displacement of oil toward recovery wells. In addition, production is increased by thermal swelling of oil, by steam distillation of light components of oil, by a gas drive resulting from the steam flush, and from a solvent dilution effect caused by the condensation of light ends in front of the steam zone (Butler, 1991). In recent years, steam injection has been identified as a promising technique for removing NAPL contaminants from the subsurface (Hunt et al., 1988), but has not yet been widely used on a commercial scale for remediation.

The groundwater remediation process differs from the petroleum recovery process in a number of ways. Essentially complete contaminant removal is required for groundwater remediation, rather than an incremental increase in oil recovery. Remobilization of NAPL leading to increased groundwater contamination is not an acceptable consequence of NAPL remediation. In addition, most NAPLs have viscosities near that of water, and many have relatively low boiling points. These factors must be considered when determining whether to remediate NAPL by steam injection.

## Mechanisms of Steam Treatment

The mechanisms operative in steam flushing include volatilization of water and organic fluids, formation of a steam zone, and hydraulic displacement of organic compounds in front of the steam zone. The relative importance of these mechanisms depends on the nature of the organic compounds present and on the distribution of the compounds in the soil.

**Volatilization of Organic Compounds.** When steam is injected into soil, the steam initially condenses, giving up latent heat to raise the temperature of the soil and pore fluids. (Radio frequency heating or electrical resistance heating, discussed later, produce a similar in situ temperature increase.) As the temperature of the soil increases, the vapor pressures of the pore fluids are increased. In the vadose zone, increasing temperatures result in an increase in the vapor phase concentrations of water and of other liquids, such as organics, present in the soil. In liquid-saturated soil, a vapor phase is formed at a particular location when the sum of the vapor pressures of the liquid phases present exceeds the in situ liquid pressure. The efficiency of thermal remediation technologies for mobilizing a particular organic compound through the volatilization mechanism is thus a function of the compound's vapor pressure.

**Formation of a Steam Zone.** When steam is injected continuously into the subsurface, it volatilizes water and organic compounds (distillation), and a steam zone is formed with a propagating condensation front. In the steam zone the amount of water remaining in the soil pores depends on the injected steam quality, temperature, and pressure. At the condensation front, a bank of condensed organic contaminant forms and moves ahead of the steam front toward a withdrawal well (see Figure 5-5). Although the organic compound may have originally been distributed at residual saturation levels and was thus immobile, the organic saturations in the bank will generally be above residual levels, and the organic bank will be mobile.

As the steam zone expands and the condensation front moves toward the extraction well, heat is transferred through the soil with fluid flow (convection) and also by conduction, as a result of temperature gradients. Heat conduction occurs in both the longitudinal and transverse directions with respect to the direction of flow. Since steam is much lighter than water, the steam zone tends to rise as it travels horizontally. This steam override, or gravity segregation, can be a significant problem in designing a steam injection program. In addition to the low density, steam also has a much lower viscosity than water, such that channeling can be a significant problem. Transverse heat conduction is an important mechanism in damping steam channeling in heterogeneous soils.

In the case of a mixture of low- and high-volatility NAPL, steam injection will result in preferential distillation of the more volatile compounds. These

Steam Injection

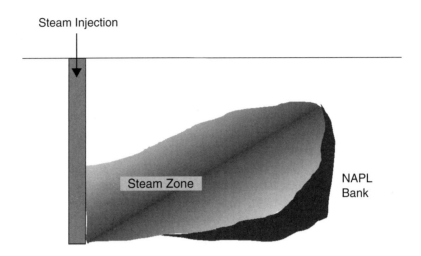

FIGURE 5-5  Steam zone and NAPL bank formation.

compounds will recondense at the front of the steam zone in the organic fluid bank. This will result in the formation of a bank of mobile organic fluid that moves ahead of the steam zone. This mobile bank will reunite stranded blobs and ganglia of NAPL and will produce a solvent drive effect. For example, in the case of a mixture of TCE and PCBs, the TCE will be preferentially stripped from the trapped blobs and ganglia and recondense in a bank at the front of the steam zone. This will result in a form of solvent drive, resulting in improved removal of the trapped PCBs as well as the TCE.

**Hydraulic Displacement of Organic Fluids.** Injection of hot water or steam can lead to hydraulic displacement of DNAPL due to the aqueous phase pressure gradients that develop. The extent of hydraulic displacement as an important mechanism during steam injection or hot water flooding depends on the DNAPL being displaced and on the nature of the porous medium, or fractures, in which the DNAPL was located. Organic fluids that are trapped as pools on low-permeability, fine-grained layers or in fractures may exist as a continuous phase at saturations above residual levels. In the case of organic fluids with low vapor pressures and high boiling points relative to steam injection temperatures, removal rates through distillation will be very low. For these fluids, such as heavy oils and PCBs, hydraulic displacement may be the major removal mechanism operative during steam injection. If the organic fluid exists in the soil at saturations *above* the residual level, then injection of steam will lead to hydraulic displacement of the organic ahead of the condensation front associated with the expanding steam zone. The efficiency of this hydraulic displacement will depend on the reduction in organic phase viscosity that occurs as the temperature is increased, and on changes in residual saturation resulting from reductions in viscosity, reductions in interfacial tension, and changes in wettability. It has been found that increasing temperature leads to decreases in residual saturation in both consolidated sands (Sinnokrot et al., 1971) and unconsolidated sands (Poston et al., 1970). Thus, steam (or hot water) flooding would be expected to remove more organic fluid than water flooding at ambient aquifer temperatures.

A major concern in the displacement of viscous liquids by less viscous liquids is viscous channelling. The less viscous displacing fluid will tend to break through the more viscous in-place fluid in a few channels, resulting in very ineffective subsequent contaminant removal. The decrease in organic fluid viscosity that occurs with the increased temperatures associated with steam or hot water flooding or electrical heating may reduce the extent of fingering that might otherwise occur in an ambient temperature hydraulic displacement process. These decreases in viscosity and interfacial tension with steam injection would be expected to result in lower remaining residual saturations than would be expected after water flooding (Poston et al., 1970; Sinnokrot et al., 1971).

Steam injection would not normally be expected to provide much additional hydraulic displacement of trapped blobs and ganglia of organic fluids left by infiltrating organic fluid, or left by steam or water flooding. This is because organic and water viscosities both decrease as temperature increases. Organic–water interfacial tensions also decrease as temperature increases. PCE–water interfacial tensions decrease by about 10 percent as the temperature is increased from 10°C to 90°C (Ma and Sleep, 1997). The ratio of viscous to capillary forces is called the capillary number. NAPL trapped by capillary forces can be displaced hydraulically if the critical capillary number, a function of the fluids and the soil structure, is exceeded. Decreasing interfacial tensions increases capillary numbers, while decreasing water viscosity decreases capillary numbers. During steam injection where one might expect water to displace NAPL, the increase in temperature is likely to decrease the water viscosity more quickly than it decreases the interfacial tensions, so that overall, raising the temperature decreases the capillary number. Thus, steam injection does not lead to any direct enhancement of hydraulic displacement of *residual* NAPL. Some swelling of the DNAPL occurs on heating, but this does not result in significant DNAPL mobilization.

**Hydrous Pyrolysis/Oxidation.** At the elevated temperatures (100°C–140°C) associated with steam flushing, it has been claimed that hydrous pyrolysis and oxidation of contaminants is a significant destruction mechanism. Knauss et al. (1999) examined the aqueous oxidation of TCE over a temperature range of 20°C–100°C, and determined Arrhenius activation energies. From this analysis, they concluded that increasing temperatures from 20°C to 100°C would increase TCE oxidation rates by a factor of 3,000. However, oxidation rates are still lower than those expected from reductive dehalogenation by anaerobic bacteria. At the Visalia field site, it was estimated that hydrous pyrolysis and oxidation accounted for 17 percent of the total removal of creosote from the subsurface, an estimate based on monitoring of carbon dioxide extracted from the subsurface (U.S. DOE, 2000).

Regardless of the operating mechanism, extraction wells are needed to recover fluids and vapors during steam flushing. In the vadose zone, vapor extraction wells may be used. In the saturated zone, the extraction wells will initially recover water, then a mixture of NAPL and water, and finally steam and organic vapors. The spacing of injection and extraction wells must be carefully chosen to ensure capture of displaced water and NAPL, steam, and organic vapors.

### Overview of Case Studies

A variety of field-scale implementations of steam injection are summarized in Table 5-4. Many of these cases involved hydrocarbons, presumably present as LNAPL. Some of the cases involve mixtures of hydrocarbons and chlorinated solvents, but it is not clear whether the mixtures are LNAPL or DNAPL. The only reported case of steam flushing for chlorinated solvents is at the Savannah River Site, but no performance data are given. Two field-scale applications of steam flushing are described in Boxes 5-7 and 5-8. The field examples indicate that steam flushing can be very effective for removing VOCs from relatively homogeneous permeable soils. As contaminants become less volatile and soils become more heterogeneous or less permeable, the effectiveness of steam flushing decreases. There is limited field experience using steam flushing for DNAPLs located below the water table and for NAPLs in fractured rock and clay. In addition, the reported performance metrics for many of the case studies are based on mass removed rather than on mass remaining, reductions in dissolved contaminant concentrations, or contaminant fluxes—all of which are better indicators of treatment efficacy than is mass removed.

### Applicability of the Technology

**Contaminants.** Steam injection is most effective, in comparison with other remediation techniques, for removing separate phase NAPL, rather than organic contaminants dissolved in the aqueous phase. Steam injection is equally suited to remediating petroleum and chlorinated hydrocarbons. The most important characteristic of the compound that should be determined is its vapor pressure over the temperature range typical of steam injection. Compounds with boiling points below that of water are readily vaporized by steam injection. As the compound's vapor pressure decreases, the mole fraction of contaminant in the vapor phase decreases, and the removal rate from vaporization decreases proportionally. For

TABLE 5-4 Field Applications of Steam Flushing

| Location | Contaminants | Hydrogeology | Volume Treated | Results |
|---|---|---|---|---|
| Former Hazardous Waste Disposal Site, Muehlacker, Germany | TCE; BTEX; Volatiles-halogenated; Volatiles-nonhalogenated. | Highly heterogeneous, weathered sandy marl in the unsaturated zone. | 3,000 cubic meters, to 49 ft (15 m) bgs | Ten months of steam injection; 2,500 kg of TCE removed, approximately 95% in gas phase, with the rest dissolved in $H_2O$. |
| Lawrence Livermore National Laboratory, Gasoline Spill Site, Livermore, CA | BTEX; Petroleum hydrocarbons; Volatiles-nonhalogenated; Gasoline (likely free-phase). | Layered system, sandy gravel and gravelly sand, clayey silts to silty clays, sandy to clayey gravels and gravelly to silty sands, etc. | 100,000 cubic yards (76,455 m$^3$) | Over 7,600 gallons (28,769 L) of gasoline removed, most as vapor over 10 weeks, conducted in phases over a 1-year period. |
| Lemoore NAS, Lemoore, CA | JP-5; Petroleum hydrocarbons; Volatiles-nonhalogenated. | Sands and silts with hydraulic conductivity of $3.9 \times 10^{-3}$ to $1.4 \times 10^{-2}$ cm/s. Water table at 4.9 m. | | 190,000 gallons (719,228 L) recovered; final vadose zone concentration of 20–50 mg/kg total petroleum hydrocarbons (TPH) |
| North Island NAS, San Diego, CA | TCE; Petroleum hydrocarbons; Volatiles-halogenated; Volatiles-nonhalogenated; JP-5; Semivolatiles-nonhalogenated; LNAPL floating on groundwater | Soil | 1,100 square yards (920 m$^2$) | 28,600 lb (12,973 kg) removed, consisting of 14,600 lb (6,622 kg) of fuel hydrocarbons and TCE in liquid phase and 14,000 lb (6,350 kg) of TCE in vapor phase. |

continued

TABLE 5-4 Field Applications of Steam Flushing

| Location | Contaminants | Hydrogeology | Volume Treated | Results |
|---|---|---|---|---|
| Rainbow Disposal, Huntington Beach, CA | Diesel Fuel; Petroleum hydrocarbons; Volatiles-nonhalogenated. | Perched water to a depth of 40 ft (12 m) bgs; sand lens at 35–40 ft (10.6–12 m) bgs. | 15,000 square yards (12,542 m$^2$) | 45,000 gallons (170,343 L) recovered; concentration reduced from 17,000 to 1,500 mg/kg (average). |
| Savannah River Site, Aiken, SC | PCE; TCE (mixed contaminants, DNAPL dominant); Volatiles-halogenated. | Interbedded clay, silt, sand; target zone bounded by clay layers. | 100 ft (30 m) × 100 ft (30 m) on surface; to depth of 165 ft (50 m); PCE at depths of 20–165 ft (6–50 m) bgs, above clay aquitard | 31,000 kg of contaminants were removed [30,000 kg of PCE (26%) and 1,000 kg of TCE (62%)]. Some removal attributed to hydrous pyrolysis, but was not quantified. |
| Site 5, Alameda Point, Alameda, CA | Volatiles-halogenated; TCE; Total extractable petroleum hydrocarbons (TEPH); Petroleum hydrocarbons; Volatiles-nonhalogenated. | Sand to silty sand, silty clays interlayered with sands and silty sands. | 180 cubic yards (138 m$^3$) of soil; 100–200 gallons (379–757 L) of NAPL; LNAPL found at depths of 3–10 ft (0.9–3 m) bgs | 600 gallons (2,271 L) of NAPL removed (84% as separate phase with 242 kg TCE). Concentrations reduced "many orders of magnitude" during treatment; TCE in soil after treatment from below detection limits (ND) to 20 mg/kg. TCE in groundwater after treatment from ND to 295 mg/L. |

| A.G. Communications Systems, Northlake, IL | TCE; cis-1,2-DCE; xylene; benzene; Dissolved TCE concentration of 45,000 mg/L. | Unsaturated and saturated zone, 330,000 cubic yards (252,303 m$^3$). | 65 steam injection wells [39 screened at depth of 35 ft (11 m) bgs, 26 wells screened at 46 ft (14 m) bgs], 186 shallow SVE wells and 76 combination groundwater/vapor extraction wells (15 to 30 gpm) | Average dissolved TCE concentrations reduced from 20,000 mg/L to 1,000 mg/L, reductions of 90% for TCE and DCE over two years. 33,000 lb (14,969 kg) of hydrocarbons removed. |
|---|---|---|---|---|
| Visalia Pole Yard NPL Site, Visalia, CA | Semivolatiles-nonhalogenated; Semivolatiles-halogenated; Polycyclic aromatic hydrocarbons (PAHs); Pesticides/herbicides; Creosote; Pentachlorophenol (PCP). | Three distinct water-bearing zones. | 75–105 ft (23–32 m) bgs | June 1997–June 1999, approximately 1,130,000 pounds (512,559 kg) of creosote removed or destroyed (50% removed as NAPL, 16% as vapors, 16% in aqueous phase, 17% destroyed by hydrous pyrolysis/oxidation in situ). |

SOURCE: http://www.clu-in.org/products/thermal. EPA (2003).

## BOX 5-7
### Visalia Pole Storage Yard

At the Visalia pole storage yard, steam flushing was used to remediate creosote and pentachlorophenol. During 25 months of operation, a total of 1,130,000 lb (141,000 gal or 533,743 L) of creosote were removed or treated (10,400 lb/wk or 4,717 kg/wk). Approximately 50 percent of the contaminants were removed in free phase—16 percent removed as vapors, 16 percent removed in an aqueous phase, and 17 percent destroyed by hydrous pyrolysis in situ (U.S. DOE, 2000). However, there were no accurate estimates of pretreatment conditions, so the removal efficiency cannot be determined. Furthermore, none of the available reports on this site discuss the impact of the remediation on contaminant concentrations, mass fluxes, or other metrics that may have been of interest.

## BOX 5-8
### Alameda Point, Site 5

Steam flushing was used at Site 5, Alameda Point, Alameda, California, to remediate a NAPL source in shallow fill soils. The NAPL source consisted of a mixture of petroleum hydrocarbons (diesel and motor oil) and TCE and was less dense than water. The site stratigraphy consisted of a surface sealed by asphalt and concrete slabs, 1.5 m of sandy silt fill, 2.5 m of fine silty sand, 0.2 m of clay, and 2–3 m of Bay mud. The water table was 1.7 m bgs.

The steam flushing implementation consisted of six injection wells located around the outside of the source zone, and an extraction well located in the source zone. To prevent NAPL condensation below the cap, steam was first injected into the vadose zone for 10 days until hot vapors were recovered in the extraction well. Steam was subsequently injected into the vadose and saturated zones for 40 days.

A total of 1,950 kg of organic liquid was recovered (83% as NAPL, the remainder as adsorbed material from water and vapors recovered) in the extraction well during the 40 days of steam flushing. The TCE was primarily in the gas leaving the last vapor/liquid separator (192 kg), with an additional 22 kg being recovered from the water phase and only 18 kg measured in recovered NAPL. VOC concentrations in the source area were reduced by two orders of magnitude, and soil concentrations of extractable hydrocarbons were reduced by an order of magnitude, with remaining compounds being mostly low-volatility compounds. Trace amounts of chlorinated compounds were found in the shallow soils just below the surface cap. Upon cooling, microbial populations in the treated soils rebounded.

SOURCE: http://www.containment.fsu.edu/cd/content/pdf/511.pdf.

compounds with very high boiling points, such as creosote and PCBs, viscosity lowering and hydraulic displacement may be the most significant removal modes operative in steam flushing.

**Hydrogeology.** The permeability and the degree of heterogeneity existing at a contaminated site will be important factors influencing the effectiveness of thermal technologies. As with most flushing technologies, very high pressure gradients are required to achieve reasonable rates of steam flow and subsequent contaminant removal from low-permeability soils. Butler (1991) recommends that steam injection not be used for petroleum recovery in formations with permeabilities below $10^{-13}$ m$^2$. The use of hydraulic and pneumatic fracturing techniques is being considered to enhance the permeability of low-permeability soil such as massive clays. In some cases parallel planes of horizontal fractures are created. Steam is injected into some fractures, and heated fluids and gases are withdrawn from adjacent fractures.

In sites with substantial contrasts in permeabilities between soil layers, significant channelling of steam along the high-permeability layers will occur. This will lead to the bypassing of contaminants trapped in the low-permeability layers. Eventually the steam zone will expand into the lower-permeability layers through transverse heat conduction. Thus, the relative efficiency of steam injection in layered systems depends on the location of the NAPL, on the permeability contrasts in the system, on the relative thickness of the layers, and on the rates of transverse heat conduction.

Steam injection can be used to access contamination under buildings provided that there is no danger of steam and organic vapors entering the building. This is most likely if the soil zone to be cleaned is an adequate depth below the bottom of the building. Deeper source zones also experience fewer problems with steam short-circuiting through permeable soil layers that are usually placed below building foundations. The building must be small enough to allow adequate spacing of injection and extraction wells, unless angle drilling technologies are used. Pavement does not pose any special problems for steam injection, and can actually help avoid breakthrough of steam and organic vapors at the ground surface.

Steam injection can theoretically be used at any depth accessible by wells. It is used at depths as great as 900 meters for tertiary petroleum recovery.

**Health, Safety, and Environmental Considerations.** The steam injection process produces heated organic vapors, which in the case of some contaminants may lead to fire or explosion hazards. It is important to control the migration of the steam and organic vapors during the remediation of near-surface NAPL contamination to avoid steam breakthrough at unexpected locations in the ground surface.

When attempting to remove DNAPL from the subsurface using thermal methods, particular attention must be paid to the potential for increasing the extent of DNAPL contamination (She and Sleep, 1999). The formation of a condensation bank of mobile DNAPL ahead of the steam zone may lead to vertical remobilization of DNAPL. It is also possible that DNAPL immobilized on fine-grained lenses could be displaced laterally off the lens by steam and could sink below the steam zone and cause contamination at deeper levels in the aquifer. Increases in temperature also decrease interfacial tensions and proportionately decrease entry pressures associated with fine-grained soil lenses. Steam injection programs must be designed so that vertical remobilization of DNAPL does not occur. This is usually accomplished by ensuring that the steam zone extends well below the source zone. When low-permeability layers are present at the bottom of the source zone, the use of electrical heating technologies may be required.

*Potential for Meeting Goals*

The expected effectiveness of steam flushing for meeting various objectives for different types of sites is summarized in Table 5-7. The entries in this table are based on available case studies and on the current understanding of the mechanisms of steam flushing. The most commonly reported performance metric for steam flushing from case studies is the amount of mass removed, as this is the easiest to measure. Removal of large amounts of mass would be expected to reduce source migration potential. Reductions in local aqueous phase concentration and contaminant mass fluxes would also be expected with reductions of mass, but the extent of these reductions would depend greatly on the hydrogeology, the initial mass present (which in many cases is not known), and the distribution of this mass. Decreases in local aqueous phase contaminant concentrations or reductions in contaminant mass fluxes do not appear to have been reported very frequently for steam flushing. Although hydrous pyrolysis has been reported as a process that may destroy some contaminants in situ at the elevated temperatures associated with steam flushing, there are no studies in which the extent of this process has been quantified from a rigorous mass balance on contaminants.

Despite these qualifications, for well-defined source zones consisting of volatile organic compounds in low-heterogeneity permeable soils, steam flushing has the potential for achieving substantial mass removal. This is expected to produce large decreases in contaminant concentrations in the treatment zone and large decreases in contaminant fluxes from the treatment zone. As soil permeability decreases, higher steam injection pressures are required to achieve the same rate of steam propagation through the soil. As the degree of heterogeneity increases, the efficiency of steam flushing decreases, with the steam channeling through high-permeability pathways. In this case, heat conduction from the steam channels to the lower-permeability zones will eventually lead to contaminant removal from these zones, but more steam, more energy for steam production, and more time will be required.

There is little laboratory or field experience with steam flushing in highly heterogeneous, low-permeability, or fractured media. However, it is clear that the risk of incomplete source removal increases with increasing subsurface heterogeneity, with the greatest risks for fractured media. In addition, as heterogeneity increases, the ability to predict and control steam movement in the subsurface decreases. Control of steam movement is important in avoiding undesired downward mobilization of DNAPLs.

Assessment of the success of steam flushing usually includes monitoring of in situ temperatures to ensure that elevated temperatures are reached throughout the source zone. As a significant amount of time (weeks) may be required for subsurface temperatures to decline to preflushing levels, temperatures and contaminant concentrations should be monitored for some time following the cessation of steam flushing.

*Cost Drivers*

Costs for a steam injection program may include a boiler for steam production, injection wells, extraction wells, vacuum extraction equipment, condenser equipment, and treatment trains for off-gas and condensate treatment. A source of feedwater for steam generation will be required, and a treatment system may be required for pretreatment of feedwater to remove dissolved solids and to prevent excessive scaling of the boiler tubes. In most cases the boiler would be fired by natural gas. The deeper the steam is injected, the greater the pressure and temperature of the steam must be, entailing higher energy costs for steam production as well as higher equipment costs. In the case of steam injection in the vadose zone, vacuum extraction is required to remove vapors, which necessitates the use of vapor extraction equipment. The vapors and condensate recovered from steam injection operations are normally passed through a condenser to condense water and organic vapors. The remaining noncondensible gas fraction will require treatment with activated carbon. The water and organic fractions of the condensate can be separated gravimetrically, with the water fraction requiring treatment to remove dissolved organics before disposal.

At the recovery wells associated with steam injection, heated organic fluids and water will be produced before steam breakthrough, and water and organic vapors will be produced after steam breakthrough. The produced fluids and condensed vapors may be separated into organic and water phases. The organic phase may be pure enough to be recycled. The water phase will be saturated with organic compounds and will require further treatment before it can be disposed of. The volume of water produced will be much lower than that produced by conventional pump-and-treat techniques since the recovery rates of DNAPL will be much higher with steam injection by virtue of the steam distillation removal mechanism.

The deeper the steam is injected, the greater must be the pressure and temperature of the steam, which entails higher costs for steam production as well as for equipment.

## Technology-Specific Prediction Tools and Models

Several models of steam injection for petroleum recovery have been developed. Unfortunately, these models are not generally applicable to modeling DNAPL remediation by steam injection because most of the models neglect dissolution of organics in the water phase, and they neglect capillary pressures. Falta et al. (1992) described a one-dimensional model for simulation of steam injection in NAPL remediation. The model, which included three-phase flow and transport of a single-component organic phase, compared favorably with laboratory steam injection experiments conducted by Hunt et al. (1988). Sleep (1993) presented a three-dimensional three-phase multicomponent model for steam injection. Modeling steam injection is very difficult and computationally demanding due to the highly nonlinear nature of the equations of nonisothermal multiphase flow and transport, and it requires considerable modeling and numerical expertise. Determination of the parameters for the various constitutive relationships involved in these models—such as how capillary pressure, saturation, and permeability are functions of temperature (e.g., see She and Sleep, 1998)—is difficult and costly.

## Research and Demonstration Needs

Little research has been conducted on the effectiveness of steam flushing in heterogeneous porous media, including fractured rock and clay. There is a need for further research in this area, including studies of NAPL displacement by hot water and steam flushing, of the impact of temperature on sorption processes, of mass transfer rates in heterogeneous systems, of the role of hydrous pyrolysis for various contaminants, of the potential for DNAPL remobilization in complex subsurface environments, and of the effect of elevated temperatures on soil properties and microbial activity (Richardson et al., 2002).

## Conductive Heating

Conductive heating, sometimes referred to as thermal conduction heating or in situ thermal desorption (ISTD), refers to the warming of the subsurface by heat conduction from electrical heating elements. Two configurations of heating elements have been used for conductive heating—thermal blankets and thermal wells (Stegemeier and Vinegar, 2001). Thermal blankets typically consist of a wire mesh woven into a ceramic cloth and may be as large as 2.4 m by 6 m. The thermal blankets are usually covered with 5–30 cm of insulation to minimize heat loss to the atmosphere. An impermeable layer placed above the insulation provides a surface seal to allow the application of a vacuum below the heater blanket to capture volatilized contaminants. Application of current to the thermal blankets increases the temperature of the blanket to as high as 800°C or 900°C. Heating of the subsurface is produced by radiation and conduction of heat downward from the thermal blankets. Because of their placement, thermal blankets are limited to shallow applications (< 1 m).

Thermal wells are used for contamination at depths greater than 1 meter. The thermal wells, oriented vertically, contain heater elements consisting of nichrome wires in ceramic insulators. As with thermal blankets, application of current to the heating elements heats the thermal wells to temperatures approaching 900°C, leading to heating of the adjacent subsurface region through heat conduction. The rate of heat transfer, or heat flux, during heat conduction is proportional to the temperature gradient in the soil according to Fourier's law of heat conduction. Water near the heaters may be vaporized, and the resulting steam will cause some convective heat transfer into the formation until all of the soil becomes dry. Thermal wells may be configured with heating elements only, or they may be configured so that a vacuum can be applied to the well to withdraw vapors and liquids from the subsurface (Stegemeier and Vinegar, 2001).

Temperatures in the vicinity of a thermal well will depend on the power of the heating element, the radiant heat transfer between the element and the soil, the thermal conductivity and the heat capacity of the soil, and the distance to neighboring heaters. The heating rate increases with thermal conductivity and decreases with heat capacity of the heated material. Both of these quantities depend on water content, but they are relatively insensitive to grain size or mineral content. As a result, temperature changes resulting from conductive heating will be relatively independent of the type of soil or rock being heated. Moreover, the change in temperature will be relatively uniform even in formations that are heterogeneous, such as interbedded sands and clays or fractured rock. At high temperatures achievable with conductive heating, soil near the heaters may become desiccated, allowing even tight clays to become permeable enough for adequate vapor extraction.

When the subsurface temperature adjacent to thermal wells or thermal blankets is raised by conductive heating, vapor pressures of water and contaminants are increased until boiling of the water and contaminants occurs. When boiling occurs there may be a significant increase in pressure due to the expansion associated with phase change from liquid to gas. This can produce a flow of gases and liquids away from the heat source, leading to convective heat transfer. In the case of thermal blankets, the movement of gases and liquids is usually controlled by application of a vacuum below the blanket. In the case of remediation using thermal wells, a vacuum is usually applied at a subset of the thermal wells in order to capture the vapors and liquids. Thus, the effectiveness of conductive heating depends particularly on the ability to recover vaporized water and contaminants produced by heating.

As conductive heating can produce very high temperatures near the heating elements, it is capable of volatilizing even very high boiling point compounds such as PCBs. The high temperatures will also accelerate desorption of contaminants from soils. Furthermore, at these high temperatures, organic contaminants may be subject to oxidation or pyrolysis (Stegemeier and Vinegar, 2001). In order to reach temperatures high enough to effectively remove PCBs (approximately 500°C), all water in the soil must be boiled off. In some cases the inflow of water into the heated area may be great enough that temperatures cannot be raised much above 100°C, severely limiting the effectiveness for low-volatility contaminants. In addition, very close well spacing (1.5–2 meters) is required to achieve adequate heating. For more volatile compounds, such as PCE and TCE, it is not necessary to reach temperatures above 100°C, and less energy input and more distantly spaced wells may be sufficient.

*Overview of Case Studies*

Laboratory treatability studies and field project experience at seven ISTD sites, summarized in Table 5-5, have confirmed that elevated temperatures applied over a period of time result in significant destruction and removal of even high boiling point contaminants such as PCBs, pesticides, PAHs, and other heavy hydrocarbons.

The effects of heating by conduction are illustrated by an example of the desiccation form of the ISTD process where 12 thermal wells were used to heat PCB-contaminated soil at Cape Giradeau, Missouri (Vinegar et al., 1998). An array of 14 temperature monitoring wells with thermocouples spaced every 0.3 m with depth was used to determine heating effectiveness. The process was operated for 42 days, and there were three distinct periods of heating. Temperatures increased from ambient conditions to 100°C as the soil and water were heated during the first nine days of the project. Boiling of pore water occurred throughout the region for the next 12–16 days, during which temperatures remained around 100°C. The temperatures increased again between days 22 and 26, after liquid water was removed completely, leaving desiccated soil throughout the treatment zone. Temperatures increased to more than 400°C during the last two weeks of the project.

*Applicability of the Technology*

**Contaminants.** Conductive heating can be used for a wide range of organic contaminants, ranging from volatile organics such as the chlorinated ethenes to low-volatility compounds such as PCBs. One of the advantages of conductive heating compared to other thermal remediation methods is the capability of generating the high subsurface temperatures conducive to the removal of very low volatility compounds. As the volatility of the contaminants decreases,

TABLE 5-5  Summary of ISTD Field Results

| Location | Treatment Zone | Treatment Method | Contaminant | Initial Concentration | Final Concentration |
|---|---|---|---|---|---|
| Cape Girardeau, MO | Vadose zone, treated to depth of 12 ft (3.7 m) bgs, weathered and unweathered loess | Thermal blanket for upper 1.5 ft (0.5 m), and 12 heater wells [to 12 ft (3.7 m) bgs on 5-ft (1.5 m) centers], 10–45 days of heating. | PCB 1260 | 20,000 ppm | < 0.033 ppm |
| Vallejo, CA | Vadose zone, 500 ft$^2$ (46 m$^2$) area, treated to depth of 12 ft (3.7 m) bgs | Thermal blanket for upper 1 ft (0.3 m), and 14 heater wells [to 14 ft (4.3 m) bgs], 10–45 days of heating. | PCB 1254/1260 | 2,200 ppm | < 0.033 ppm |
| Portland, IN | Vadose zone, silty clay soil | 130 heater wells on 7.5-ft (2.3-m) centers, to depth of 19 ft (5.8 m) bgs, 9 weeks of treatment. | 1,1-DCE / PCE / TCE | 0.65 ppm / 3,500 ppm / 79 ppm | 0.053 ppm / < 0.5 ppm / 0.02 ppm |
| Eugene, OR | Vadose zone, gravel [1–4 ft (0.3–1.2 m) bgs], silt [11–16 ft (3.4–4.9 m) bgs], gravel/sand/clay (below silt) | 277 vacuum/heater wells and 484 heater-only wells [7 ft (2.1 m) centers to 10–12 ft (3–3.7 m) bgs] within and outside a building and adjacent to residences. | Benzene / Gasoline / Diesel | 33 ppm / 3,500 ppm / 9,300 ppm + free product | < 0.044 ppm / 250,000 lb (113,398 kg) free product removed |
| Ferndale, CA | Vadose zone, 40 × 30 × 15 ft (12 × 9.1 × 4.6 m) deep, silty and clayey colluvial soils | Heater-only and vacuum/heater wells in hexagonal pattern, with 6-ft (1.8-m) spacing. | PCB 1254 | 800 ppm | < 0.17 ppm |

SOURCE: Stegemeier and Vinegar (2001).

increasingly higher subsurface temperatures must be reached, requiring closer well spacing, limited water inflow, and greater energy inputs.

**Hydrogeology.** Conductive heating has the potential to be effective over a fairly wide range of geologic conditions, including saturated zones. Thermal wells have no depth limitations other than those of typical well drilling. However, most applications of conductive heating to date have been in the vadose zone. For contamination in the saturated zone, water recharge can be a concern if it is necessary to desiccate the soil to achieve high temperatures, since much more heat is required to boil large quantities of water. More experience is needed to better evaluate the potential of conductive heating in saturated zones and to develop methods of stopping or minimizing water recharge, if necessary. In addition, the extent to which subsurface heterogeneity impacts the ability to capture vapors and control contaminant migration needs to be further explored. This method will be infeasible at some sites, however, where sensitive structures at the ground surface preclude the installation of either thermal wells or blankets.

**Health, Safety, and Environmental Considerations.** Conductive heating involves the use of large quantities of electric power and produces high temperatures, requiring very stringent safety measures. The potential for underground fires is not expected to be significant. Studies indicate that the formation of dioxins and furans in the soil is not significant (Stegemeier and Vinegar, 2001). Stack gas testing is required to ensure that hazardous gas emissions do not occur. At most sites recovered gas streams are further treated with thermal oxidizers and carbon adsorbers. The presence of buried drums in the vicinity of the source zone would be problematic, as the high temperatures may cause explosion of these drums; measures should be taken to ensure that sealed drums are not present in the treatment area.

The migration of contaminants away from the treatment area due to pressure increases associated with heating and liquid vaporization is controlled through the application of a vacuum at thermal wells. If complete capture is not achieved, there is potential for outward spread of contaminant vapors and contaminated water and for downward mobilization of DNAPLs. As soil heterogeneity increases, achievement of complete capture could be more problematic.

*Potential for Meeting Goals*

Conductive heating is a very aggressive technology, typically involving close well spacing, high energy inputs, and high subsurface temperatures. Most applications have been in the vadose zone for removal of low-volatility contaminants. Under these conditions for almost any type of geology, it would be expected to be very effective at mass removal and at achieving reductions in contaminant concentrations, fluxes, and source migration potential.

There is limited field experience applying conductive heating below the water table. If water inflow can be limited, then conductive heating would be expected to be effective in all granular media. However, achieving adequate capture of vapors and liquids and limiting water inflow may be more difficult as heterogeneity increases. There is no experience with conductive heating in saturated fractured media or karst. As control of water inflow may be problematic in fractured media and karst, and capture of contaminants may be difficult, effectiveness is expected to be limited in these settings.

Assessment of the success of conductive heating usually includes monitoring of in situ temperatures to ensure that design temperatures are reached throughout the source zone. As a significant amount of time (weeks) may be required for subsurface temperatures to decline to pretreatment levels, temperatures and contaminant concentrations should be monitored for some time following the cessation of heating.

### Cost Drivers

Costs for conductive heating include the costs of thermal blankets or thermal wells, temperature monitoring, vacuum extraction equipment, treatment trains for off-gas and condensate treatment, and electrical power for heating. As very close well spacing is typically used, well installation costs will increase significantly as the depth to the source zone increases. As contaminant volatility decreases, higher temperatures must be generated, increasing power input requirements. Similarly, if there is significant water inflow, energy consumption increases.

### Technology-Specific Prediction Tools and Models

Subsurface heating due to conductive heating can be simulated with simple models when phase change and fluid flow do not significantly impact subsurface heating. Comprehensive modeling of conductive heating, including phase change in a multiphase (gas–water–NAPL) system is a complex nonlinear process and requires the use of sophisticated numerical models such as those of Falta et al. (1992). Elliott et al. (2004) used a multiphase flow and transport simulator to investigate the design of conductive heating in the saturated zone.

### Research and Demonstration Needs

Compared to many other remediation technologies, there is very limited experience with conductive heating, and very little has been published in the refereed scientific and engineering literature. In particular, there has been little application of conductive heating below the water table. Little is known about the permeability increases that occur when low-permeability soils are desiccated by heating, and there has been no significant evaluation of the potential for increased contaminant spreading due to application of conductive heating.

## Electrical Resistance Heating

Electrical resistance heating (ERH) was originally developed as an enhanced oil recovery technique (Harvey et al., 1979; Wattenbarger and McDougal, 1988). In recent years it has been proposed as a method for remediation of subsurface contamination (Buettner et al., 1992; Gauglitz et al., 1994; McGee et al., 1994). Application of ERH at a site involves installation of electrodes into the ground in hexagonal or triangular arrays. Typically, three-phase or six-phase electricity is applied to the electrodes. The resulting electric field set-up results in heating of the subsurface. The heating rate is equivalent to the power dissipated in the subsurface, so heating will be greatest where the current flow is greatest (McGee et al., 1994). The applied voltage, rather than the current, is adjusted in the field to produce the current that is needed to induce resistive heating at whatever rate is required. Electrical resistance heating can raise the temperature of the subsurface to the boiling point of water, which creates an in situ source of steam to strip contaminants from the subsurface. As the contaminants are converted to vapors, they are captured and removed using soil vapor extraction, which is applied at the electrodes and through additional wells in the vadose zone.

The ability to produce steam in situ between electrodes can produce a more uniform distribution of temperatures than steam flooding and conductive heating, where heat moves outward from wells. The configuration of electrodes is critical to creating a uniform distribution of heat. The six-phase configuration uses six metal electrodes placed in a circle around a central neutral electrode (Gauglitz et al., 1994). The six metal electrodes are connected in a spatially phase-sequenced pattern so that each electrode conducts to every other electrode in the array—a configuration that can produce a fairly uniform heating pattern. Electrical resistance heating can also be employed using three-phase heating or even by using a single phase. The most commonly employed configuration of electrical resistance heating for treatment of large areas is three-phase heating, in which each of the electrodes of a repeating triangular pattern is wired to one of three phases (McGee et al., 2000).

Because electrical conductivities of soils vary much less than soil permeabilities, ERH is able to produce more uniform heating of heterogeneous soils than steam flushing. In particular, low-permeability clay layers may be preferentially heated due to the presence of ions in clay that increase the current flow and thus increase the heat deposition in the clay zones. However, as soils dry out due to conversion of liquid water to steam, the electrical conductivity decreases. Thus, it is necessary to retain some liquid water in the soil to maintain conduction of current and heating (McGee et al., 2000). For this reason, in contrast to conductive heating, ERH is limited to temperatures in the 100°C range (depending on depth). Dry-out around electrodes and electrode overheating are particular problems that are usually dealt with through the use of a water drip at the electrodes.

*Overview of Case Studies*

A number of case studies of ERH application for source zone remediation are summarized in Table 5-6. Most of these applications are for shallow contamination by volatile chlorinated solvents with soil types ranging from sands to clays. 6-phase heating is more common than 3-phase heating, most likely due to vendor specialization rather than to technical considerations. Remedial objectives, which were not reached at two of the sites, ranged from removal of contaminant mass (e.g., Cape Canaveral, Portland) to reductions in soil and groundwater contaminant levels. At all sites it is not clear whether posttreatment contaminant concentration rebound occurred. Many of the applications were plagued by problems with electrodes and by inadequate heating.

*Applicability of the Technology*

**Contaminants.** The primary means of contaminant removal with ERH is through volatilization and recovery by soil vapor extraction. As it is necessary to maintain liquid water in the soil to allow current conduction, ERH temperatures are limited to the boiling point of water (McGee et al., 2000), which limits the effectiveness of ERH to volatile contaminants (boiling points below 150°C), similar to steam flushing. As contaminant volatility decreases, removal rates will decrease correspondingly, and the risks of DNAPL remobilization increase.

**Hydrogeology.** Most applications of ERH have been to unsaturated and shallow saturated zones, although there are no technical limitations to how deep electrodes can be placed. Although ERH shows promise for removal of contaminants from low-permeability clays due to the preferential heating of these electrically conductive soils, there has been limited application of ERH to these soils. In particular, the difficulties with contaminant vapor recovery in low-permeability soils and the consequences for SVE well spacing have not been very thoroughly investigated. ERH has the potential to be more effective than steam flushing in moderately heterogeneous soils due to more uniform heating than is possible with steam flushing. In addition, through control of the power supplied to electrodes, it is possible to focus the heating to make efficient use of energy input. There have not been any applications of ERH in fractured rock or karst systems. ERH is not expected to be particularly effective for these settings due to the low conductivity of low-porosity rocks and the difficulty in maintaining control of fluid migration.

**Health, Safety, and Environmental Considerations.** The ERH process involves high-voltage electrical systems, requiring extensive safety precautions

TABLE 5-6 Case Studies Using Electrical Resistance Heating

| Location | Media and Contaminants | Application Method | Results |
|---|---|---|---|
| Savannah River, GA | Dissolved PCE and TCE (100–200 ppm) in a 10-ft (3-m)-thick clay layer 40 ft (12 m) bgs. | 6-phase ERH. Electrodes placed in 30-ft (9.1-m)-diameter circle, 25 days of heating. | 100°C reached in 10 days. After 25 days, 99.99% removal of contaminants in treatment zone. (EPA, 1995a) |
| Avery Dennison Site, Waukegan, IL | Methylene chloride (MeCl) source (16,000 yd³ or 12,233 m³) in the vadose zone. | 6-phase ERH. 20 treatment cells with perimeter electrodes to a depth of 24 ft (7.3 m). Total of 95 copper electrodes, including 6 installed below an active street and 16 installed inside the existing building. Treatment goals: 24 mg/kg MeCl in soil. | After four weeks of operation, there was inadequate heating due to electrode deterioration. 1-inch (2.5-cm) galvanized steel pipes were installed around each electrode. Concentrations of MeCl were reduced to below the treatment goals (24 mg/kg in soil) after 5 months, except for 4 treatment cells. Addition of extra galvanized steel pipe electrodes and another month of operation met remedial goals. Average MeCl concentrations in soil were reduced to 2.51 mg/kg. (EPA, 2003) |
| Skokie, IL, Site | TCE (130 mg/L maximum; 54.4 mg/L average), TCA (150 mg/L maximum; 52.3 mg/L average), and DCE (160 mg/L maximum; 37.6 mg/L average). DNAPL present. Heterogeneous silty sands with clay lenses to 18 ft (5.5 m) bgs ($10^{-5}$–$10^{-4}$ cm/s); underlain by dense clay till aquitard ($10^{-8}$ cm/s). Depth to groundwater 7 ft (2.1 m) bgs. | 6-phase ERH. ERH (13.8 kV local service at 1,250 kW) combined with soil vapor extraction. Tier III remedial goals were TCE (17.5 mg/L); TCA (8.85 mg/L); and DCE (35.5 mg/L). 23,100 cubic yards (17,661 m³) treated in 6 months. Additional 11,500 cubic yards (8,792 m³) treated. | In five to six months of operation, Tier III cleanup goals were achieved for TCE, TCA, and DCE in all wells in the initial area of contamination. Average groundwater concentrations reduced by more than 99% for TCE (54.4 mg/L to 0.4 mg/L); more than 99% for TCA (52.3 mg/L to 0.2 mg/L), and more than 97% for DCE (37.6 mg/L to 0.8 mg/L). (EPA, 2000) |

244

| Site | Contaminants/Site Conditions | System Description | Results |
|---|---|---|---|
| ICN Pharmaceuticals Incorporated Site, Portland, OR | TCE, cis-1,2-DCE, and VC. DNAPL suspected based on presence of contaminants in groundwater at >1% of solubility. Source zone (saturated and unsaturated) of 48,000–65,000 yd³ (36,699–49,696 m³) at a depth of 40 ft (12 m). | 6-phase ERH. 60 electrodes (hexagonal arrays of 6 electrodes each, with a 7th neutral electrode in the middle of each array) directing power to 3 zones: 20–30 ft (6–9 m) bgs, 34–44 ft (10–13 m) bgs, and 48–58 ft (15–18 m) bgs. Initial heating limited to bottom interval to create a hot floor to prevent downward contaminant migration. 53 vapor extraction wells installed in unsaturated region above heated region (5–10 ft or 1.5–3 m bgs). Remedial objectives: prevent and contain migration of separate-phase DNAPL during treatment and reduce contaminant groundwater concentrations to levels that indicate DNAPL has been removed or treated. | System expanded after steam and hot water outside treatment area. As of December 2001, maximum groundwater contaminant concentrations in one layer had been reduced to 100 µg/L for TCE, 1,300 µg/L for DCE, and 50 µg/L for VC with concentrations above Oregon MCLs as of June 2002. In another, lower layer, concentrations of VC were reduced, but concentrations of DCE and benzene increased as a result of a possible compromise in the well casings; these wells were abandoned in April 2002 because dissolved phase VOCs remained above the Department of Environmental Quality's generic risk-based screening levels at various locations at the site. (EPA, 2003) |
| Poleline Road Disposal Area, Arrays 4, 5, and 6, Fort Richardson, AK | Soil: PCE - 120 mg/kg; TCE - 640 mg/kg. Pentachloroethane (PCA) - 12,000 mg/kg. Groundwater: PCE - 0.30 mg/L; TCE - 7.8 mg/L; PCA - 18 mg/L. DNAPL present. 7,300 yd³ (5,581 m³) treatment area 36 ft (11 m) deep. | 6-phase ERH. Electrode arrays (7 electrodes installed to a depth of 38 ft or 11.6 m), three SVE wells installed in each of 3 phases. Remedial goals: PCE - 0.005 mg/L groundwater, 4 mg/kg soil; TCE - 0.005 mg/L groundwater, 0.015 mg/kg soil; PCA - 0.052 mg/L groundwater and 0.1 mg/kg soil. | Estimated mass of PCE, TCE, and PCA removed in the off-gas was 1,385 lb (628 kg) • The ERH system reduced groundwater concentrations of PCA, PCE, and TCE an average of 49%, 75% and 56%, respectively; however, at the end of the field demonstration, concentrations of PCA, PCE, TCE, and cis-1,2-DCE were above the remedial action objectives. • The ERH system reduced soil concentrations of PCA and PCE to below the remedial action objectives; however, TCE concentrations remained above the remedial action objective. (EPA, 2003) |

continued

246

TABLE 5-6 Continued

| Location | Media and Contaminants | Application Method | Results |
|---|---|---|---|
| Launch Complex 34, Cape Canaveral Air Force Station, FL | TCE - Estimated mass of 11,313 kg in test plot, with 10,490 kg as DNAPL. Source zone test plot was 75 ft by 50 ft by 45 ft (23 m by 15 m by 14 m) deep. | 6-phase ERH. 13 electrodes, each with two conductive intervals [25–30 ft (7.6–9 m) bgs and 38–45 ft (11.6–13.7 m) bgs]. Lower heating interval configured to provide a hot floor. 12 SVE wells installed with 2-ft (0.6-m) screens to depth of 4–6 ft (1.2–1.8 m) bgs. Remedial objective of 90% TCE mass removal. | Excessive rainfall from a hurricane caused water table rise, resulting in insufficient heating of the shallow portion of the test plot. Ground rods were installed near the electrodes to heat the 5–10 ft (1.5–3 m) bgs interval. TCE and DNAPL in the soil in the test plot were reduced by 90% and 97%, respectively. Heating was more efficient in deeper portion of the aquifer. Sampling hot cores of soil (90°C) may have resulted in some losses of chlorinated VOCs. (EPA, 2003). |
| Navy Base in Charleston, SC | Total VOCs (PCE, TCE, cis-DCE) in groundwater of 70,000 mg/l in silty sand with a thin clay layer at approximately 10 ft (3 m) bgs. Groundwater at 5 ft (1.5 m) bgs. Treatment area of 18,000 ft² (1,672 m²), volume of 8,000 yd³ (6,116 m³). | 90 electrodes and co-located vapor recovery (VR) wells. The cleanup objective was to reduce the total chlorinated VOCs (CVOCs) in groundwater by an average of 95%. Heated interval was from 2 to 11 ft (0.6–3.4 m) bgs, 1 ft (0.3 m) into the clay layer. | After 9 months of ERH operations, average total CVOC concentrations in groundwater reduced by approximately 86%. (Source: Thermal Remediation Services), http://www.thermalrs.com/TRSPages/Projects/CProj3_NavalBase.html#) |

| Location | Description | Treatment | Results |
|---|---|---|---|
| Lowry Landfill, Aurora, CO | PCE DNAPL and xylene LNAPL in landfill waste pit. Large amount of metallic debris in the form of buried 55-gallon drums and miscellaneous metal including automobile bodies and mattress springs. Also wood debris, car tires, and municipal waste in the landfill. A plenum consisting of a layer of gravel covered by a clay cap is at the surface. Soil lithology consists of clay, silt, sand, and bedrock with groundwater at about 20 ft (6 m) bgs. | ERH combined with multiphase extraction. 107 electrodes, 7 multiphase extraction wells. Performance criteria: heat treatment zone to an average 90°C for 120 days, maintain vapor capture and control of ambient air emissions, and reduce xylene concentrations by 90%. The heating interval was to be 9 ft (2.7 m) and 24 ft (7.3 m) bgs, with hot floor in the lower heated interval and sweeping the heat up toward the top of the interval in two phases, capturing vapors, steam, and liquids at the surface. | After four months of ERH and multiphase extraction operations, treatment zone temperature was > 75°C. 15,000 kg of total VOCs were recovered (70% average reduction). 4,000 kg of total xylenes were recovered (80% average reduction). 2,500,000 kW-hrs of electricity input. (Source: Thermal Remediation Services) |
| Air Force Plant 4, Fort Worth, TX | TCE 95 mg/L in groundwater and 91 mg/kg in heterogeneous interbedded silt, clay, gravel under building. | ERH and SVE. 60 electrodes and co-located vapor recovery wells covering an area of about 0.5 acre (0.2 hectares) inside a building. Remediation objective: 90% reduction of TCE in groundwater and soil. | Soil temperatures in 60°C–90°C range. Final average TCE concentration in soil was 391 $\mu$g/kg. TCE concentrations in groundwater reduced by an average 93%. |
| Chicago, IL | PCE, at maximum of 13,600 mg/kg and average of 5,424 mg/kg in silt and clay under building. | ERH and SVE. 17 electrodes, vertical and horizontal vapor recovery wells. Cleanup objective: 529 mg PCE/kg in soil. | 77% to 99.6% reduction in PCE in soil, all below cleanup goals. |

and specialized expertise for design and operation. The heating process produces heated organic vapors, which in the case of some contaminants could lead to fire or explosion hazards. It is important to control the migration of the steam and organic vapors in the remediation of near-surface NAPL contamination to avoid steam breakthrough at unexpected locations in the ground surface. As with steam flushing, attention must be paid to the potential for increasing the extent of DNAPL contamination (She and Sleep, 1999). With the ability to focus heating with ERH, hot floors below the source zones have been created to volatilize and capture any vertically remobilized DNAPL. The effectiveness of this strategy has not been very widely investigated for the range of contaminants and site conditions that could be encountered.

*Potential for Meeting Goals*

At ERH sites, measures of remedial effectiveness have typically been contaminant mass removal or groundwater or soil contaminant concentrations in the treatment zone. The monitoring of decreases in aqueous phase contaminant concentration and of reductions in contaminant mass flux does not appear to have been reported very frequently for ERH. For volatile DNAPLs, ERH has the potential to remove significant quantities of mass from unconsolidated media. The relatively small variation in subsurface electrical conductivities, compared to soil permeabilities, can produce more uniform heating than steam flushing, and ERH therefore has the greater potential to remove contaminants from heterogeneous soils, provided vaporized contaminants can be captured. As soil permeability decreases, steam zone expansion and recovery of vaporized contaminants will become more difficult, increasing remediation times and potentially reducing effectiveness. In highly permeable soils, influx of water may lead to problems in attaining target temperatures, thereby limiting effectiveness of mass removal.

Characterization of the subsurface and of the DNAPL distribution is required for effective design and implementation of ERH at a site. Problems with ERH typically have occurred due to electrode corrosion or poor design of the ERH system, resulting in inadequate or uneven heating. Buried metal objects can also distort heating patterns. Although hydrous pyrolysis has been reported as a process that may destroy some contaminants in situ at the elevated temperatures associated with ERH, there are no studies in which the extent of this process has been quantified from a rigorous mass balance on contaminants.

There is limited experience with ERH in fractured media, particularly consolidated media. The difficulties of installation of electrodes in consolidated media, and the low water contents and therefore low electrical conductivities of many consolidated media, would be problematic. There has not been any application of ERH in karst media. Given the difficulty in characterizing karst and of installing an ERH system in karst, ERH has little potential in such a hydrogeologic setting.

Assessment of the success of ERH usually includes monitoring of in situ temperatures to ensure that steam temperatures are reached throughout the source zone and monitoring of vapor composition in the SVE system. Monitoring should include subsurface gases or water pressures to ensure adequate contaminant capture. Provision of adequate drip water at electrodes to prevent soil desiccation around the electrodes and monitoring of electrodes to ensure there is no overheating are required. A significant amount of time (weeks) may be required for subsurface temperatures to decline to preflushing levels, so temperatures and contaminant concentrations should be monitored for some time following the cessation of ERH.

## Cost Drivers

The costs for ERH are associated with the elements of the electrical system (electrodes, electrical network, power controls), SVE system, electrode and SVE well installation, power provision, and off-gas treatment.

## Technology-Specific Prediction Tools and Models

Prediction of ERH processes requires modeling of current flow and the resulting heat generation in situ. This requires knowledge of both electrical and hydraulic soil properties and of how these properties change as steam forms and as moisture content changes. There have been some publications in the petroleum industry of simplified models for modeling current flow from ERH application, based on assumptions of constant electrical and hydraulic properties (Vermuelen et al., 1979; Vinsome et al., 1994). There are no published applications of these models for DNAPL remediation. In order to be comprehensive, these ERH models should be coupled to thermal models such as those of Falta et al. (1992).

## Research and Demonstration Needs

There have been very few refereed journal publications on the ERH process. The variability of soil heating due to variations in soil types and to changes in water content needs to be investigated. The potential for microcrack formation of low-permeability materials during heating has been anecdotally reported but never fully analyzed. The effectiveness of vapor and liquid recovery with ERH in heterogeneous soils also should be studied. There is a need for laboratory and field studies of ERH in fractured rock to evaluate its potential effectiveness. Improved modeling capability is also needed, particularly the capability to predict changes in heating as a result of changes in moisture content.

# BIOLOGICAL TECHNOLOGIES

The final two source remediation technologies either directly or indirectly invoke biological processes to degrade contaminants in situ. Air sparging accomplishes contaminant removal primarily by stripping volatile compounds from the subsurface while simultaneously supporting in situ biodegradation of contaminants. Enhanced bioremediation refers to any in situ treatment in which chemicals are introduced into the subsurface with the goal of stimulating microorganisms that can degrade or transform the contaminants of concern.

## Air Sparging

Air sparging is an in situ remedial technology for volatile solvents that utilizes injection wells to pump air below the water table, stripping contaminants from the dissolved, sorbed, and nonaqueous phases by volatilization. Commonly, contaminant-laden air is simultaneously extracted from the unsaturated zone in a process equivalent to soil vapor extraction. Although the primary removal mechanism for this technology is physical, the introduction of oxygen into the contaminated zone associated with the injection of air often promotes substantial removal of contaminants by biodegradation in both the saturated and unsaturated zones of the aquifer (EPA, 1995b). This technology is also sometimes referred to as *in situ air stripping* or *in situ volatilization*. It must be followed by capture and treatment of the waste vapor stream. Air sparging is less aggressive than chemical transformation or thermal treatment options, and may be especially suitable for pairing with bioremediation options. It is considered a mature technology when it is applied to dissolved contaminants and innovative when it is applied to source zones.

Air sparging is based upon the principle that injected air moving through saturated porous media will volatilize contaminants that are present in NAPLs, dissolved in the aqueous phase, and sorbed onto solids (Figure 5-6). Partitioning of contaminants into the vapor phase is a complex function of vapor pressures, Henry's Law constants, and sorption equilibrium constants (NRC, 1999). Understanding how injected air is distributed within the aquifer and how this affects partitioning of contaminants is critical to the success of this technology. A large number of flow visualization and characterization studies have been conducted to improve our understanding of air distribution in saturated porous media. The effects of flow rate, injection pressure, and pulsing schemes have been studied in laboratory and field studies (summarized in Johnson et al., 2001). Consensus is that (1) air flow is irregular in shape and is sensitive to very subtle changes in soil structure (see Figure 5-7), (2) increased airflow rates generally produce more dense flow field patterns, (3) vertical wells in homogeneous soils generally result in an airflow distribution radius of less than 3 m, and (4) heterogeneous soils may have either a positive or negative effect on air distribution. Contaminant removal

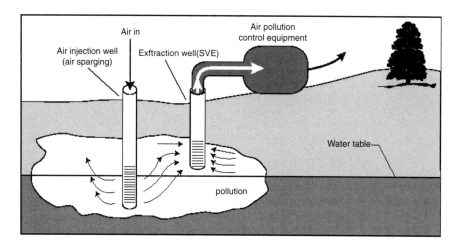

FIGURE 5-6 Typical application of in situ air sparging coupled with soil vapor extraction. SOURCE: EPA (2001).

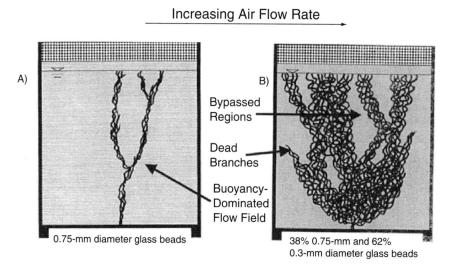

FIGURE 5-7 Effects of flow rate and grain size on channelization produced by air injected into saturated porous media. Panels (A) and (B) show columns packed with different grains sizes and experiencing different air flow rates. SOURCE: Reprinted, with permission, from Ji et al. (1993). © 1993 National Ground Water Association.

during air sparging is initially dominated (over a timeframe of months) by volatilization into the discrete air channels generated by the injection wells. Subsequently, contaminant removal is controlled by mass transfer through the liquid phase surrounding the air channels (Leeson et al., 2002). Biodegradation may also play a role in subsequent removal; however, for most chlorinated solvents aerobic biodegradation will not be a significant removal process.

*Overview of Case Studies*

To date, air sparging has been chosen as a component of remediation strategies at 48 Superfund sites (EPA, 2001). Fluor Daniel GTI compiled information on 32 field applications of air sparging to remediate chlorinated solvents or petroleum hydrocarbons (Brown, 1998). Of these 32, four were designed to treat dissolved contaminants alone, while the remainder were designed to address source areas. Seven of the sites were chlorinated solvent sites and 25 were hydrocarbon sites. Of the solvent sites, none were thought to contain mobile DNAPL. The field sites represented a wide range of hydrogeologic conditions and were distributed across 13 states. Injection well spacing ranged from 3.6 to 24 m, flow rate per well ranged from 85 to 1,000 standard liters per minute, and injection networks ranged from 1 to 16 wells. Both horizontal and vertical wells were represented; some were continuous injection systems and others were pulsed injection. Sparging duration ranged from several months to over four years. Remediation success was measured in terms of percent reduction in contaminant mass. Mass reductions were estimated from rebound concentrations in groundwater samples taken months after sparging was discontinued. Of the 28 source area sites, 20 achieved estimated reductions of $\geq$ 80 percent, while 12 were estimated at $\geq$ 95 percent. In general, sparging at solvent sites was more successful than at hydrocarbon sites, and dissolved phase contaminants were more effectively removed than sorbed contaminants. Not surprisingly, sites with closely spaced wells (average spacing of 8 m) performed better than sites with widely spaced wells (average spacing of 13 m). Box 5-9 describes a site where air sparging was used in combination with soil vapor extraction for source remediation of DNAPLs and other compounds.

*Applicability of the Technology*

**Contaminants.** Air sparging was developed in the late 1980s for in situ remediation of volatile contaminants. It has most commonly been used to treat petroleum hydrocarbons and chlorinated solvents. In fact, it can be applied to any contaminant that is sufficiently volatile, and is most effective with contaminants that have dimensionless Henry's constants much greater than 0.01.

---

**BOX 5-9**
**Air Sparging with Soil Vapor Extraction for**
**DNAPL Source Remediation**

A large-scale application of air sparging coupled with soil vapor extraction (AS/ SVE) was performed to treat a chlorinated solvent site in Burlington County, New Jersey (Gordon, 1998). The contamination covered 1.7 acres (0.7 hectares) in a coastal plain, with medium to coarse sand and an unconfined aquifer with a water table 1–9 ft (0.3–2.7 m) below ground surface. The contamination was bounded underneath by a clay layer. A variety of contaminants were present at the site, including TCE, 1,1,1-TCA, and 1,1-DCA in both DNAPL and dissolved phase. Following an extensive site characterization and development of a site conceptual model, an AS/SVE pilot test was conducted to estimate the radii of influence of air sparging and vapor extraction wells. Pilot test results were used to design the full-scale system, including the number and spacing of wells and the optimal airflow and vacuum rates. The full-scale system included 134 air sparging wells and 58 SVE wells distributed across 1.7 acres (0.7 hectares), designed for full coverage in the presumed source area and for plume interception downstream. To avoid hydraulic mounding and reduce channelization of air bubbles, 15 individual sections of the system were operated in sequential pulsing, with 30 minutes of operation followed by 60 minutes of down time. Pulsing cycles were coordinated across the entire site. During the first two years of operation, more than 500 kg of vapor phase VOCs were removed, and VOC concentrations in downstream wells declined to below detection limits for most solvents. TCE concentrations in the downstream wells declined by a factor of 30–500. However, in the source area wells, solvent concentrations remained high throughout the two years, suggesting that AS/SVE was continually volatilizing DNAPL masses. Direct recovery of DNAPL by pumping in the source area was stimulated by the AS/SVE process for the first seven months, and subsequently declined. In addition, vapor phase solvent removal declined during the first eight months with a characteristic tailing effect thereafter. Potential contaminant rebound subsequent to discontinuation of the AS/SVE system was not reported.

---

**Hydrogeology.** Aquifer heterogeneities can significantly hinder contaminant transport and the effective zone of influence of the sparging vapors via plugging and the formation of preferential pathways. These conditions can be difficult to predict or monitor, making the performance of the technology highly empirical in nature (Leeson et al., 2002). Nonetheless, given a fairly homogenous aquifer with a high hydraulic conductivity, air sparging is expected to be effective for removing significant mass within a three- to five-year time span. The level of site characterization required to implement this technology successfully, and the depth limitations of the technology, are similar to that required for application of aqueous pump-and-treat (i.e., hydraulic containment).

**Health, Safety, and Environmental Considerations.** Since air sparging does not involve the active pumping of groundwater, human health risks associated with the contaminants are limited to the potential for exposure to sparging vapors. Further, since no additional chemicals are involved in the application of air sparging, chemical exposure is not an issue. If sparging is accomplished with compressed air, the presence of a compressed gas cylinder may pose a slight hazard. However, the major exposure route would be with extracted vapors containing the contaminants as they move through the aboveground treatment train.

*Potential for Meeting Goals*

Designing the zone of influence to target the entire volume of the contaminant source is extremely important in order to attain cleanup goals. The zone of influence is a function of the number and placement of injection wells, the airflow to injection wells, and the hydrogeology of the site. Precipitation or biofouling can lead to the plugging of injection wells, resulting in decreased airflow rates and a decreased the zone of influence. Increased injection pressures and redevelopment of wells are actions applied to counteract plugging. Sparging vapors generated by this technology may be laden with sufficiently high concentrations of contaminants to necessitate vapor extraction with associated gas treatment. Also, control of sparging vapors may be necessary to limit further contaminant migration in the subsurface. Heterogeneities in the hydrologic condition of the subsurface can cause short-circuiting of the sparge vapors, decreasing the effectiveness of the technology (Leeson et al., 2002). Thus, due to the large number of potential variables associated with the application of this technology in the subsurface, achieving success with in situ air sparging requires significant engineering judgment and expertise. Consequently, even within homogenous media with high transmissivity, it is problematic to predict effectiveness, as the values for likely effectiveness in Table 5-7 reflect. Indeed, in a recent review of air sparging advances, Johnson et al. (2001) voice the opinion that many sparging systems are operated inefficiently and that conventional monitoring techniques are inadequate to assess effectiveness. They go further to state, "In brief, [in situ air sparging] system design remains largely empirical with an apparent lack of appreciation for the complexity of the phenomena and the sensitivity of the technology to design and operating conditions." Since there have been few well-documented applications of air sparging for treatment of DNAPL source zones in fractured, heterogeneous, or impermeable media, little is known about its effectiveness in those settings; consequently, likely effectiveness ratings in Table 5-7 have been listed as low.

## Cost Drivers

The major costs of air sparging are associated with the injection of air under the water table, possibly coupled with the extraction and treatment of subsurface vapors. The cost of the oxygen delivery will primarily be affected by the depth to the contaminant zone (well drilling cost and energy cost), the areal distribution of contaminant, the hydraulic conductivity and heterogeneities of the subsurface (which control the number of wells required to provide adequate coverage), and the mass and volatility of the contaminant (which control the amount of time required for site closure). The cost of the soil vapor extraction will be driven by the mass and volatility of the contaminant (which control the extraction flow rate and duration) and the hydraulic conductivity (which controls the number of extraction wells). The cost of vapor treatment at the surface will be driven by the nature and mass of contaminant (which control the treatment method, loading rate, and duration). Generated vapors are generally dilute, resulting in high per unit mass costs for treatment.

## Technology-Specific Prediction Tools and Models

A great number of numerical models have been developed to predict air sparging effectiveness and to aid in the design of air sparging systems (e.g., Marley et al., 1992; van Dijke et al., 1995; Lundegard and Andersen, 1996; McCray and Falta, 1997; Philip, 1998; Rabideau and Blayden, 1998; van Dijke et al., 1998; Elder et al., 1999). Despite this, the intense complexity of the process still necessitates a strong reliance on pilot-scale and feasibility testing for effective application (Johnson et al., 2001). Therefore, the technology is not highly predictive, and field-testing is essential for evaluating the potential for success.

## Research and Development Needs

A recent review of air sparging summarizes the key research needs associated with this technology (Johnson et al., 2001):

• Improved understanding of air flow distributions and the effects of geology and injection flow rate.
• Development of better characterization methods for air flow distributions at the pilot and field scales.
• Improved predictions of how transient operating conditions such as pulsing can affect performance and reduce equipment costs.
• Development of innovative monitoring approaches that are capable of accurately assessing system performance.

## Enhanced Bioremediation

Many contaminants can be transformed in some fashion by subsurface micro-organisms. Indeed, these processes take place during natural attenuation, which is considered a suitable remediation technology when it results in contaminants disappearing more rapidly than they migrate, resulting in a stable or shrinking contaminant plume. When natural attenuation occurs too slowly or is inhibited by a lack of substrates or nutrients or by some other condition, *enhanced bioremediation* may be an appropriate technology to pursue. Enhanced bioremediation involves the stimulation of contaminant-degrading microorganisms within a subsurface aquifer or vadose zone by delivering chemical amendments to the contamination zone. Subsurface microorganisms are stimulated by delivery of substrates, electron acceptors, and/or nutrients by means of subsurface injection or surface infiltration (Figure 5-8). The major advantage associated with in situ bioremediation is that the contaminants are destroyed largely in place, minimizing contaminant transport to the surface and preventing transfer of the contaminant to a new medium for subsequent treatment or disposal. It should be noted that the amount of time required to completely remediate a source zone using biological processes

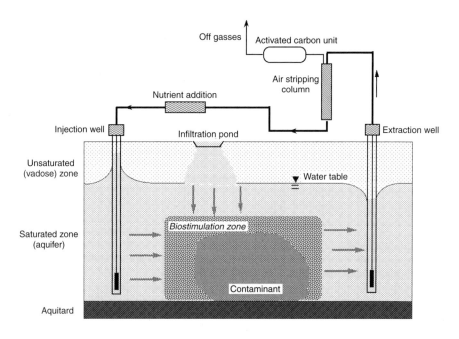

FIGURE 5-8 Enhanced bioremediation employing an injection well and infiltration pond for nutrient delivery and air stripping with activated carbon for treatment of extracted water.

will likely be greater than that required for more aggressive options such as thermal or (to a lesser extent) chemical treatments.

For remediation of chlorinated solvents, enhanced bioremediation can be achieved either by metabolic reactions, with the contaminant serving as either an electron donor or electron acceptor for energy generation, or by cometabolic reactions, with the contaminant degrading fortuitously due to the presence of an alternate substrate. Under strictly anaerobic conditions and in the presence of a reduced electron donor, chlorinated solvents will undergo a metabolic reaction known as reductive dechlorination (McCarty, 1997). Reductive dechlorination generally occurs in a series of reaction steps, with transient production of lesser-chlorinated intermediates. For example, reductive dechlorination of PCE results in the generation of TCE followed sequentially by DCE and VC prior to complete dechlorination to ethene. In some cases, dechlorination does not proceed to completion, and intermediates such as DCE build up.

Substrates that are suitable for promoting reductive dechlorination include hydrogen gas, a wide variety of defined organics such as lactate, methanol, butyrate, and sugars, as well as complex organics such as molasses and vegetable oil (Lee et al., 1998; Yang and McCarty, 1998). Slow-release polymers that result in a long-term source of lactate are commercially available for this application. The subsurface introduction of these reduced substrates results first in microbial consumption of all available electron acceptors; this can occur within days of the initial application. Subsequently, the reduced organics are fermented, generating fermentation products that include hydrogen gas. The hydrogen gas is then used by reductive dechlorinating bacteria to degrade the chlorinated solvents. Observable reductive dechlorination may require months to achieve (e.g., see the case study described in Box 5-10).

Conversely, *cometabolic* reactions for the degradation of chlorinated solvents are generally aerobic reactions that involve the delivery of cosubstrates such as methane, propane, toluene, or butane to the subsurface along with a source of oxygen (reviewed by Alvarez-Cohen and Speitel, 2001). For chlorinated solvent applications, the cometabolic reactions have a number of disadvantages compared to the metabolic reactions, including the difficulties associated with achieving proper mixing of the cosubstrate, oxygen, and contaminants, competition between cosubstrate and contaminant for the active enzyme, and potential for product toxicity. However, observable degradation can be achieved within months of application, and with careful control of injection strategies, long-term effectiveness is possible (McCarty et al., 1998). In some cases, it may be advantageous to create anaerobic zones followed by aerobic zones to ensure more complete removal of contaminants and their daughter products.

Bioremediation of a contaminant that provides energy to microorganisms for growth can be highly efficient since it involves a built-in termination mechanism: when the contaminant is consumed, growth of the microbial population ceases. On the other hand, bioremediation of contaminants that do not provide growth or energy to microorganisms (e.g., cometabolized compounds) requires an alternate growth substrate to stimulate microbial activity. In addition, achieving the appropriate mixing and transport of substrates to the zone of contamination to promote cometabolism can be highly challenging, and generally requires more hydrologic control than bioremediation of an energy-yielding substrate.

*Overview of Case Studies*

Although enhanced bioremediation has primarily been applied to dissolved plumes of contaminants rather than to source zones, there have been a number of recent laboratory studies suggesting that source zones could potentially be treated with this technology (Isalou et al., 1998; Nielsen and Keasling, 1999; Carr et al., 2000; Cope and Hughes, 2001; Yang and McCarty, 2000). For example, Nielsen and Keasling (1999) reported that a mixed microbial culture enriched on TCE was capable of degrading PCE and TCE in the presence of DNAPL in a batch system and that degradation kinetics increased at high solvent concentrations. Isalou et al. (1998) showed that rapid degradation of high concentrations of PCE could be achieved in continuous-flow columns. Demonstrations have shown that active microbial communities could lead to the enhanced dissolution of PCE DNAPL in continuous-flow liquid reactors (Carr et al., 2000) and columns (Cope and Hughes, 2001). In fact, Seagren et al. (1994) used a modeling approach to demonstrate that biodegradation reactions could increase the concentration gradient near NAPL sources, resulting in increased dissolution rates. Further, Yang and McCarty (2000) demonstrated that saturation concentrations of PCE inhibit the activity of cells within the dechlorinating community that compete for hydrogen, thereby increasing the utilization efficiency of the delivered electron donor (e.g., lactate). As a whole, these laboratory studies suggest that enhanced bioremediation is a promising technology for application to chlorinated solvent source zones. However, well-characterized field demonstrations and a better understanding of the specific relationship between DNAPL dissolution and biodegradation is needed in order to fully exploit this promise. Two field case studies are presented in Boxes 5-10 and 5-11.

*Applicability of the Technology*

The feasibility of applying in situ bioremediation at a specific site depends upon a number of factors, including aquifer hydrologic and geochemical characteristics, the indigenous microbial population, and the nature and distribution of the contaminants. The fundamental requirement for bioremediation is that the contaminant, the microorganisms, and any other required reactant are brought into contact so that the biodegradation reaction can proceed. For contaminants that exert microbial toxicity at high concentrations, the application of in situ bioremediation in source zones would involve additional challenges.

**Contaminants.** A wide variety of contaminants are amenable to bioremediation, including chlorinated solvents, hydrocarbons, creosote, polychlorinated phenols, nitrotoluenes, and PCBs. Nonetheless, there are a number of potential limitations associated with the application of enhanced bioremediation to source zones containing these compounds. The high concentrations of solvents associated with source zones may inhibit robust microbial growth. However, as discussed above for the case of chlorinated ethenes, there is some evidence that reductive dechlorinating communities may be capable of surviving and indeed flourishing under source zone conditions. Chemical explosives can be the target of enhanced bioremediation, as discussed in Chapter 2 (for Badger AP) and in a subsequent section.

---

**BOX 5-10**
**Case Study of Enhanced Bioremediation at INEEL**

In 1999, a field pilot study was performed to evaluate the potential for enhanced in situ bioremediation to treat a TCE-contaminated source zone at the Idaho National Environment and Engineering Laboratory (INEEL) Test Area North (TAN) site (Song et al., 2002). A mixture of waste materials, including low-level radioactive isotopes, sewage, and chlorinated solvents, was injected into the aquifer from the 1950s to the 1970s, resulting in a 2-km plume of TCE with concentrations as high as 2.3 mM (300 ppm). Lesser amounts of *cis*-1,2-DCE and *trans*-1,2-DCE were also observed prior to the pilot study. The aquifer consists of permeable basalts with a 61-m-thick saturated zone under a 64-m-thick vadose zone (Figure 5-9).

The pilot study began with the injection of 76 L/min. of clean water into TSF-05 (day 0–day 25). Lactate was introduced into the aquifer through TSF-05 via pulsed injections of 907 kg of sodium lactate dissolved in 1,140 L water at 38 L/min. (day 52–day 77), in 2,270 L at 76 L/min. (day 78–day 105), in 11,400 L at 95 L/min. (day 106–day 204), and in 22,700 L at 95 L/min. (day 205–day 296). Lactate was not injected between days 296 and 449, but resumed on day 450. Chlorinated solvents, organic acids, and ions were all monitored during the study. In addition, stable carbon isotope ratios of the chlorinated organics were measured to track treatment progress.

Concentration data indicated that lactate injection promoted the degradation of TCE to ethene with transient generation of *cis*-DCE and VC. Compound-specific stable isotope monitoring of the solvents was used to differentiate the effects of groundwater transport, dissolution of DNAPL at the source, and enhanced bioremediation. Within the zone of lactate influence, the carbon isotope ratio of ethene at the end of the study matched the ratio of the initial dissolved TCE, confirming the complete conversion of dissolved TCE to ethene (see Figure 5-10). Observed shifts in TCE isotope ratios in the source zone indicated that dissolution of DNAPL was promoted during the lactate injection.

*continued*

**BOX 5-10 Continued**

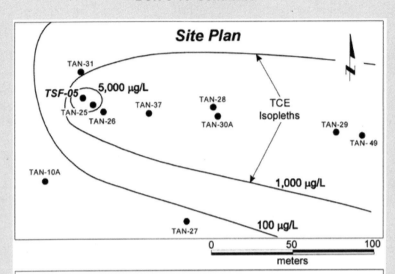

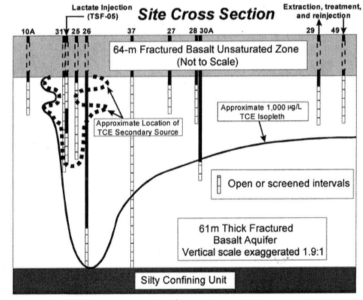

FIGURE 5-9 Site plan and cross section of a fractured basalt aquifer containing a TCE source zone at which a field-scale demonstration of biostimulation using lactate injection was performed. SOURCE: Reprinted, with permission, from Song et al. (2002). © 2002 American Chemical Society.

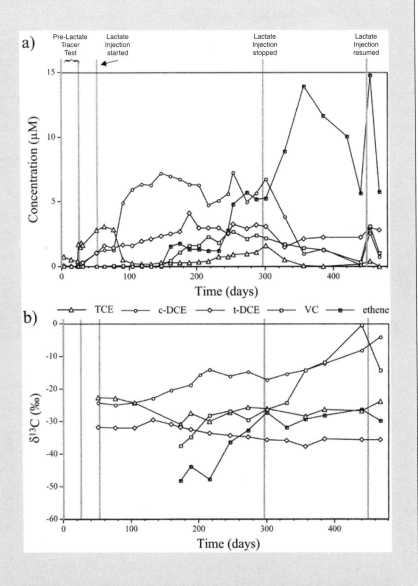

FIGURE 5-10 Solvent concentration (a) and isotope data (b) at a downgradient shallow well (TAN-25) located within the zone of high lactate exposure. SOURCE: Reprinted, with permission, from Song et al. (2002). © 2002 American Chemical Society.

An important factor that can limit the feasibility of in situ bioremediation is the availability of the contaminant for microbial attack. That is, contaminants that have extremely low solubilities (e.g., PCBs), are present in NAPL phases (e.g., PCE and TCE), or are otherwise physically inaccessible will be more difficult to degrade than dissolved phase contaminants, and they will tend to persist in the environment. Interestingly, recent studies with chlorinated solvent NAPLs have shown that microorganisms are capable of facilitating dissolution of the contaminants into the aqueous phase. Further, since the aqueous solubility increases as the chlorinated ethenes become more reduced, the daughter products partition more favorably into the aqueous phase, potentially increasing removal rates (Carr et al., 2000). Methods to enhance the bioavailability of contaminants by employing surfactants or elevated temperatures to increase solubilities are currently under study.

**Hydrogeology.** Aquifer heterogeneities can significantly hinder transport of contaminants and reactants to the microorganisms, severely limiting remediation rates. In general, aquifers with extremely low hydraulic conductivities ($\leq 10^{-6}$ cm/s) are not suitable for this technology. The level of site characterization required to implement this technology successfully is similar to that required for application of pump-and-treat, but bioremediation also requires information on the physical-chemical properties of the groundwater (e.g., ions, organics, potential electron donors and acceptors, redox potential, and pH). As with many other technologies discussed in this chapter, there are no depth limitations associated with enhanced bioremediation other than those associated with well drilling.

**Health, Safety, and Environmental Considerations.** One of the major advantages of enhanced in situ bioremediation is that most of the reaction takes place in the subsurface, so human exposure to the contaminants is minimized. In fact, contaminants are only transported to the surface if groundwater extraction is required to hydrologically control the plume or to deliver substrates. The chemicals used for reductive dechlorination reactions are generally benign, while some of those used for cometabolic applications may be somewhat toxic (e.g., toluene, phenol). Finally, application of enhanced bioremediation is gentle to the environment, utilizing naturally occurring microorganisms and biological reactions. Contamination of additional quantities of groundwater, as would be caused by pump-and-treat operations, are avoided with this technology.

*Potential for Meeting Goals*

Proving that contaminant destruction by microbial degradation is occurring can be challenging due to the inaccessibility of the subsurface, aquifer heterogeneities, the complexity of differentiating biological from nonbiological processes, and the wide range of potential contaminant fates. Overlapping lines of evidence

## BOX 5-11
## In Situ Bioremediation of Chlorinated Solvents
## Using Bioaugmentation

The Caldwell Trucking Superfund Site is an 11-acre (4.5-hectare) property of a former sewage-hauling firm in Fairfield, Essex County, New Jersey. From the 1950s until 1973 the owners hauled industrial waste and discharged it into unlined lagoons. The sludge in the lagoons contained, among other things, TCE, chloroform, and lead. Groundwater contaminants include chlorinated ethenes, ethanes, and methanes up to 1,200 m downgradient of the source zone. Groundwater flows through a glacial sand and gravel aquifer that overlies a fractured basalt aquifer. Over 50 wells have been closed in the area because of contamination.

Prior source remediation activities included removal of underground storage tanks and of highly contaminated soils, soil vapor extraction (SVE) for VOCs, and stabilization/solidification of metal-contaminated soils. An SVE system was operated for six months and removed about 12,000 kg of VOCs. SVE was discontinued due to sewage odors even though significant mass removal was occurring.

Recent analyses in the source zone show TCE levels up to 700 mg/L (about 60% of TCE solubility) due to the presence of residual DNAPL in the basalt bedrock. Low levels of natural TCE biodegradation had been occurring over much of the site. A full-scale field test of source zone bacterial reductive dechlorination was started in 2001 (Finn et al., 2004). The test goals were to accelerate the dissolution and treatment of source material and reduce the overall lifetime and impact of the source, rather than to achieve specific concentrations of parent and daughter compounds in groundwater.

The test design included six nutrient injection wells screened in glacial deposits and bedrock, and seven monitoring wells. The initial substrate feed was a mixture of lactate, acetate, and methanol, later modified to lactate, methanol, and ethanol. The injection wells were also bioaugmented with a microbial consortium that included *Dehalococcoides ethenogenes*. Gene probes were used to verify initial and continued survival and propagation of *D. ethenogenes*. Figure 5-11 shows concentration vs. time for chlorinated ethenes in selected wells. Well MW-B23 is an overburden monitoring well that showed complete disappearance of PCE and TCE concomitant with ethene production. Well MW-C22 is the bedrock well that exhibited the highest initial concentrations of contaminants. Analyses show that PCE and TCE in bedrock have been degraded to a mixture of *cis*-1,2-DCE, VC, and ethene.

Overall results of the test include average net reductions in PCE and TCE concentrations across the treatment zone of 95% and 93%, respectively, over a 30-month monitoring period. After 30 months, two monitoring wells contained no PCE, and one well had no detectable TCE. Significant solvent reductions occurred in both injection wells and monitoring wells accompanied by large increases in ethene concentrations, indicating that a continuous treatment zone was present across the test area. The average observed ethene concentration was 723 µg/L, which exceeds the average PCE concentration (131 µg/L) and is similar to the average TCE concentration (790 µg/L).

The test results indicate that in situ bioremediation is a viable source treatment/control technology for Caldwell Trucking. A formal Record of Decision (ROD) change from pump-and-treat to in situ bioremediation is currently under consideration by EPA.

*continued*

# BOX 5-11 Continued

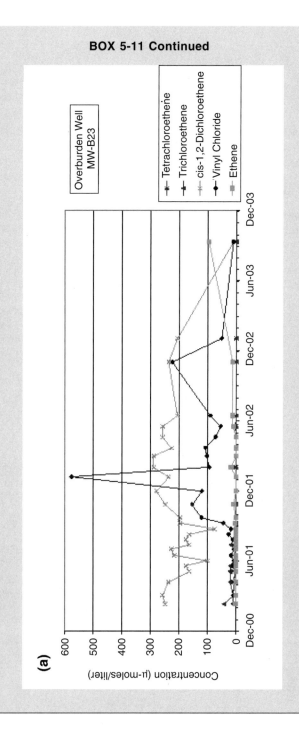

(a)

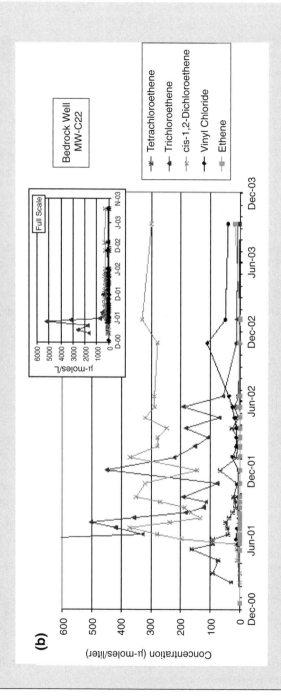

FIGURE 5-11 Solvent concentrations in a well screened in the overburden (a) and a well screened in the bedrock (b) during a full-scale study of enhanced bioremediation using a mixture of organics coupled with bioaugmentation. SOURCE: Reprinted, with permission, from Finn et al. (2004). © 2004 Battelle Press.

from a range of field monitoring techniques are required for indication of successful in situ bioremediation (NRC, 1993). Examples of monitoring observations that provide evidence for in situ bioremediation include contaminant disappearance in the bioactive zone, increased biological activity, generation of degradation intermediates, depletion of electron acceptors, and changes in stable isotope ratios of degradation products.

In some cases, it is possible that the physical–chemical and biological conditions of the subsurface are not conducive to enhanced bioremediation. Conditions that may limit microbial growth rates such as low permeability, low temperature, or high metals concentrations may also hinder remediation rates. Since indigenous microbial populations in the majority of aquifers are capable of degrading a wide range of contaminants including chlorinated solvents, there generally is no need for the introduction of exogenous microorganisms for bioremediation applications. However, in cases where the appropriate indigenous strains are not present, injection of laboratory-enriched microbial populations to bioaugment the site may be possible. Successful bioaugmentation for treatment of chlorinated solvents has been documented at a number of field sites (Ellis et al., 2000; Major et al., 2002).

Microbial metabolism is significantly affected by temperature, that is, the microbial reactions tend to decelerate with decreasing subsurface temperatures. Although temperatures within the top ten meters of the subsurface may fluctuate seasonally, subsurface temperatures down to 100 m typically remain within 1°C–2°C of the mean annual surface temperature (Freeze and Cherry, 1979), suggesting that bioremediation within the subsurface would occur more quickly in temperate climates. Additional factors that may limit microbial activity include pH values outside the range of neutral (pH < 6, pH > 8), desiccating moisture conditions, and extreme redox potentials.

Experience with the use of enhanced bioremediation for the treatment of source zones containing DNAPLs is extremely limited. More often, enhanced bioremediation is applied as part of a permeable reactive barrier to intercept and treat a contaminant plume. Although there have recently been studies that suggest bioremediation in the vicinity of DNAPLs is possible (Isalou et al., 1998; Nielsen and Keasling, 1999; Carr et al., 2000; Cope and Hughes, 2001; Yang and McCarty, 2000), there have been limited numbers of well-documented field studies to demonstrate this. For the few demonstrations that have been published, success is generally reported in concentration reduction rather than mass removal or mass flux reduction. Further, the long-term effectiveness of source zone bioremediation in heterogeneous or low permeability media has not been demonstrated. Consequently, the likely effectiveness values in Table 5-7 are primarily "low" or "low-medium" in low-permeability and heterogeneous settings. Entries in the table for change in toxicity reflect the strong probability that significant biodegradation of chlorinated solvents to intermediates having extremely different toxicities (e.g., the carcinogenic vinyl chloride vs. the relatively harmless ethane) would result in toxicity changes. Finally, entries for reduction in source migration potential have a specific notation reflecting the fact that microorganisms are known to generate surfactants that could potentially increase the mobility of DNAPLs in the subsurface (Carr et al., 2000; Cope and Hughes, 2001).

*Cost Drivers*

Since enhanced bioremediation involves the use of injection and/or extraction of subsurface fluids, it has higher costs associated with it than natural attenuation. However, enhanced bioremediation involves less reliance upon naturally occurring subsurface conditions and growth factors than does natural attenuation and is therefore applicable over a broader range of site and contaminant characteristics. In addition, enhanced bioremediation may be the more appropriate option when time constraints or liability constraints are a concern, or when it is necessary to show good faith to the surrounding community or regulators by "engineering" a solution.

The costs of enhanced bioremediation are driven primarily by delivery of the stimulants to the subsurface, cost of the stimulants themselves, and the monitoring required to demonstrate the treatment's effectiveness. The cost of the stimulant delivery will primarily be affected by the depth to the contaminant zone (well drilling cost and energy cost) and by the areal distribution of contaminant and the hydraulic conductivity (which affect the number of wells required to provide adequate coverage). The cost of the stimulants will be driven by the nature of the contaminant (which determines the type of stimulant) and the mass of contaminant (which controls the amount of stimulant required). Finally, the cost of monitoring is also driven by the mass of contaminant, as well as by all of the factors listed above for delivery of stimulants.

It should be noted that because the contaminants are not brought to the surface for treatment during bioremediation, the costs associated with pumping large quantities of water and with treating that water at the surface, which are common to other remedies, are avoided.

*Technology-Specific Prediction Tools and Models*

Numerous reactive transport models have been developed to simulate in situ bioremediation of *plumes* downgradient of DNAPL sources. Semprini and McCarty developed a single-phase two-dimensional reactive transport model to simulate TCE degradation by methane-stimulated bacteria at Moffett Field, California; the model incorporated microbial growth, cometabolic transformation, and competitive inhibition (Semprini and McCarty, 1991, 1992). Subsequently, the TRAMPP model was developed to simulate the same process at Savannah River, Tennessee; this two-dimensional model built upon the Semprini and McCarty model by integrating gas phase transport and predation of the methane-oxidizing bacteria

(Travis and Rosenberg, 1997). More recently, Gandhi et al. (2002) modeled the cometabolic biodegradation of TCE by toluene oxidizers that were stimulated at Edwards Air Force Base in California using two recirculating wells. This reactive transport model incorporated oxygen outgassing as well as hydrogen peroxide inhibition into the microbial growth kinetics. Cirpka and Kitanidis (2001) modeled the same system using a simplified one-dimensional travel-time approach to evaluate effective substrate pulsing methodologies. Hossain and Corapcioglu (1996) developed a reactive transport model to simulate the sequential reductive dechlorination of PCE in the presence of excess stimulating substrate, and predicted pulsing methodologies that would be more effective than continuous injection. While each of these models could be calibrated and adequately fit to the field data for bioremediation within chlorinated solvent plumes, none of them incorporated bioremediation within the source zone or the effects of DNAPLs.

Several reactive transport models that have been developed explicitly incorporate source zones containing NAPLs into simulations of bioremediation. For example, Gallo and Manzini (2001) developed a one-dimensional model that incorporates dual-phase transport with biodegradation of a single contaminant that is degraded as a growth substrate. This single-substrate model has not yet been calibrated and validated with field data. Malone et al. (1993) also developed a one-dimensional dual-phase model applicable to aerobic bioremediation in the presence of LNAPLs that differentiated between oil fractions with high and low mass transfer potential. SEAM3D is a three-dimensional reactive transport model that incorporates biodegradation of multiple substrates and electron acceptors along with NAPL dissolution (Brauner and Widdowson, 2001). To date, this model has been applied only to hydrocarbon mixtures and has been shown to be especially sensitive to NAPL dissolution rates.

Reactive transport models that explicitly incorporate DNAPL sources include RT3D, a three-dimensional multisubstrate model that is capable of integrating DNAPL dissolution with contaminant transport and first-order sequential solvent degradation (http://bioprocess.pnl.gov/rt3d.htm). This model has been calibrated and validated at the Area-6 site at Dover Air Force Base in Delaware (Clement et al., 2000), among others. Not surprisingly, a sensitivity analysis of this model showed that the shape of the solvent plume was most sensitive to the aquifer transmissivity and that the mass of TCE in the resultant plume was most sensitive to biodegradation estimates.

Although a wide variety of comprehensive reactive transport models have been developed to predict in situ bioremediation in both single and multiple dimensions, few of them incorporate DNAPL presence and dissolution, and all of them are plagued by imprecise hydrologic information and by in situ biodegradation rates that are difficult to estimate. There is limited information available on microbial kinetics in the subsurface, especially with respect to dehalogenating communities, and this hinders the predictive abilities of even the most comprehensive models. Further, the limited information on the effects of DNAPL on subsurface microbial ecology hinders the development of appropriate microbial kinetic expressions.

*Research and Demonstration Needs*

There are two extremely important research needs associated with the application of enhanced bioremediation for the treatment of source zones. The first is a better understanding of the specific relationship between DNAPL dissolution and biodegradation. That is, it is important to better understand the potential toxic effects of DNAPLs on microbial communities, the potential for biosurfactants or microbial uptake to enhance DNAPL dissolution, and the interactions between cell community dynamics and DNAPL presence. The second critical research need is to evaluate the microbial kinetics associated with DNAPL and dissolved solvent concentrations in the field. Improving understanding of field-associated microbial kinetics will allow improved reactive transport models to better design and predict the performance of enhanced in situ bioremediation processes. The major demonstration needs follow on directly from the research needs, with well-characterized field-scale demonstrations of enhanced bioremediation within source zones containing DNAPLs topping the list.

## INTEGRATION OF TECHNOLOGIES

Integration of technologies can be critical to effectively treating multiple contaminants (e.g., organics and heavy metals), to treating contaminants in multiple compartments (e.g., vadose and saturated zones), and to optimizing the treatment of a given contaminant in a single compartment. This discussion focuses on technology integration for optimizing remediation of a given source zone, as well as for optimizing the overall treatment of the source zone and the dissolved plume. Integration of technologies is most effective when the weakness of one technology is mitigated by the strength of another technology, thus producing a more efficient and cost-effective solution. This topic is considered briefly here, with more information available in a recent NRC report (NRC, 2003).

An example of optimizing source zone remediation includes coupling of thermal and surfactant/cosolvent technologies for addressing an extremely challenging NAPL (e.g., coal tar). Dwarakanath et al. (2000) evaluated surfactant-enhanced remediation of a highly viscous (1,000 cP) heating oil, and determined that it would be more efficient to integrate surfactant and thermal technologies. A field study conducted at $50^{\circ}C$ demonstrated that this approach was able to achieve 88 percent removal of the NAPL.

Research has also evaluated a combination of surfactant and chemical oxidation processes. The concept is that surfactants can be used to remove the bulk of the residual contamination, followed by low-concentration chemical oxidation as a polishing step. By removing the bulk of the mass with the surfactants, a much lower contaminant load remains to be attacked by the chemical oxidation, allowing the implementation of a much lower-concentration chemical oxidation process (Shiau et al., 2003). Using lower concentrations of chemical oxidants has the advantages of being safer to implement, generating lower levels of heat and gas, and being more economical.

In an interesting innovation on the classical application of air sparging, Jeong et al. (2002) have proposed the concurrent injection of air and cosolvent for remediation of chlorinated solvent DNAPLs. Their proposed strategy is to use the preferential flow paths produced by air injection to enhance contact between the injected cosolvent and the DNAPL, resulting in enhanced contaminant dissolution into the cosolvent. Although they demonstrated promising results with this technique in laboratory studies, it has not yet been tested in the field. Similarly, Kim et al. (2004) investigated the use of low-concentration (below the CMC) surfactants to improve the size of source zones swept by air sparging. In laboratory experiments with homogeneous sand, they found that the swept zone created by air sparging was 5.2 times larger in the presence of sodium dodecyl benzene sulfonate than in its absence.

In situ thermal processes such as steam flushing or soil heating generally promote bulk removal of contaminants. Therefore, it would be useful to integrate a cost-effective follow-on technology capable of remediating the residual contaminants. For example, a logical combination would have in situ bioremediation following thermal treatment. A series of bench- and field-scale studies have been conducted with creosote- and solvent-contaminated soils to evaluate the effects of steam-enhanced extraction on soil microbial activity (Richardson et al., 2002). Results demonstrated that mesophilic subsurface microorganisms survive in situ steam treatment and that the surviving community has the potential for post-steam bioremediation.

Abriola et al. (2003) also coupled "biopolishing" to a source zone remediation activity, in this case surfactant-enhanced aquifer remediation. The study was conducted at a former dry cleaning facility in Oscoda, Michigan, and the target contaminants were chlorinated solvents and BTEX. Acetate production in the formation was consistent with fermentation of the surfactant, Tween 80, which was believed to have stimulated reductive dechlorination microorganisms in the aquifer. This was also consistent with substantial degradation of the low-level PCE concentrations remaining after the surfactant-based remedial efforts. Ongoing activities are seeking to isolate microbial populations at the site responsible for this reductive dechlorination.

The above discussion has focused on coupling technologies within the source zone. It is also important to make sure that the technologies implemented in the source zone do not negatively impact processes occurring in the downgradient dissolved plume. At the same time, it is possible that downgradient processes may be enhanced by activities taking place in the source zone. One example of the latter case is where amendments added to the source zone may enhance downgradient biodegradation processes. For example, when alcohol is used to flush the source zone, residual alcohol remaining after source zone remediation may serve as an electron donor to promote natural attenuation within the dissolved plume (Rao et al., 2001). This is analogous to the processes reported above by Abriola et al. (2003) and Richardson et al. (2002) within the source zone. At the same time, there is a danger that the source zone remediation technique may actually hinder the ensuing natural attenuation (e.g., by changing redox conditions due to chemical oxidation). Thus, the integrated technologies must be carefully selected to both maximize the synergism while avoiding undesirable side effects.

One specific question the committee was asked to address was whether and how monitored natural attenuation (MNA) should be used following more aggressive source zone remediation. Indeed, a commonly expressed goal at sites the committee is familiar with is to reduce concentrations of source zone contaminants enough to "allow MNA to take over." There is very little scientific evidence to address this question.

The scientific understanding of and experience with MNA of chlorinated compounds is growing and changing very quickly (see NRC, 2000, and the more up-to-date summary in Major et al., 2002). MNA relies on a variety of processes that transform or reduce contaminant concentrations in the subsurface, although MNA is most closely associated with microbial degradation. If source zone treatments are successful in lowering contaminant mass and concentration, several conditions are required for using MNA as a follow-on activity: presence of the necessary bacteria, electron donors and acceptors, and the necessary macro- and trace nutrients.

Because there have been no documented cases of using MNA as a follow-on to source remediation, one can only surmise about its potential, which is done here using reductive dechlorination as an example. Bacterial reductive dechlorination requires a reducing aquifer environment, which is maintained by fermentative bacteria that degrade carbon compounds and release hydrogen. There are three main posttreatment possibilities. If after a source treatment there are no carbon sources remaining that can support fermentation and maintain reduced aquifer conditions, infiltrating groundwater will carry oxygen into the source zone and the plume. This oxygen will kill dechlorinating bacteria and halt the primary natural attenuation processes. Alternatively, there could be enough biodegradable carbon sources remaining after source remediation to maintain anaerobic conditions, but the required bacteria may not be present if the source remedy has reduced their numbers. It is not known whether dechlorinating bacteria routinely survive source zone treatments. They would be least likely to survive treatments that rely on aggressive oxidation (see Box 5-12 for a description of recent work in this area). Treatments that introduce air, heat, or surfactants are also likely to kill bacteria. If the bacteria do not survive in numbers large enough to support MNA, or if they were not naturally present before treatment, bioaugmentation (adding bacterial cultures to the subsurface, also see Box 5-12) can be used to assure that the contaminant-biodegrading capacity is present when needed (Ellis et al., 2000; Major et al., 2002; He et al., 2003). In this case, MNA would be less effective and a plume of partially dechlorinated compounds would continue to migrate. The best-case scenario for MNA is if both the required

---

**BOX 5-12**
**Laboratory Study of Sequential**
**Chemical Oxidation and Bioaugmentation**

Soon-to-be-published work describes the use of in situ chemical oxidation using permanganate to rapidly remove DNAPL mass in a source zone. Because VOC rebound can occur after treatment, a secondary polishing technology such as enhanced bioremediation was considered. Bioremediation is potentially a highly effective technology for containing the residual groundwater plume, although it can be limited by the slow rate of microbially mediated reductive dechlorination, microbial inhibition by high chloroethene concentrations, and accumulation of more toxic degradation products (i.e., *cis*-1,2-dichloroethene and vinyl chloride). Furthermore, the prior addition of an oxidant may have severe impacts on the indigenous microbial population and may result in an ambient redox potential inhibitory to reductive dechlorination. To date, while some work has evaluated the sequential application of in situ chemical oxidation using Fenton's reagent with aerobic degradation, no published studies have assessed coupling permanganate addition with anaerobic dechlorination.

The potential consequences of sequencing these technologies have been evaluated in an ongoing laboratory optimization study prior to implementing sequential chemical oxidation/bioremediation at a pilot site in the United States. TCE was placed in six columns packed with soil collected from a site with an existing VOC plume, and the columns were flushed with distilled water. The study then evaluated the effects of short-term permanganate addition on the indigenous microbial activity and the groundwater geochemistry. The permanganate flushing was followed by biostimulation of four columns, and then bioaugmentation of two of these biostimulated columns with an enriched PCE/TCE dechlorinating consortium.

Residual TCE concentrations following permanganate flushing were approximately 1 mg/L in the flushed columns. Initial biostimulation was conducted with distilled water containing acetate and ethanol. Two of these biostimulated columns were also bioaugmented. No biodegradation of acetate or ethanol, and no transformation products of TCE, were detected in any of the biostimulated or bioaugmented columns. Three weeks after switching from distilled water to site groundwater, the column became gradually reduced due to degradation of acetate and ethanol, followed by commencement of TCE dechlorination to *cis*-1,2-DCE and methanogenesis. Under reduced conditions significant dissolution of $MnO_2$ was observed, both visually and through increases in dissolved Mn. Subsequent rebioaugmentation of one column led to dechlorination to ethene. The study thus illustrates the feasibility of sequential in situ chemical oxidation and bioaugmentation.

bacteria and biodegradable carbon sources are present following the source treatment. In that case, the essential question will be whether a sufficient supply of electron donors remains after treatment to support bacteria until the contaminants of interest are completely destroyed. Accurate posttreatment site assessment and mass estimates of both carbon sources and DNAPLs are essential and are required to address this issue.

At this time no general statement can be made about the ability of MNA to contain plumes after mass removal due to the very site-specific physical and chemical nature of individual releases and the uncertain presence or absence of bacteria. Site-specific analyses of natural attenuation potential—done by qualified and experienced experts—will be required in all cases.

* * *

The limited work done to date on integrating source remediation technologies with one another and with MNA suggests that coupling of technologies has the potential to significantly improve the overall treatment of source zone systems. The obvious challenge is to find an optimal combination that both maximizes remediation efficiency while minimizing remedial costs. Integration of technologies may require modification to one or both technologies. For example, coupling of thermal and bioremediation technologies may require use of lower-temperature thermal processes to prevent inactivation of the microbial population. Nonetheless, some of the greatest advances yet to be realized may well result from the integration of source zone remedial technologies, which ongoing research should continue to explore.

## COMPARISON OF TECHNOLOGIES

Table 5-7 summarizes the DNAPL source remediation technologies covered in this chapter. Since a detailed comparison of the technologies depends on a complex integration of a wide range of site and contaminant properties, the table provides a qualitative comparison only. It lists the types of contaminants for which each technology is applicable. The table then provides a rank of "high," "medium," "low," or "not applicable" relative to each technology's ability to achieve (1) mass removal in the source zone, (2) local aqueous concentration reduction in the source zone, (3) mass flux reduction from the source zone, (4) reduction of source migration potential, and (5) changes in the toxicity of the source area contaminants. "High," "medium," and "low" are qualitative terms that describe the likelihood that a given technology would be effective at achieving the listed objective. While a score of high suggests that a given technology is likely to achieve the source zone remedial objective, a score of medium suggests that progress might be realized, depending on specific circumstances. A score of

TABLE 5-7 Comparison of DNAPL Source Remediation Technologies

| Technology | Applicable Contaminant Types | Media Settings[a] | Likely Effectiveness at Appropriate Sites | | | | | Limitations | Comments |
|---|---|---|---|---|---|---|---|---|---|
| | | | Mass Removal | Local Aqueous Concentration Reduction | Mass Flux Reduction | Reduction of Source Migration Potential | Change in Toxicity | | |
| Physical Containment | All compounds | I | Not Applicable | Not Applicable | High | High | Not Applicable | Usually used with sources < 200 feet (61 m) deep. Difficult to use in karst. | The most commonly approved remedy for source areas containing DNAPL. Failure rates are low for properly constructed systems, but all projects should be monitored long-term. |
| | | II | Not Applicable | Not Applicable | High | High | Not Applicable | | |
| | | III | Not Applicable | Not Applicable | High | High | Not Applicable | | |
| | | IV | Not Applicable | Not Applicable | Low–High* | Low–High* | Not Applicable | | |
| | | V | Not Applicable | Not Applicable | Low–High* | Low–High* | Not Applicable | *Effectiveness depends on the fracture network | |

| Technology | Contaminants | Type | | | | | | | |
|---|---|---|---|---|---|---|---|---|---|
| Excavation | All compounds | I | High | High | High | High | Not Applicable | Site assessment must clearly define depth and area. Contaminant mass and concentration is not a factor. Technology is difficult when source is in bedrock (Type IV, V), but see discussion for exempting circumstances. | The most aggressive of all source remediation methods. Source materials must be treated or disposed of after excavation. |
| | | II | High | High | High | High | Not Applicable | | |
| | | III | High | High | High | High | Not Applicable | | |
| | | IV | Not Applicable | Not Applicable | Not Applicable | Not Applicable | Not Applicable | | |
| | | V | Not Applicable | Not Applicable | Not Applicable | Not Applicable | Not Applicable | | |
| Hydraulic Containment (Pump-and-Treat) | All organics | I | Low | Low | High | Low | Low | Hydraulic containment effective for managing plumes, but it must be assumed to operate continually. | Very commonly used, design requires good site characterization, not well suited to highly permeable or low permeability sites. |
| | | II | Low | Low | Low | Low | Low | | |
| | | III | Low | Low | Medium-High | Medium-High | Low | | |
| | | IV | Low | Low | Low | Low | Low | | |
| | | V | Low | Low | Low | Low | Low | | |

*continued*

TABLE 5-7 Continued

| Technology | Applicable Contaminant Types | Media Settings[a] | Likely Effectiveness at Appropriate Sites | | | | | | | Comments |
|---|---|---|---|---|---|---|---|---|---|---|
| | | | Mass Removal | Local Aqueous Concentration Reduction | Mass Flux Reduction | Reduction of Source Migration Potential | Change in Toxicity | Limitations | |
| Multiphase Extraction | Organics with low to moderate viscosity | I | Low-Medium | Low | Low-Medium | Low | Low-Medium | Difficult to find DNAPL pools precisely enough for extraction. Residual NAPL will not be recovered. This technology only applies to shallow source zones. | While commonly used for LNAPL recovery, has not been successfully demonstrated on DNAPL. |
| | | II | Low | Low | Low | Low | Low | | |
| | | III | Low | Low | Low-Medium | Low | Low | | |
| | | IV | Low | Low | Low | Low | Low | | |
| | | V | Low | Low | Low | Low | Low | | |

| Surfactant/ Cosolvent Flushing | All organics | I | High | High | High | High | Low-Medium | Must avoid undesirable downward migration of DNAPL. | Limited experience in fractured media. Foam may be needed if heterogeneity is high. |
|---|---|---|---|---|---|---|---|---|---|
| | | II | Low | Low | Low | Low | Low-Medium | | |
| | | III | Medium-High | Low-Medium | Medium-High | High | Low-Medium | | |
| | | IV | Low-Medium | Low-Medium | Low-Medium | Low-Medium | Low-Medium | | |
| | | V | Low-Medium | Low-Medium | Medium | Low-Medium | Low-Medium | | |

*continued*

278

TABLE 5-7 Continued

| Technology | Applicable Contaminant Types | Media Settings[a] | Likely Effectiveness at Appropriate Sites | | | | | Limitations | Comments |
|---|---|---|---|---|---|---|---|---|---|
| | | | Mass Removal | Local Aqueous Concentration Reduction | Mass Flux Reduction | Reduction of Source Migration Potential | Change in Toxicity | | |
| Chemical Oxidation | Halogenated ethenes and ethanes | I | Medium-High | Medium | Medium | Low | Medium-High | May be large heat release, soil fouling ($MnO_2$ ppt from $KMnO_4$), or metals released due to pH changes. Delivery of chemical oxidants will be poor in all but high-permeability media. Significant natural organic matter will limit efficacy. | Only applicable to immobilized sources (low NAPL saturation, or sorbed). Limited experience in fractured media, most failures attributed to channeling in heterogeneous media. May require multiple injections. |
| | | II | Low | Low | Low | Low | Low | | |
| | | III | Low-Medium | Low-Medium | Medium-High | Low | Low-Medium | | |
| | | IV | Low | Low-Medium | Low-Medium | Low-Medium | Low | | |
| | | V | Low | Low | Low | Low | Low | | |

| Technology | Contaminant | | | | | | | | Comments |
|---|---|---|---|---|---|---|---|---|---|
| Soil Mixing/ Chemical Reduction | Chlorinated and fluorinated compounds, primarily chlorinated ethanes | I | High | High | High | High | Medium-High | The source area has to be defined, and be located in a soil that is amenable to mixing. Has not been used in bedrock. | Rapid cleanup and site reuse, as little as one day for a small source, several weeks for a larger one. Developing technology with extremely limited experience to date. |
| | | II | High | High | High | High | Medium-High | | |
| | | III | High | High | High | High | Medium-High | | |
| | | IV | Not Applicable | Not Applicable | Not Applicable | Not Applicable | Not Applicable | | |
| | | V | Not Applicable | Not Applicable | Not Applicable | Not Applicable | Not Applicable | | |
| Steam Flushing | Volatile organic compounds and the more volatile end of the semivolatile organic compounds | I | High | High | High | High | Medium-High | Steam over-ride, vertical remobilization may be problems. Steam generation, vapor–liquid capture and treatment require significant infrastructure. | Few examples in fractured rock/ clay. No research on rate limitations. Some evidence that limited contaminant transformation may occur due to hydrous pyrolysis. |
| | | II | Low | Low | Low | Low | Low | | |
| | | III | Medium | Medium | Medium-High | Medium | Medium-High | | |
| | | IV | Low | Low | Low | Low | Low | | |
| | | V | Low | Low | Low | Low | Low | | |

*continued*

TABLE 5-7 Continued

| Technology | Applicable Contaminant Types | Media Settings[a] | Likely Effectiveness at Appropriate Sites | | | | | Limitations | Comments |
|---|---|---|---|---|---|---|---|---|---|
| | | | Mass Removal | Local Aqueous Concentration Reduction | Mass Flux Reduction | Reduction of Source Migration Potential | Change in Toxicity | | |
| Conductive Heating (ISTD) | All organics and some metals | I | High | High | High | High | Medium-High | High water flux below water table increases energy requirements. Migration into interval below heating is a concern. Close well spacing required. | Little documentation on demonstrations in saturated media or fractured media. |
| | | II | Medium-High | Medium-High | High | High | Medium-High | | |
| | | III | Medium-High | Medium-High | High | High | Medium-High | | |
| | | IV | Low-Medium | Medium | Medium | Medium | Low-Medium | | |
| | | V | Low-Medium | Medium | Medium | Medium | Low-Medium | | |

| Technology | Compounds | | | | | | | | Comments | |
|---|---|---|---|---|---|---|---|---|---|---|
| Electrical Resistance Heating | Compounds with boiling points less than that of water | I | Medium-High | Medium-High | High | Medium-High | Medium-High | Medium-High | Little documentation on demonstrations in saturated media or fractured media. | Cold temperatures increase energy requirements. Otherwise, limitations are the same as ISTD |
| | | II | Low-Medium | Low-Medium | Medium | Low-Medium | Low-Medium | Low-Medium | | |
| | | III | Medium | Medium | Medium-High | Medium | Medium | Medium-High | | |
| | | IV | Low | Low | Low | Low | Low | Low | | |
| | | V | Low-Medium | Low-Medium | Low-Medium | Low-Medium | Low-Medium | Low-Medium | | |
| Air Sparging (sometimes with SVE) | Volatile compounds such as organic solvents and gasoline aromatics | I | Low-Medium | Low-Medium | Low-Medium | Low | Low | Low-Medium | Good potential to promote concurrent aerobic bioremediation. Only applicable to immobilized sources (low NAPL saturation, or sorbed). | High potential for rebound following cessation of pumping. Heterogeneities result in channeling and decreased effectiveness. |
| | | II | Low | Low | Low | Low | Low | Low | | |
| | | III | Low | Low | Low-Medium | Low | Low | Low-Medium | | |
| | | IV | Low | Low | Low | Low | Low | Low | | |
| | | V | Low | Low | Low | Low | Low | Low | | |

continued

TABLE 5-7 Continued

| Technology | Applicable Contaminant Types | Media Settings[a] | Mass Removal | Likely Effectiveness at Appropriate Sites | | | | | Limitations | Comments |
|---|---|---|---|---|---|---|---|---|---|---|
| | | | | Local Aqueous Concentration Reduction | Mass Flux Reduction | Reduction of Source Migration Potential | Change in Toxicity | | | |
| Enhanced Bioremediation | Most volatile, semivolatile and nonvolatile organics, some metals, some inorganic ions, some explosives | I | Low-Medium | High | High | Low** | Medium-High | Good potential for contaminant destruction. Difficult to completely predict/control. May take a long time to see effects. May require multiple treatments.<br><br>**In cases where microbes make biosurfactants, DNAPL mobility can be enhanced. | Performance will be limited by dissolution rates and can be a function of geochemical conditions and indigenous microbial population. Limited experience in DNAPL proximity or in fractured media. |
| | | II | Low | Low | Low | Low** | Low | | |
| | | III | Low-Medium | Medium | Medium | Low** | Medium-High | | |
| | | IV | Low-Medium | Low-Medium | Low-Medium | Low** | Low-Medium | | |
| | | V | Low | Low-Medium | Low-Medium | Low** | Low-Medium | | |

[a]Media settings are as follows:

I Granular media with low heterogeneity and moderate to high permeability
II Granular media with low heterogeneity and low permeability
III Granular media with moderate to high heterogeneity
IV Fractured media with low matrix porosity
V Fractured media with high matrix porosity

low suggests that the technology is unlikely to be effective in meeting the specific objective. "Not applicable" is used in cases where the technology cannot, by design, accomplish the particular objective. For example, containment purposely has no effect on mass removal. Similarly, excavation is unable to bring about any change in toxicity. It should be kept in mind that the performance of a given technology is extremely site specific, as are the objectives associated with any remediation strategy. Thus, the scores of high, medium, and low are somewhat subjective, and are intended for guidance purposes rather than as an absolute ranking.

To try to capture some of the site-dependent performance of the technologies, a rank is given for each of the five hydrogeologic settings presented in Chapter 2. While these five hydrogeologic settings are intended to provide structure to the discussions, they have limitations. At times a single site may encompass several media settings, or it may not clearly fall into one of the five settings. In addition, inadequate site characterization may place the site into one type of setting whereas in actuality, the site may better fit into another setting. All of this points to the importance of successive source characterization by experienced professionals at each stage of the remedial process.

The five physical objectives listed in Table 5-7 (mass removal, local aqueous concentration reduction, mass flux reduction, reduction of source migration potential, and change in toxicity) correspond to those discussed in detail in Chapter 4, and each are briefly reviewed here. The time scale for evaluating progress toward meeting the objective is immediately after the source zone remedial effort ceases. The one exception is pump-and-treat, which is considered as a continuous process, since it is so inefficient for source zone removal.

The first physical objective is mass removal, which refers to destruction or extraction of contaminant mass from the source zone. For example, while containment isolates the mass from contact with the groundwater, it does not extract the mass from the subsurface; therefore, containment is deemed to be "not applicable" for mass removal.

The second physical objective is local aqueous concentration reduction. This parameter refers a weighted average of local contaminant concentrations in the zone of interest, and can thus deviate from concentrations measured a well sample, especially in a heterogeneous system (Type III media). In a heterogeneous system, during well purging the removed water will preferentially flow through the high conductivity zones, resulting in sample concentrations that are skewed toward the concentration found in this high conductivity zone. Thus, the resulting sample concentration is a function of the well itself (screened length and screened interval) and the way in which the well is sampled (amount and rate of purging, etc.). In addition, since the high conductivity zone tends to be well swept during flushing-based remedial activities, the resulting sample will have lower contaminant concentrations than expected from the contaminant remaining in the system. For these reasons, we choose instead to define the objective of reducing "local aqueous concentration"—that is, the concentration that would be realized if point sampling could be conducted at the pore level and could be averaged over a zone of interest. While this parameter cannot be measured at the field scale, it removes the ambiguity of how the well is constructed and/or sampled. Furthermore, as defined here, this objective provides insight into the efficiency of each technology because it can be clearly distinguished from the third physical objective (mass flux reduction), which reflects groundwater flow-averaged concentrations. Mass flux reduction and volume averaged concentration reduction would differ in cases in which there is significant heterogeneity in the source zone and a significant fraction of the contamination is contained in the low-permeability portions. Cases where the scores on these two parameters differ are cases where postremediation rebound in contaminant concentrations may be more likely to occur.

The third physical objective in Table 5-7 is mass flux reduction, which refers to the mean mass of contaminant that is flowing past an imaginary plane at the downgradient side of the source zone. This parameter can be determined by integrating concentration measurements from several wells located along this imaginary plane, as discussed further in Chapter 4. Thus, while similar to concentrations commonly measured in a downgradient monitoring well, mass flux provides a more complete picture by accounting for spatial variability in contamination and flow fields.

The fourth physical objective is reduction of source migration potential. This parameter refers to an actual reduction in DNAPL migration potential, which primarily results from removal of source zone mass. Selective removal of components from the NAPL might also impact the source migration potential by changing properties of the NAPL (e.g., viscosity). Nevertheless, some technologies have the potential to inadvertently promote downward mobility of the DNAPL (e.g., by reducing the oil–water interfacial tension), and these effects are noted in the "limitations" column.

The final physical objective in Table 5-7 is potential for change in toxicity, which comes from the Superfund criteria. Since this report focuses on source zone contamination, toxicity here refers to aqueous phase contaminants in the source zone area (as opposed to contamination at a downstream receptor). For common DNAPLs, change in toxicity can result from changes in the compound itself or in the composition of the DNAPL. For example, enhanced biodegradation of chlorinated ethenes generally occurs by a progressive series of dechlorinations—PCE to TCE to 1,1-DCE to vinyl chloride to ethene—although the toxicity of these compounds does not follow the same order. When a complex mixture is present in a source zone, and the constituent chemicals are differentially affected by the treatment, there may be a change in the effective toxicity of the mixture. These changes may *increase* or *decrease* the overall toxicity. Such effects would be extremely difficult to predict without detailed information on subsurface conditions during treatment. Because of this inherent uncertainty, the entries are provided as a range (e.g., low–medium) of potential for change in toxicity (either increasing or decreasing), and should not be considered as an absolute indication of the toxicity after remediation.

The final two columns in Table 5-7 document limitations and provide additional comments for each technology. It should be noted that plume size reduction is not listed as an objective in Table 5-7 even though it was discussed in Chapter 4 along with the other five physical objectives. Plume size reduction was omitted here because during the course of source zone remediation, contaminant concentrations and the plume size may initially increase prior to a long-term decrease—a subtlety that is not well reflected in such a simple table.

One common absolute objective of source zone remediation is risk reduction, which is not explicitly listed in Table 5-7. Instead, the objectives of mass removal, concentration reduction, and mass flux reduction attempt to capture the temporal and spatial nature of exposure. Which of these three objectives is best related to risk reduction is dependent on site type and other factors. For example, in spite of significant mass removal, heterogeneities may still cause there to be locally high concentrations. However, this can be accounted for by considering mass flux rather than discrete concentrations. Ultimately, risk assessment requires integration of these parameters with fate and transport simulators to determine the downgradient exposure level. Risk assessment also allows consideration of the temporal nature of the exposure. For example, in certain cases the long-term exposure may be more significantly impacted than the near-term exposures, as when mass removal does not result in significant concentration reduction but does greatly reduce the duration of exposure.

The goal of this table is to help guide the professional considering source zone remediation at a given site. Under no circumstance should this table be used to make the final selection of a technology (as discussed in Chapter 6). Rather, this table can help identify a list of the most viable technologies that should be thoroughly evaluated for use under site-specific conditions. In the end, the scoring in Table 5-7 should be considered more relative (one technology compared to another) than absolute.

The Table 5-7 entries are based on results from reported case studies as well as the committee's best professional judgment (where there is a lack of comprehensive full-scale demonstrations). Few of the objectives in Table 5-7 have been measured in the field for the technologies. Where full-scale demonstrations have been done, success is most frequently reported in terms of mass removal, such that many of the entries under this column reflect field data. However, the committee' best professional judgment was also critical to determining this column's entries because statements about the percent of mass removed during remediation rely on estimates of the original mass in place, which are subject to considerable uncertainty. With regards to the concentration column, there have occasionally been successes reported using concentration data from wells, but these are very hard to interpret and may not represent local groundwater concentrations. Virtually no field data exist for the third, fourth, and fifth objectives listed in Table 5-7, necessitating that the committee judge the technologies' likely effectiveness based the known operative mechanisms and the characteristics of the hydrogeologic settings. The lack of full-scale studies in all of the five hydrogeologic settings makes it very difficult to make generalizations about source remediation technologies. Nonetheless, in comparing the technologies in Table 5-7 and their individual descriptions in previous sections, it becomes apparent that certain site limitations apply to most of the technologies. For example, low-permeability materials pose serious challenges to almost all of the technologies. Most, if not all, of the technologies will struggle in certain hydrogeologic settings, such as karst (a variant of a Type V setting). Karst formations are characterized by large fractures and solution cavities, such that characterization and remediation of source zones in karst are extremely challenging and involve a large risk of reaching erroneous conclusions.

In lieu of full-scale demonstration projects, our ability to predict the effectiveness of source remediation is dependent on technology-specific mathematical models, bench-scale tests, and field pilot projects. The successful use of mathematical models is highly dependent on the quality of the site assessment and on the level of sophistication of the models and the model users. Flow models that encompass the movement of groundwater to carry the active agents of remediation are well developed and frequently used. When adequate site assessment and hydrogeologic information is available, these models can be quite accurate in their predictions (e.g., Brown et al., 1999). A typical use of modeling is to discover which subsurface properties are likely to have the largest influence on the remediation outcome, and then to investigate these properties in the field in sufficient detail to understand whether the proposed remedial action could succeed. One significant challenge is that many of the processes that are important for source remediation are highly nonlinear and thus difficult to model. The available models that incorporate this type of behavior are generally difficult for the nonspecialist to use successfully. At the present time, surfactant flushing appears to have the most successful predictive models available.

For those technologies that lack adequate mathematical models, bench studies and field-scale pilot studies can be used to evaluate their potential for success and to obtain key design parameters. Bench-scales test are helpful for evaluating chemical oxidation, chemical reduction, thermal techniques, and enhanced bioremediation. In these cases one must choose samples to be evaluated from a representative part of the source zone, and collect, transport, and handle those samples in ways that preserve their essential characteristics.

Field-scale pilot studies are defined as those done at a scale large enough to encompass likely variation in the hydrogeologic and chemical properties of the source zone. It was the committee's collective judgment, based on experience as well as ITRC (1998) and Morse et al. (1998), that a useful size for a pilot study at most sites is between 5 percent and 10 percent of the volume of the source zone.

At sites with exceptional complexity this rule-of-thumb may not hold, and a larger pilot test or multiple pilot tests may be necessary. The pre- and posttreatment contaminant masses and the mass and fate of treated or removed contaminants should be well documented during a field-scale pilot study. A successful pilot test also must be run long enough for problems that could be caused by variability in the subsurface or by unanticipated reactions—for example, poor reaction rates, fouling, or contaminant rebound—to be clearly observed. Three to six months should be long enough for most pilot studies, though pilot studies in source zones with exceptional volume or contaminant mass may have to be operated for longer times. Enhanced bioremediation studies may need to be conducted for longer periods due to the relatively slow growth rates of some bacteria. Even with pilot testing results in hand, substantial care should be taken to ensure that treatment systems are scaled up in a manner that is both practical and appropriate for the individual site in question.

Even at sites where mathematical models and bench-scale tests have been conducted, field-scale pilot studies are still very helpful before full-scale source remediation projects are started. They can help refine the full-scale activity in ways that save time and money.

## EXPLOSIVES REMOVAL TECHNOLOGIES

Historically, source remediation for explosives contaminated sites has focused on near-surface soils because of the tendency of these compounds to readily precipitate out of solution following disposal (see Chapter 2). Excavation to remove highly contaminated soils followed by ex situ treatment has been most common. Initially, incineration was the preferred ex situ treatment; however, composting has emerged as the standard for large volumes of explosives contaminated soils. Alternative technologies to composting (e.g., bioslurry) have been evaluated in pilot tests, but have not shown a competitive advantage (Craig et al., 1999). Among Army sites, incineration was the chosen ex situ treatment at Louisiana Army Ammunition Plant (AAP), Cornhusker AAP, Savanna Army Depot (AD), and Alabama AAP, while composting was performed at Newport AD, Hawthorne AAP, Iowa AAP, Milan AAP, Camp Navajo AD, Joliet AAP, and Umatilla AD. Lesser used technologies include capping where contaminated vadose zone soils extend deeper than conventional excavation depth, in situ flushing (e.g., water flooding), and enhanced bioremediation.

For the most part, other source area treatment alternatives for explosives have not been evaluated. Most of the aggressive technologies discussed for DNAPLs (such as chemical oxidation and electrical resistance heating) have been avoided due to concerns that localized temperature increases could initiate a detonation. Cosolvent or steam flushing appears attractive because explosives are highly soluble in the typical cosolvents used for DNAPLs (alcohols) and in hot water, but these technologies have not yet been considered.

Table 5-8 shows the limited number of Army sites contaminated with explosives where source area remedies have been completed or are in progress. In situ flushing at Umatilla has shown good progress, as evidenced by decreased concentrations of RDX and HMX in the regional groundwater. Remedial project managers (RPMs) at the site have considered enhancing the desorption of bound nitroaromatic residues from soil using induced cation exchange (calcium chloride injections); however, field tests have not been funded. At Badger, the treatment of the DNT source area via enhanced bioremediation is being optimized in field tests. RPMs have found that controlling soil moisture (low-grade in situ flushing), nutrients (phosphate), and waste products (nitrite) is important to optimizing treatment kinetics (Fortner et al., 2003; Rubingh, 2003).

Volunteer AAP is unique in that it is located in a hydrogeologically complex environment (karst bedrock with fractured residuum, a Type V setting). Extensive site characterization has been completed, yet there is still a lack of understanding regarding the nature of the explosive material released and the distribution of source areas within the residuum and/or bedrock. Site environmental managers

TABLE 5-8 Source Area Remedies for Explosives Contaminated Sites at Army Installations

| Site | Operable Unit | Contaminants | Source Area Remedy | Plume Remedy |
|------|---------------|--------------|--------------------|--------------|
| Umatilla, OR | Washout lagoons | TNT, RDX | Excavation, composting, in situ flushing | Pump-and-treat |
| Badger, WI | Propellant burning ground waste pits | DNT | Excavation, incineration, in situ flushing/enhanced bioremediation | Pump-and-treat |
| Volunteer, TN | North TNT manufacturing valley | TNT, DNT | Monitored natural attenuation | Monitored natural attenuation |
| Louisiana | Lagoons | TNT, RDX | Excavation, incineration, capping | Pump-and-treat |
| Milan, CO | Ditch E/ Wolf Creek | TNT, RDX, HMX | Source area not identified | ISCO-Fenton's |
| Pueblo, CO | TNT washout facility and discharge system | TNT, RDX, TNB | Excavation, composting | ISCO-Fenton's |

are considering the benefits of additional site characterization and extensive excavation. Pump-and-treat has been attempted, but is not now considered a viable alternative for containment of either the source area or the plume. Given the extreme hydrogeologic complexity at this site, MNA is being considered as an attractive remedy. Biodegradation in the residuum, dilution in the bedrock, and mass flux will be used to measure the extent of attenuation. An explosives-specific MNA protocol has been developed by the U.S. Army Corps of Engineers Engineer Research and Development Center (Pennington et al., 1999).

Development and deployment of technologies for treating explosives-contaminated plumes have been much more extensive than for source areas. Bioremediation, chemical oxidation, and chemical reduction have been the principal treatment alternatives to pump-and-treat for managing plumes. For the principal explosive compounds in this study (TNT, DNT, RDX, and HMX), there are several key chemical features that influence the efficacy of these different types of plume treatment. For example, the electrophilic nitro groups of TNT are susceptible to microbial reduction under both oxidative and reductive conditions (Rieger and Knackmuss, 1995). The cyclic nitramine RDX structure, however, is more easily biodegraded under strong reducing conditions than under aerobic conditions (McCormick et al., 1981, 1985; Speitel et al., 2001). HMX follows a similar pattern as RDX, but at a much slower rate. In contrast to the cometabolism of TNT and RDX, DNT appears to be oxidatively metabolized and used as a carbon, nitrogen, and energy source (Fortner et al., 2003).

Research on in situ chemical oxidation of TNT and RDX with Fenton's reagent has shown good results in laboratory studies (Li et al., 1997; Bier et al., 1999); however, much less work has been completed for HMX (Zoh and Stenstrom, 2002). In one study, a systematic evaluation of potassium permanganate oxidation of RDX was done by determining the degradation and mineralization kinetics using $^{14}$C-labeled RDX (Comfort, 2003); results showed that permanganate produces slow but sustained rates of RDX destruction and mineralization in the presence of aquifer solids.

Abiotic reduction of RDX has been explored using both zero valent iron and in situ geochemical reduction. Zero valent iron (ZVI) has been shown to effectively destroy RDX in aqueous solution and in soil slurries (Hundal et al., 1997), and recent work has shown that ZVI may be used in excavated soil windrows as an alternative to composting (Comfort et al., 2003). In situ geochemical reduction uses a chemical reductant (sodium dithionite) to reduce the iron present in aquifer solids (Fruchter et al., 2000). The structural ferrous iron in the reduced zone then reacts with RDX, dissolved oxygen, and other electron acceptors. Laboratory tests have shown that in the presence of dithionite-reduced sediments, RDX transformation occurs rapidly; however, mineralization is small. Follow-on studies showed that secondary biotreatment of the residuals emerging from a permeable treatment barrier significantly improved the total mineralization (Comfort, 2003).

\* \* \*

Because the characterization of explosive source materials and their interactions with geologic media lag far behind the knowledge base that exists for DNAPLs, fewer innovative source remediation technologies have been developed for explosives. The potential for dangerous explosions during remediation of explosives source areas has also been a major impediment. Indeed, whether for characterization or treatment technology development, laboratory and field assessment of explosives source zones must be performed with utmost care within specialized facilities.

## TECHNOLOGY COST CONSIDERATIONS

Although anecdotal cost data are available for some source remediation technologies (particularly surfactant flooding and thermal technologies), actual cleanup costs are highly dependent on site-specific hydrogeologic, geochemical, and contaminant conditions (NRC, 1997), such that absolute statements regarding the relative costs of different technologies are of limited utility. For example, some technology costs are independent of depth, while others (like soil mixing) go up exponentially with depth. The cost of in situ chemical oxidation is controlled primarily by the amount of contaminant in the subsurface and has little to do with the size of the site, unlike the cost of surfactant flooding. Furthermore, different analysts often use different assumptions to estimate costs, which can lead to different conclusions regarding the relative financial merits of competing technologies. For example, estimates obtained from technology vendors may include vendor costs, but not the additional costs that must be incurred by the site in order to utilize the technology.

Of the wide variety of cost measures reported in the literature, the life cycle cost metric is recommended because it represents the total cost resulting from a course of action over the entire period of time affected by the action, and it therefore avoids the problems of suboptimization presented by other cost metrics. It is recommended that life cycle cost analyses be performed at individual sites, taking into consideration all of the costs associated with deployment of the specific technology at the specific site. For example, life cycle cost estimates should include research, development, test, and evaluation costs; preparation and mobilization costs; capital costs; operation and maintenance costs; site restoration costs; and long-term management costs including potential costs associated with future liabilities. Finally, cost estimates should be probabilistic to reflect the uncertainties inherent in the estimates. Chapter 4 presents a general life cycle cost analysis approach that can be used to compare the costs of competing technologies at a particular site.

To the committee's knowledge, the only example of a comprehensive life cycle cost analysis for a site undergoing source remediation is from Hill AFB (see

Box 4-3). The decision-making framework and commensurate life cycle cost estimate used at Hill will include an updated geosystem model of the OU2 DNAPL source zone and a numerical model that can assist in the quantitative evaluation of the fate of the contaminant given a variety of possible management strategies. The goal is to use the site conceptual model, quantitative tools, and predictive contaminant transport modeling to provide the technical basis for a life cycle cost analysis that measures the effect of aggressive DNAPL removal operations on the dissolved phase plume and predicts plume dynamics into the future. Furthermore, these tools will be used to analyze various modifications to the current remedial strategy, including such possibilities as further source zone treatment (e.g., biopolishing) and/or decommissioning of existing remedial systems, and to aid in the determination of life cycle costs associated with each scenario. Life cycle cost estimates will most likely be probabilistic and will incorporate both capital and operation costs along with certain externalities. Examples of externalities that may be included are potential costs associated with mitigation of indoor air contamination, natural resource damage liability, and changes in regulatory standards.

## CONCLUSIONS AND RECOMMENDATIONS

The committee's review of source remediation technologies, both within the Army and at other sites, is based primarily on pilot-scale tests because the number of full-scale demonstrations across the country is quite small (unlike more established technologies such as pump-and-treat, for which an evaluation of 77 full-scale sites is summarized in NRC, 1994). Only thermal technologies and surfactant-flushing technologies have undergone multiple, carefully documented full-scale tests. The results of pilot-scale projects typically do not provide quantitative information on the ability of the various technologies to meet most remediation objectives. For example, although it is relatively easy to measure mass removal during a pilot test, such tests are rarely if ever designed to enable measurement of the objectives likely to be important when a remedy has gone to full-scale, such as concentration reduction at a downstream compliance point, mass flux reduction, or risk reduction. Furthermore, for the limited performance data available, usually only positive results are reported.

This lack of data and information upon which to make definitive statements about source remediation is echoed in a recently completed survey of 53 source remediation projects conduct-ed for the Navy (GeoSyntec, 2004). At 36 of the sites (68 percent) the RPM reported "success," even though mass removal was not estimated at 33 of these sites (63 percent), mass flux was unknown at more than half the sites, and rebound was unevaluated at 60 percent of the sites. Thus, it was not clear for the majority of cases what the perception of success was based on. Indeed, it is due to the lack of adequate, full-scale performance data that many of the entries in Table 5-7 are based on the committee's best professional judgment regarding the underlying physical processes of certain treatment technologies rather than on documented case studies.

The following conclusions and recommendations are made regarding current technologies for source remediation.

Some source remediation technologies have been demonstrated to achieve substantial mass removal across a range of sites and contaminants. A number of these studies have also demonstrated concentration reductions (at only one or a few wells), but the meaning of these measurements is highly debatable and few of these cases include long-term data on postremediation concentrations. Mass flux reduction, reduced migration of the source, and changes in toxicity have not yet been demonstrated at any of the source remediation case studies reviewed. This is partly because of the difficulty in making such measurements. Furthermore, there are few field data to support both the hypothesis and existing laboratory data that suggest that partial mass removal can affect local concentration and downgradient mass flux (see Box 4-1). **Thus, available data from field studies do not demonstrate what effect source remediation is likely to have on water quality.**

Although ongoing research is developing a depletion profile approach for assessing mass flux reduction, to date these profiles have only been developed for surfactant/cosolvent technologies. A substantial body of theoretical work suggests that mass removal could result in a mass flux reduction, but field evidence to support or contradict this is sparse. This and related approaches should be further developed in future research.

**Performance of *most* technologies is highly dependent on site heterogeneities.** All remediation technologies are affected by site-specific heterogeneities that should be taken into account for effective use of the technology, but some technologies are much more sensitive to heterogeneities than others. In general, the efficiency of flushing methods decreases as the heterogeneity increases, although the degree of impact depends on the specific site characteristics and on the operative processes. In the case of surfactant flushing, foam generated by air injection has emerged as a viable way to mitigate heterogeneities. Steam flushing is affected by preferential flow of the steam, but conduction mitigates this impact to some degree. Soil heating by conduction is least sensitive to heterogeneities because thermal conductivity varies very little with media properties. Chemical oxidation and enhanced bioremediation are more sensitive to heterogeneities than are thermal methods, and air sparging is the most sensitive to heterogeneity because there are no mitigating factors preventing the preferential flow of air and the bypassing of the target DNAPL. Heterogeneities are more likely to affect one's ability to achieve mass removal and local aqueous concentration reductions compared to mass flux from the source zone.

**Most of the technologies are not applicable in, are negatively impacted by, or have not been adequately demonstrated in low-permeability or frac- tured materials.** The effectiveness of flushing technologies in low-permeability settings (Type II) is limited due to the difficulty in moving flushing solutions (surfactants, oxidants, reductants, steam) through low permeability formations. Technologies which do not use fluid flow as a delivery mechanism, such as conductive heating and electrical resistance heating, have greater potential in Type II settings. Applications of source remediation technologies in fractured media (Types IV and V) have been limited due to difficulties in and cost of characterizing the fracture networks and delineating the source zone. Design and control of most source remediation technologies in these difficult-to-characterize systems is problematic. In addition, channeling along high-permeability fractures results in poor removal of mass from lower-permeability matrix zones for most technologies, with the possible exception of conductive heating since heat can be conducted efficiently through the rock matrix.

**Each technology has the potential to produce negative side effects that need to be accounted for in the design and implementation of that technology.** Examples of potential side effects include surfactant/cosolvent/steam-induced vertical migration of DNAPL, alteration of the redox potential by chemical oxidants or reductants (potentially serving to release previously bound nontarget compounds into the groundwater), and changes in the indigenous microbial popu- lation due to chemical or thermal treatment. These side effects can at times be avoided by an experienced design/implementation team. In other cases, the nega- tive side effects should be factored into the design/implementation process.

**Additional research is needed to determine how different source remediation technologies can be combined to achieve greater overall effec- tiveness.** Examples of potential synergism include combining surfactants and low-level thermal processes to solubilize high-viscosity oils, following contami- nant extraction with low-level chemical oxidation as a polishing step, allowing posttreatment levels of surfactants or alcohols to promote biotransformation of remaining contaminants, and allowing the elevated temperatures characteristic of thermal processes to promote the rate of biodegradation.

**Almost all of the source remediation technologies evaluated require more systematic field-scale testing to better understand their technical and eco- nomic performance.** Of the innovative technologies reviewed, only surfactant flooding has amassed a substantial number of field-scale studies in the peer- reviewed literature. Because full-scale applications of source remediation tech- nologies are scarce, there is insufficient information to thoroughly evaluate most technologies, especially with regard to the long-term impact of mass reduction in the source zone. Furthermore, due to insufficient economic data and the site-specific nature of both performance and cost, it is not possible to generically predict the impact of source remediation technologies on life cycle costs.

**The level and type of source zone characterization required to design, implement, and monitor the performance of remedies is dependent on the chosen objectives and the remediation technology.** For example, in situ chemical oxidation requires accurate estimates of source zone mass and composition and matrix oxygen demand, or else the remedy could be plagued by stoichiometric limitations or by the consumption of oxidant by unidentified co-contaminants. The properties of the source material (e.g., composition, viscosity, density, interfacial tension) should be determined for field-weathered samples in order to assess such remedies as surfactant-enhanced flushing. The location and geometry of source zone materials should be known to some level of certainty in order to design containment systems. For example, the most effectively designed slurry wall will have less effect on downstream mass flux if it is placed across the source zone rather than around it. With respect to performance monitoring, judging the effectiveness of in situ chemical oxidation by monitoring mineralization products or by monitoring the consumption of oxidant could overestimate treatment effectiveness in cases where alternate contaminants are present.

**Development of treatment technologies for explosives source zones is in its infancy because the characterization of explosive source materials and of their interactions with geologic media lags far behind the knowledge base that exists for DNAPLs.** Before one can understand the utility or performance characteristics of treatment technologies for explosives contamination, one should understand the chemical and physical nature of the explosives source zones. Furthermore, source areas containing high concentrations of explosives have the potential for dangerous explosions during remediation, which will necessitate laboratory and field assessment of explosives source zones within specialized facilities.

## REFERENCES

Abriola, L. M., T. J. Dekker, and K. D. Pennell. 1993. Surfactant-enhanced solubilization of residual dodecane in soil columns: 2—mathematical modeling. Environ. Sci. Technol. 27(12):2341–2351.

Abriola, L. M., C. A. Ramsburg, K. D. Pennell, F. E. Loeffler, M. Gamache, and E. A. Petrovskis. 2003. Post-treatment monitoring and Biological Activity at the Bachman Road Surfactant-enhanced Aquifer Remediation Site. 43(1) Extended Abstract, American Chemical Society Meeting, New Orleans, LA, March 23–27, 2003.

Abston, S. 2002. U.S. Army. Presentation to the NRC Committee on Source Removal of Contaminants in the Subsurface. August 22, 2002.

Alvarez-Cohen, L., and G. E. Speitel. 2001. Kinetics of aerobic cometabolism of chlorinated solvents. Biodegradation 12(2):105–126.

AATDF. 1997. Technology Practices Manual for Surfactants and Cosolvents. CH2MHILL. http:// clu-in.org/PRODUCTS/AATDF/Toc.htm.

Baker, R., D. Groher, and D. Becker. 1999. Minimal desaturation found during multi-phase extraction of low permeability soils. Ground Water Currents, 33, EPA 542-N-99-006.

Bier, E. L., J. Singh, Z. Li, S. D. Comfort, and P. J. Shea. 1999. Remediating RDX-contaminated water and soil by Fenton Oxidation. Environ. Toxicol. Chem. 18:1078–1084.

Brauner, J. S., and M. A. Widdowson. 2001. Numerical Simulation of a Natural Attenuation Experiment with a Petroleum Hydrocarbon NAPL Source. Groundwater 39:939–952.

Brown, C. L., M. Delshad, V. Dwarakanath, R. E. Jackson, J. T. Londergan, H. W. Meinardus, D. C. McKinney, T. Oolman, G. A. Pope, and W. H. Wade. 1999. Demonstration of surfactant flooding of an alluvial aquifer contaminated with dense nonaqueous phase liquid. Pp. 64–85 *In:* Innovative Subsurface Remediation: Field Testing of Physical, Chemical and Characterization Technologies. M. L. Brusseau, D. A. Sabatini, J. S. Gierke, and M. D. Annable (eds.). American Chemical Society Symposium Series 725. Washington, DC: ACS.

Brown, R. A. 1998. An analysis of air sparging for chlorinated solvent sites. Pp. C1.5:285–291 *In:* Physical, Chemical and Thermal Technologies. G. B. Wickramanayake and R. E. Hinchee (eds.). Columbus, OH: Battelle Press.

Buettner, H. M., W. D. Daily, and A. L. Ramirez. 1992. Enhancing cyclic steam injection and vapor extraction of volatile organic compounds in soils with electrical heating. Proc., Nuclear and Hazardous Waste Mgmt., Spectrum 1992:1321–1324.

Butler, R. M. 1991. Thermal Recovery of Oil and Bitumen. Englewood Cliffs, NJ: Prentice Hall.

Carr, C. S., S. Garg, and J. B. Hughes. 2000. Effect of dechlorinating bacteria on the longevity and composition of PCE-containing nonaqueous phase liquids under equilibrium dissolution conditions. Environ. Sci. Technol. 34:1088–1094.

Chown, J. C., B. H. Kueper, and D. B. McWhorter. 1997. The use of upward hydraulic gradients to arrest downward DNAPL migration in rock fractures. Journal of Ground Water 35(3):483–491.

Cirpka, O. A., and P. K. Kitanidis. 2001. Travel-time based model of bioremediation using circulation wells. Groundwater 39(3):422–432.

Clement, T. B., C. D. Johnson, Y. Sun, G. M. Klecka, and C. Bartlett. 2000. Natural attenuation of chlorinated ethene compounds: model development and field-scale application at the Dover site. Journal of Contaminant Hydrology 42:113–140.

Comfort, S. D. 2003. Evaluating in-situ permanganate oxidation and biodegradation of RDX in a perched aquifer. Project Report, University of Nebraska, October 2003.

Comfort, S. D., P. J. Shea, T .A. Machacek, and T. Satapanajaru. 2003. Pilot-scale treatment of RDX-contaminated soil with zerovalent iron. J. Environ. Qual. 32:1717–1725.

Cope, N., and J. B. Hughes. 2001. Biologically-enhanced removal of PCE from NAPL source zones. Environ. Sci. Technol. 35:2014–2021.

Craig, H. D., W. E. Sisk, M. D. Nelson, and W. H. Dana. 1999. Bioremediation of explosives-contaminated soils: a status review. Proceedings of the 10th Annual Conference on Hazardous Waste Research. Great Plains-Rocky Mountain Hazardous Substance Research Center, Kansas State University, Manhattan, Kansas. May 23–24, 1995.

Delshad, M., G. A. Pope, and K. Sepehrnoori. 1996. A compositional simulator for modeling surfactant enhanced aquifer remediation: 1—formulation. Journal of Contaminant Hydrology 23(4):303–327.

Delshad, M., G. A. Pope, L. Yeh, and F. J. Holzmer. 2000. Design of the surfactant flood at Camp Lejeune. The Second International Conference on Remediation of Chlorinated and Recalcitrant Compounds, Monterey, CA, May 22–25, 2000.

Dicksen, T., G. J. Hirasaki, and C.A. Miller. 2002. Mobility of foam in heterogeneous media: flow parallel and perpendicular to stratification. SPE Journal, June 2002.

Dwarakanath, V., and G. A. Pope. 2000. Surfactant phase behavior with field degreasing solvent. Environ. Sci. Technol. 34(22):4842.

Dwarakanath, V., K. Kostarelos, G. A. Pope, D. Shotts, and W. H. Wade. 1999. Anionic surfactant remediation of soil columns contaminated by nonaqueous phase liquids. J. Contaminant Hydrology 38(4):465–488.

Dwarakanath, V., L. Britton, S. Jayanti, G. A. Pope, and V. Weerasooriya. 2000. Thermally enhanced surfactant remediation. *In:* Proceedings of the Seventh Annual International Petroleum Environmental Conference. Albuquerque, NM, November 7–10, 2000.

Elder, C. R., C. H. Benson, and G. R. Eykholt. 1999. Modeling mass removal during in situ air sparging. J. Geotech. Geoenviron. Eng. 125(11):947–958.

Elliott, L. J., G. A. Pope, and R. T. Johns. 2004. Multi-dimensional numerical reservoir simulation of thermal remediation of contaminants below the water table. *In:* Proceedings of the Fourth International Conference on Remediation of Chlorinated and Recalcitrant Compounds, Monterey, CA, May 24–27, 2004.

Ellis, D. E., E. J. Lutz, J. M. Odom, R. J. Buchanan, C. L. Bartlett, M. D. Lee, M. R. Harkness, and K. A. Deweerd. 2000. Bioaugmentation for accelerated in situ anaerobic bioremediation. Environ. Sci. Technol. 34:2254–2260.

Enfield, C. G., A. L. Wood, M. C. Brooks, and M. D. and Annable. 2002. Interpreting tracer data to forecast remedial performance. Pp. 11–16 *In:* Groundwater Quality: Natural and Enhanced Restoration of Groundwater Pollution. S. Thornton and S. Oswald (eds.). Sheffield, UK: IAHS.

Environmental Protection Agency (EPA). 1995a. In Situ Remediation Technology Status Report: Thermal Enhancements. EPA/542-K-94-009. Washington, DC: EPA Office of Solid Waste and Emergency Response and Technology Innovation Office.

EPA. 1995b. How to Evaluate Alternative Cleanup Technologies for Underground Storage Tank Sites: A Guide for Corrective Action Plan Reviewers. EPA 510-B-95-007. Washington, DC: EPA.

EPA. 1996. Pump-and-Treat Ground-Water Remediation, A Guide for Decision Makers and Practitioners. EPA/625/R-95/005. Washington, DC: EPA Office of Research and Development.

EPA. 1997. Presumptive Remedy: Supplemental Bulletin Multi-Phase Extraction Technology for VOCs in Soil and Groundwater. Directive No. 9355.0-68FS, EPA 540-F-97-004 PB97-963501. Washington, DC: EPA.

EPA. 1998a. Abstracts of Remediation Case Studies, Volume 3. EPA 542-98-010. Washington, DC: EPA Federal Remediation Technologies Roundtable.

EPA. 1998b. Field Applications of In Situ Remediation Technologies: Chemical Oxidation. EPA 542-R-98-008. Washington, DC: EPA.

EPA. 1998c. Permeable Reactive Barrier Technologies for Contaminant Remediation. EPA/600/R-98/125. Washington, DC: EPA Office of Research and Development and Office of Solid Waste and Emergency Response.

EPA. 1999. Multi-Phase Extraction: State-of-the-Practice. EPA 542-R-99-004. Washington, DC: EPA.

EPA. 2000. Abstracts of Remediation Case Studies, Volume 4. EPA 542-R00-006. Washington, DC: EPA Federal Remediation Technologies Roundtable.

EPA. 2001. A Citizen's Guide to Soil Vapor Extraction and Air Sparging. EPA 542-F-01-006. Washington, DC: EPA.

EPA. 2003. Abstracts of Remediation Case Studies, Volume 7. Washington, DC: EPA Federal Remediation Technologies Roundtable.

Falta, R. W., K. Pruess, I. Javandel, and P. A. Witherspoon. 1992. Numerical modelling of steam injection for the removal of nonaqueous phase liquids from the subsurface. I. Numerical formulation. Water Resources Research 28(2):433–449.

Falta, R. W., C. M. Lee, S. E. Brame, E. Roeder, J. T. Coates, C. Wright, A. L. Wood, and C. G. Enfield. 1999. Field test of high molecular weight alcohol flushing for subsurface nonaqueous phase liquid remediation. Water Resources Research 35(7):2095–2108.

Finn, P. S., A. Kane, J. Vidumsky, D. W. Major, and N. Bauer. 2004. In-situ bioremediation of chlorinated solvents in overburden and bedrock using bioaugmentation. In: Bioremediation of Halogenated Compounds. V. S. Magar and M. E. Kelley (eds.). Columbus, OH: Battelle Press.

Fortner, J. D, C. Zhang, J. C. Spain, and J. B. Hughes. 2003. Soil column evaluation of factors controlling biodegradation of DNT in the vadose zone. Environ. Sci. Technol. 37(15):3382–3391.

Fountain, J. C., R. C. Starr, T. Middleton, M. Beikirch, C. Taylor, and D. Hodge. 1996. A controlled field test of surfactant-enhanced aquifer remediation. Ground Water 34(5):910–916.

Freeze, R. A., and J. A. Cherry. 1979. Groundwater. Englewood Cliffs, NJ: Prentice Hall.

Fruchter, J., C. Cole, M. Williams, V. Vermeul, J. Amonette, J. Szecsody, J. Istok, and M. Humphrey. 2000. Creation of a subsurface permeable treatment barrier using in situ redox manipulation. Ground Water Monitor. Rev. 66–77.

Gallo, C., and G. Manzini. 2001. A fully coupled numerical model for two-phase flow with contaminant transport and biodegradation kinetics. Communications in Numerical Methods in Engineering 17:325–336.

Gandhi, R. K., G. D. Hopkins, M. N. Goltz, S. M. Gorelick, and P. L. McCarty. 2002. Full-Scale demonstration of in situ cometabolic biodegradation of trichloroethylene in groundwater. II. Comprehensive analysis of field data using reactive transport modeling. Water Resources Research 38:1040.

Gates, D. D., and R. L. Siegrist. 1995. In situ chemical oxidation of trichloroethylene using hydrogen peroxide. J. Environ. Eng. September:639–644.

Gates-Anderson, D. D., R. L. Siegrist, and R. L. Cline. 2001. Comparison of potassium permanganate and hydrogen peroxide as chemical oxidants for organically contaminated soils. ASCE Journal of Environmental Engineering 127(4):337–347.

Gauglitz, R. A., J. S. Roberts, T. M. Bergsman, R. Schalla, S. M.Caley, M. H. Schlender, W. O. Heath, T. R. Jarosch, M. C. Miller, C. A. Eddy Dilek, R. W. Moss, and B. B. Looney. 1994. Six-phase soil heating for enhanced removal of contaminants: volatile organic compounds in non-arid soils integrated demonstration, Savannah River Site, PNL-101 84. Battelle Pacific Northwest Laboratory.

GeoSyntec Consultants. 2004. Assessing the Feasibility of DNAPL Source Zone Remediation: Review of Case Studies. Port Hueneme, CA: Naval Facilities Engineering Service Center.

Glaze, W. H., and J. W. Kang. 1988. Advanced oxidation processes for treating groundwater contaminated with TCE and PCE: laboratory studies. J. Amer. Water Works 80(5):57–63.

Gordon, M. J. 1998. Case history of a large-scale air sparging soil vapor extraction system for remediation of chlorinated volatile organic compounds in ground water. Ground Water Monitoring and Remediation 18(2):137–149.

Ground-Water Remediation Technology Analysis Center (GWRTAC). 1999. In situ chemical oxidation. Technology Evaluation Report TE-99-01. http://www.gwrtac.org.

Harvey, A. H., M. D. Arnold, and S. A. El-Feky. 1979. Selective electric reservoir heating. J. Cdn. Pet. Tech. (July–Sept. 1979):45–47.

Hasegawa, M. H., B. J. Shiau, D. A. Sabatini, R. C. Knox, J. H. Harwell, R. Lago, and L. Yeh. 2000. Surfactant-enhanced subsurface remediation of DNAPLs at the former Naval Air Station Alameda, California. Pp. 219–226 In: Treating Dense Nonaqueous-Phase Liquids (DNAPLs). G. B. Wickramanayake, A. R. Gavaskar, and N. Gupta (eds.). Columbus, OH: Battelle Press.

He, J., K. M. Ritalahti, M. R. Aiello, and F. E. Loeffler. 2003. Complete detoxification of vinyl chloride by an anaerobic enrichment culture and identification of the reductively dechlorinating population as a Dehalococcoides species. Appl. Environ. Microbiol. 69(2):996–1003.

Hirasaki, G. J., C. A. Miller, R. Szafranski, D. Tanzil, J. B. Lawson, H. Meinardus, M. Jin, R. E. Jackson, G. Pope, and W. H. Wade. 1997. Field demonstration of the surfactant/foam process for aquifer remediation. In: Proceedings - SPE Annual Technical Conference, San Antonio, TX, SPE 39292.

Holmberg, K., B. Jonsson, B. Kronberg, and B. Lindman. 2003. Surfactants and Polymers in Aqueous Solution. 2nd ed. New York: Wiley.

Holzmer, F. J., G. A. Pope, and L. Yeh. 2000. Surfactant-enhanced aquifer remediation of PCE-DNAPL in low permeability sands. Pp. 187–193 *In* Treating Dense Nonaqueous-Phase Liquids (DNAPLs). G. B. Wickramanayake, A. R. Gavaskar, and N. Gupta (eds.). Columbus, OH: Battelle Press.

Hood, E. D., and N. R. Thomson. 2000. Numerical simulation of in situ chemical oxidation. *In:* Proceedings from the Second International Conference on Remediation of Chlorinated and Recalcitrant Compounds, Monterey, CA, May 22-25, 2000.

Hood, E. D., N. R. Thomson, D. Grossi, and G. J. Farquhar. 1999. Experimental determination of the kinetic rate law for oxidation of perchloroethylene by potassium permanganate. Chemosphere 40(12):1383–1388.

Hossain, M. A., and M. Y. Corapcioglu. 1996. Modeling primary substrate controlled biotransformation and transport of halogenated aliphatics in porous media. Transport in Porous Media 24(2):203–220.

Huang, K. C., G. E. Hoag, P. Chheda, B.A. Woody, and G. M. Dobbs. 1999. Kinetic study of oxidation of trichloroethylene by potassium permanganate. Environ. Eng. Sci. 16(4):265–274.

Huang, K. C., G. E. Hoag, P. Chheda, B. A. Woody, and G. M. Dobbs. 2002. Kinetics and mechanism of oxidation of tetrachloroethylene with permanganate. Chemosphere 46(6):815–825.

Hundal L., J. Singh, E. L. Bier, P. J. Shea, S. D. Comfort, and W. L. Powers. 1997. Removal of TNT and RDX from water and soil using iron metal. Environ. Pollut. 97:55–64.

Hunt, J. R., N. Sitar, and K. S. Udell. 1988. Nonaqueous phase liquid transport and cleanup. I. Analysis of mechanisms. Water Resources Research 24(8):1247.

Illangasekare, T. H., and D. D. Reible. 2001. Pump-and-treat remediation and plume containment: applications, limitations, and relevant processes. Pp. 79–119 *In:* Groundwater Contamination by Organic Pollutants: Analysis and Remediation. Reston, VA: American Society of Civil Engineers.

Interstate Technology and Regulatory Council (ITRC). 2001. Technical and regulatory guidance for in situ chemical oxidation of contaminated soil and groundwater. Washington, DC: ITRC In Situ Chemical Oxidation Work Team.

ITRC. 1998. Technical and regulatory requirements for enhanced *in situ* bioremediation of chlorinated solvents. Washington, DC: ITRC In Situ Bioremediation Subgroup.

Isalou, M., B. E. Sleep, and S. N. Liss. 1998. Biodegradation of high concentrations of tetrachloroethene in a continuous column system. Environ. Sci. Technol. 32(22):3579–3585.

Jackson, R. E.; V. Dwarakanath, H. W. Meinardus, C. M. Young. 2003. Mobility control: how injected surfactants and biostiumulants may be forced into low-permeability units. Remediation, Summer:59–66.

Jawitz, J. W., M. D. Annable, P. S. C. Rao, and R. D. Rhue. 1998. Field implementation of a Winsor Type I surfactant/alcohol mixture for in situ solubilization of a complex LNAPL as a single-phase microemulsion. Environ. Sci. Technol. 32(4):523–530.

Jayanti, S., and G. A. Pope. 2004. Modeling the benefits of partial mass reduction in DNAPL source zones. *In:* Proceedings of the Fourth International Conference on Remediation of Chlorinated and Recalcitrant Compounds, Monterey, CA, May 24-27, 2004.

Jayanti, S., L. N. Britton, V. Dwarakanath, and G. A. Pope. 2002. Laboratory evaluation of custom designed surfactants to remediate NAPL source zones. Environ. Sci. Technol. 36(24):5491–5497.

Jeong, S.-W., A. L. Wood, and T. R. Lee. 2002. Enhanced contact of cosolvent and DNAPL in porous media by concurrent injection of cosolvent and air. Environ. Sci. Technol. 36:5238–5244.

Ji, W. A., A. Dahmani, D. Ahlfeld, J. D. Lin, and E. Hill. 1993. Laboratory study of air sparging: air flow visualization. Ground Water Monitor. Remed., Fall:115–126.

Johnson, P. C., R. L. Johnson, C. L. Bruce, and A. Leeson. 2001. Advances in in situ air sparging/
    biosparging. Bioremediation Journal 5(4):251–266.
Kastner, J. R., J. A. Domingo, M. Denham, M. Molina, and R. Brigmon. 2000. Effect of chemical
    oxidation on subsurface microbiology and trichloroethene (TCE) biodegradation. Bioremediation
    Journal 4(3):219–236.
Kim, H., H.-E. Soh, M. D. Annable, D.-J. Kim. 2004. Surfactant-enhanced air sparging in saturated
    sand. Environ. Sci. Technol. 38(4):1170–1175.
Knauss, K. G., M. J. Dibley, R. N. Leif, D. A. Mew, and R. D. Aines. 1999. Aqueous oxidation of
    trichloroethene (TCE): a kinetic analysis. Applied Geochemistry 14:531–541.
Knox, R. C., B. J. Shiau, D. A. Sabatini, and J. H. Harwell. 1999. Field Demonstration of Surfactant
    Enhanced Solubilization and Mobilization at Hill Air Force Base, UT. Pp. 49–63 In: Innovative
    Subsurface Remediation: Field Testing of Physical, Chemical and Characterization Technolo-
    gies. M. L. Brusseau, D. A. Sabatini, J. S. Gierke and M. D. Annable (eds.). American Chemical
    Society Symposium Series 725. Washington, D.C.: ACS.
Lee, E. S., Y. Seol, Y. C. Fang, and F. W. Schwartz. 2003. Destruction efficiencies and dynamics of
    reaction fronts associated with the permanganate oxidation of trichloroethylene. Environ. Sci.
    Technol. 37:2540–2546.
Lee, M. D., J. M. Odom, and R. J. Buchanan Jr. 1998. New perspectives on microbial dehalogenation
    of chlorinated solvents. Annu. Rev. Microbiol. 52:423–425.
Leeson, A., P. C. Johnson, R. L. Johnson, C. M. Vogel, R. E. Hinchee, M. Marley, T. Peargin, C. L.
    Bruce, I. L. Amerson, C. T. Coonfare, R. D. Gillespie, and D. B. McWhorter. 2002. Air
    sparging design paradigm. ESTCP technical document CU-9808. Washington, DC: ESTCP.
Li, Z. M., S. D. Comfort, and P .J. Shea. 1997. Destruction of 2,4,6-Trinitrotoluene (TNT) by Fenton
    Oxidation. J. Environ. Qual. 26:480–487.
Liang, S., L. S. Palencia, R. S. Yates, M. K. Davis, J.-M. Bruno, and R. L. Wolfe. 1999. Oxidation of
    MTBE by ozone and peroxone processes. J. Amer. Water Works Assoc. 91(6):104–114.
Liang, S., R. S. Yates, D. V. Davis, S. J. Pastor, L. S. Palencia, and J. M. Bruno. 2001. Treatability of
    MTBE contaminated groundwater by ozone and peroxone. J. Amer. Water Works Assoc.
    93:110–120.
Liang, C., C. J. Bruell, M. C. Marley, and K. L. Sperry. 2003. Thermally activated persulfate oxida-
    tion of trichloroethylene (TCE) and 1,1,1-trichloroethane (TCA) in aqueous systems and soil
    slurries. Soil & Sediment Contamination 12(2):207–228.
Londergan, J. T., H. W. Meinardus, P. E. Mariner, R. E. Jackson, C. L. Brown, V. Dwarakanath, G.
    A. Pope, J. S. Ginn, and S. Taffinder. 2001. DNAPL Removal from a Heterogeneous Alluvial
    Aquifer by Surfactant-Enhanced Aquifer Remediation. Ground Water Monitoring and Remedia-
    tion, Fall:57–67.
Lundegard, P. D., and G. Andersen. 1996. Multi-phase numerical simulation of air sparging perfor-
    mance. Groundwater 34 (3):451–460.
Ma, Y., and B. E. Sleep. 1997. Thermal variation of organic fluid properties and impact on thermal
    remediation feasibility. J. Soil Contamination 6(3):281–306.
MacKinnon, L. K., and N. R. Thomson. 2002. Laboratory-scale in situ chemical oxidation of a
    perchloroethylene pool using permanganate. J. Contam. Hydrol. 56:49–74.
Major, D. W., M. L. McMaster, E. E. Cox, E. A. Edwards, S. M. Dworatzek, E. R. Hendrickson, M.
    G. Starr, J. A. Payne, and L. W. Buonamici. 2002. Field demonstration of successful bio-
    augmentation to achieve dechlorination of tetrachloroethene to ethene. Environ. Sci. Technol.
    36:5106–5116.
Malone, D. R., C. M. Kao, and R. C. Borden. 1993. Dissolution and biorestoration of nonaqueous
    phase hydrocarbons—model development and laboratory evaluation. Water Resources Research
    29:2203–2213.
Marley, M. C., L. Fengming, and S. Magee. 1992. The application of a 3-D model in the design of air
    sparging systems. Ground Water Manag. 14:377–392.

Martel, R., and P. Gelinas. 1996. Surfactant solutions developed for NAPL recovery in contaminated aquifers. Ground Water 34:143–154.

Martel, R., P. J. Gelinas, and L. Saumure. 1998. Aquifer washing by micellar solutions: field test at the Thouin Sand Pit, L'Assomption, Quebec, Canada. Journal of Contaminant Hydrology 30:33–48.

Mason, A. R., and B. H. Kueper. 1996. Numerical simulation of surfactant-enhanced solubilization of pooled DNAPL. Environ. Sci. Technol. 30(11):3205–3215.

May, I. 2003. Army Environmental Center. Presentation to the Committee on Source Removal of Contaminants in the Subsurface. January 30, 2003.

McCarty, P. L. 1997. Breathing with chlorinated solvents. Science 276:1521–1522.

McCarty, P. L., M. N. Goltz, G. D. Hopkins, M. E. Dolan, J. P. Allan, B. T. Kawakami, and T. J. Carrothers. 1998. Full-scale application of in situ cometabolic degradation of trichloroethylene in groundwater through toluene injection. Environ. Sci. Technol. 32:88–100.

McClure, P., and B. E. Sleep. 1996. Simulation of bioventing for remediation of organics in groundwater. ASCE. J. Env. Eng. 122(11):1003–1012.

McCormick, N. G., J. H. Cornell, and A. M. Kaplan. 1981. Biodegradation of hexahydro-1,3,5-trinitro-1,3,5-triazine. Appl. Environ. Microbiol. 42:817–823.

McCormick, N. G., J. H. Cornell, and A. M. Kaplan. 1985. The anaerobic biotransformation of RDX, HMX and their acetylated derivatives. AD Report A149464 (TR85-007). Natick, MA: U.S. Army Natick Research and Development Center.

McCray, J. E., and R. W. Falta. 1997. Numerical Simulation of Air Sparging for Remediation of NAPL. Ground Water 35(1):99–110.

McGee, B. C. W., F. E. Vermeulen, P. K. W. Vinsome, M. R. Buettner, and F. S. Chute. 1994. In-situ decontamination of soil. The Journal of Canadian Petroleum Technology, October:15–22.

McGee, B. C. W., B. Nevokshonoff, and R. J. Warren. 2000. Electrical heating for the removal of recalcitrant organic compounds. In: Proceedings of the Second International Conference on Remediation of Chlorinated and Recalcitrant Compounds, Monterey, CA, May 22–25, 2000.

Meinardus, H. W., V. Dwarakanath, J. Ewing, K. D. Gordon, G. J. Hirasaki, C. Holbert, and J. S. Ginn. 2002. Full-scale field application of surfactant-foam process for aquifer remediation. In: Proceedings of the Third International Conference on Remediation of Chlorinated and Recalcitrant Compounds, Monterey, CA, May 20–23, 2002.

Morgan, D., D. Bryant, and K. Coleman. 2002. Permanganate chemical oxidation used in fractured rock. Ground Water Currents Issue 43.

Morse, J.J., B. C. Alleman, J. M. Gossett, S. H. Zinder, D. E. Fennell, G. W. Sewell, and C. M. Vogel. 1998. Draft technical protocol—a treatability test for evaluating the potential applicability of reductive biological in situ treatment technology to remediate chloroethenes. Washington, DC: DOD Environmental Security Technology Certification Program.

Myers, D. 1999. Surfaces, Interfaces, and Colloids, Principles and Applications. 2nd ed. New York: Wiley-VCH.

Nash, J. H. 1987. Field Studies of In-Situ Soil Washing. EPA/600/2-87/110. Cincinnati, OH: U.S. Environmental Protection Agency.

National Research Council (NRC). 1993. In Situ Bioremediation: When Does it Work? Washington, DC: National Academy Press.

NRC. 1994. Alternatives for Ground Water Cleanup. Washington, DC: National Academy Press.

NRC. 1997. Innovations in Ground Water and Soil Cleanup: From Concept to Commercialization. Washington, DC: National Academy Press.

NRC. 1999. Groundwater Soil Cleanup. Washington, DC: National Academy Press.

NRC. 2000. Natural Attenuation for Groundwater Remediation. Washington, DC: National Academy Press.

NRC. 2003. Environmental Cleanup at Navy Facilities: Adaptive Site Management. Washington, DC. National Academies Press.

NAVFAC. 1999. In-situ chemical oxidation of organic contaminants in soil and groundwater using Fenton's Reagent. TDS-2071-ENV. Port Hueneme, CA: Naval Facilities Engineering Service Center.

Naval Facilities Engineering Service Center (NFESC). 2002. Surfactant-Enhanced Aquifer Remediation (SEAR) Design Manual. NFESC Technical Report TR-2206-ENV. Battelle and Duke Engineering and Services. http://enviro.nfesc.navy.mil/erb/erb_a/restoration/technologies/remed/phys_chem/sear/tr-2206-sear.pdf

Nielsen, R. B., and J. D. Keasling. 1999. Reductive dechlorination of chlorinated ethene DNAPLs by a culture enriched from contaminated groundwater. Biotechnology and Bioengineering 62:162–165.

Nyer, L. K., and D. Vance. 1999. Hydrogen peroxide treatment: the good, the bad, the ugly. Ground Water Monitoring and Remediation 19(3):54–57.

O'Haver, J. H., R. Walk, B. Kitiyanan, J. H. Harwell, and D. A. Sabatini. 2004. Packed column and hollow fiber air stripping of a contaminant-surfactant stream. Journal of Environmental Engineering ASCE 130(1):4–11.

Pennington, J. C., M. Zakikhani, and D. W. Harrelson. 1999. Monitored natural attenuation of explosives in groundwater—Environmental Security Technology Certification Program Completion Report. Technical Report EL-99-7. Vicksburg, MS: U.S. Army Corps of Engineers, Waterways Experiment Station.

Philip, J. R. 1998. Full and boundary-layer solutions of the steady air sparging problem. J. Contam. Hydrol. 33(3–4):337–345.

Poston, S. W., S. Ysrael, A. K. M. S. Hossain, E. F. Montgomery III, and H. J. Ramey, Jr. 1970. The effect of temperature on irreducible water saturation and relative permeability of unconsolidated sands. Soc. Pet. Eng. J., June:171–180.

Rabideau, A. J., and J. M. Blayden. 1998. Analytical model for contaminant mass removal by air sparging. Ground Water Monitor. Remed. 18(4):120–130.

Ramsburg, C. A., and K. D. Pennell. 2002. Density-modified displacement for DNAPL source zone remediation: density conversion and recovery in heterogeneous aquifer cells. Environ. Sci. Technol. 36 (14):3176–3187.

Rao, P. S. C., A. G. Hornsby, D. P. Kilcrease, and P. Nkedi-Kizza. 1985. Sorption and transport of hydrophobic organic chemicals in aqueous and mixed solvent systems: model development and preliminary evaluation. Journal of Environmental Quality 14(3):376–383.

Rao, P. S. C., M. D. Annable, R. K. Sillan, D. Dai, K. Hatfield, W. D. Graham, A. L. Wood, and C. G. Enfield. 1997. Field-scale evaluation of in situ cosolvent flushing for enhanced aquifer remediation. Water Resources Research 33(12):2673–2686.

Rao, P. S. C., H. W. Jawitz, C. G. Enfield. R. W. Falta, Jr., M. D. Annable, and A. L. Wood. 2001. Technology integration for contaminated site remediation: cleanup goals and performance criteria. Presented at Groundwater Quality 2001, Sheffield, UK, June 18–21, 2001.

Rathfelder, K. M., L. M. Abriola, T. P. Taylor, and K. D. Pennell. 2001. Surfactant enhanced recovery of tetrachloroethylene from a porous medium containing low permeability lenses. II. Numerical simulation. Journal of Contaminant Hydrology 48(3–4):351–374.

Ravikumar, J. X., and M. D. Gurol. 1994. Chemical oxidation of chlorinated organics by hydrogen peroxide in the presence of sand. Environ. Sci. Technol. 28(3):395–400.

Reitsma, S., and Q. Dai. 2000. Reaction-enhanced mass transfer from NAPL pools. In: Proceedings from the Second International Conference on Remediation of Chlorinated and Recalcitrant Compounds, Monterey, CA, May 22–25, 2000.

Rice, B. N., and R. F. Weston. 2000. Lessons learned from a dual-phase extraction field application. Pp. 93–100 In: Physical and Thermal Technologies: Remediation of Chlorinated and Recalcitrant Compounds. G. B. Wickramanayake and A. R. Gavaskar (eds.). Proceedings of the Second International Conference on Remediation of Chlorinated and Recalcitrant Compounds. Columbus, OH: Battelle Press.

Richardson, R. E., C. A. James, V. K. Bhupathiraju, and L. Alvarez-Cohen. 2002. Microbial activity in soils following steam exposure. Biodegradation 13(4):285–295.

Rieger, P. G., and H. J. Knackmuss. 1995. Basic knowledge and perspectives on biodegradation of 2,4,6-trinitrotoluene and related nitroaromatic compounds in soil. In: Biodegradation of Nitro-aromatic Compounds. J. C. Spain (ed.). New York: Plenum Press.

Rosen, M. J. 1989. Surfactants and Interfacial Phenomena. 2nd ed. New York: Wiley.

Rubingh, D. 2003. Presentation to the NRC Committee on Source Removal of Contaminants in the Subsurface. April 14, 2004. Washington, DC.

Sabatini, D. A., J. H. Harwell, M. Hasegawa, and R. C. Knox. 1998. Membrane processes and surfactant-enhanced subsurface remediation: results of a field demonstration. Journal of Membrane Science 151(1):89–100.

Sabatini, D. A., J. H. Harwell, and R. C. Knox. 1999. Surfactant selection criteria for enhanced subsurface remediation. In: Innovative Subsurface Remediation: Field Testing of Physical, Chemical and Characterization Technologies. M. L. Brusseau, D. A. Sabatini, J. S. Gierke and M. D. Annable (eds.). ACS Symposium Series 725. Washington, DC: American Chemical Society.

Sabatini, D. A., R. C. Knox, J. H. Harwell, and B. Wu. 2000. Integrated design of surfactant enhanced DNAPL remediation: effective supersolubilization and gradient systems. Journal of Contaminant Hydrology 45(1):99–121.

Schnarr, M., C. Truax, G. Farquhar, E. Hood, T. Gonullu, and B. Stickney. 1998. Laboratory and controlled field experiments using potassium permanganate to remediate trichloroethylene and perchloroethylene DNAPLs in porous media. J. Contam. Hydrol. 29:205–224.

Schroth, M. H., M. Oostrom, T. W. Wietsma, and J. D. Istok. 2001. In-situ oxidation of trichloro-ethene by permanganate: effects on porous medium properties. J. Contam. Hydrol. 50:79–98.

Seagren, E. A., B. E. Rittmann, and A. J. Valocchi. 1994. Quantitative evaluation of the enhance-ment of NAPL-pool dissolution by flushing and biodegradation. Environ. Sci. Technol. 28:833–839.

Semprini, L., and P. L. McCarty. 1991. Comparison between model simulations and field results for in-situ biorestoration of chlorinated aliphatics. I. Biostimulation of methanotrophic bacteria. Ground Water 29:365–374.

Semprini, L., and P. L. McCarty. 1992. Comparison between model simulations and field results for in-situ biorestoration of chlorinated aliphatics. II. Cometabolic transformations. Ground Water 30:37–44.

She, Y., and B. E. Sleep. 1998. The effect of temperature on capillary pressure-saturation relation-ships for air-water and perchloroethylene-water systems. Water Resources Research 34(10):2587–2597.

She, H., and B. E. Sleep. 1999. Removal of PCE from soil by steam flushing in a two-dimensional laboratory cell. Ground Water Monitoring and Remediation 19(2):70–77.

Shiau, B. J., J. M. Brammer, D. A. Sabatini, J. H. Harwell, and R. C. Knox. 2003. Recent Develop-ment of In Situ Surfactant Flushing and Case Studies for NAPL-impacted Site Closure and Remediation. Pp. 92–106 In: Proceedings of 2003 Petroleum Hydrocarbon and Organic Chemi-cals in Ground Water: Prevention, Assessment and Remediation. National Ground Water Association, August 19-22, 2003, Costa Mesa, CA.

Siegrist, R. L., K. S. Lowe, L. C. Murdoch, T. L. Case, and D. A. Pickering. 1999. In situ oxidation by fracture emplaced reactive solids. ASCE J. Env. Eng. 125(5):429–440.

Sinnokrot, A. A., H. J. Ramey Jr., and S. S. Marsden Jr. 1971. Effect of temperature level upon capillary pressure curves. Soc. Petroleum Engrs. J. 11:13–22.

Sleep, B. E. 1993. Modelling and laboratory investigations of steam flushing below the water table. Groundwater 31(5):831.

Sleep, B. E., L. Sehayek, and, C. Chien. 2000. A modeling and experimental study of LNAPL accumulation in wells and LNAPL recovery from wells. Water Resources Research 36(12):3535–3546.

Song, D. L., M. E. Conrad, K. S. Sorenson, and L. Alvarez-Cohen. 2002. Stable carbon isotopic fractionation during enhanced in-situ bioremediation of trichloroethylene. Environ. Sci. Technol. 36(10):2262–2268.

Speitel, G. E., T. L. Engels, and D. C. McKinney. 2001. Biodegredation of RDX in unsaturated soil. Biorem. J. 5:1–11.

Stegemeier, G. L., and H. J. Vinegar. 2001. Thermal conduction heating for in-situ thermal desorption of soils. Pp. 1–37 *In:* Hazardous and Radioactive Waste Treatment Technologies Handbook. C. H. Oh (ed.). Boca Raton, FL: CRC Press.

Stroo, H. F., M. Unger, C. H. Ward, M. C. Kavanaugh, C. Vogel, A. Leeson, J. Marqusee, and B. Smith. 2004. Remediating chlorinated solvent source zones. Environ. Sci. Technol. 37(11):225A–230A.

Struse, A. M., R. L. Siegrist, H. E. Dawson, and M. A. Urynowicz. 2002. Diffusive transport of permanganate during in situ oxidation. J. Environ. Eng. 128(4):327–334.

Szafranski, R., J. B. Lawson, G. J. Hirasaki, C. A. Miller, N. Akiya, S. King, R. E. Jackson, H. Meinardus, and J. Londergan. 1998. Surfactant/foam process for improved efficiency of aquifer remediation. Progr. Colloid Polym. Sci. 111:162–167.

Tarr, M. A., M. E. Lindsey, J. Lu, and G. Xu. 2000. Fenton oxidation: bringing pollutants and hydroxyl radicals together. *In:* Proceedings from the Second International Conference on Remediation of Chlorinated and Recalcitrant Compounds, Monterey, CA, May 22-25, 2000.

Travis, B. J., and N. D. Rosenberg. 1997. Modeling in situ bioremediation of TCE at Savannah River: effects of product toxicity and microbial interactions on TCE degradation. Environ. Sci. Technol. 31:3093–3102.

U.S. Department of Energy (DOE). 1999. In Situ Chemical Oxidation Using Potassium Permanganate. Innovative Technology Summary Report. DOE/EM-0496. Washington, DC: DOE Subsurface Contaminants Focus Area, Office of Environmental Management, Office of Science and Technology.

U.S. Department of Energy (DOE). 2000. Hydrous pyrolysis Oxidation/Dynamic Underground Stripping. Innovative Technology Summary Report, DOE/EM-054. Washington, DC: DOE.

van-Dijke, M. I. J., and S. E. A. T. M. Van der Zee. 1998. Modeling of air sparging in a layered soil: numerical and analytical approximations. Water Resources Research 34 (3):341–353.

van-Dijke, M. I. J., S. E. A. T. M. Van der Zee, and C. J. Van Duijn. 1995. Multi-phase flow modeling of air sparging. Advan. Water Resour. 18(6):319–333.

Vermeulen, F. E., F. S. Chute, and M. R. Cervenan. 1979. Physical modelling of the electromagnetic heating of oil sand and other earth-type and biological materials. Canadian Electrical Engineering Journal 4(4):19–28.

Vinegar, H. J., E. P. de Rouffignac, G. L. Stegemeier, J. M. Hirsch, and F. G. Carl. 1998. In situ thermal desorption using thermal wells and blankets. *In:* The Proceedings of the First International Conference on Remediation of Chlorinated and Recalcitrant Compounds. G. B. Wickramanayake and R. E. Hinchee (eds.). Physical, Chemical and Thermal Technologies. Monterey, California, May 18–21, 1998.

Vinsome, P. K. W., B. C. W. Mcgee, F. E. Vermeulen, and F. S. Chute. 1994. Electrical heating. Journal of Canadian Petroleum Technology 33(9):29–35.

Wattenbarger, R. A., and F. McDougal. 1988. Oil response to in-situ electrical resistance heating (ERH). Journal of Canadian Petroleum Technology 27(6):45–50.

Watts, R. J., M. D. Udell, S. Kong, and S. W. Leung. 1999. Fenton-like soil remediation catalyzed by naturally occurring iron minerals. Environ. Eng. Sci. 16(1):93–103.

White, P. D., and J. T. Moss. 1983. Thermal Recovery Methods. Oklahoma: PennWell Books.

Yan, Y. E, and F. Schwartz. 1999. Oxidative degradation and kinetics of chlorinated ethylenes by potassium permanganate. J. Contam. Hydrol. 37(3):343–365.

Yang, Y. R., and P. L. McCarty. 1998. Competition for hydrogen within a chlorinated solvent dehalogenating anaerobic mixed culture. Environ. Sci. Technol. 32:3591–3597.

Yang, Y., and McCarty, P. L. 2000. Biologically enhanced dissolution of tetrachloroethene DNAPL. Environ. Sci. Technol 34:2979–2984.

Zhang, H., and R W. Schwartz. 2000. Simulation of oxidative treatment of chlorinated compounds by permanganate. *In:* Proceedings from the Second International Conference on Remediation of Chlorinated and Recalcitrant Compounds, Monterey, CA, May 22–25, 2000.

Zoh, K., and K. D. Stenstrom. 2002. Fenton oxidation of hexahydro-1,3,5-trinitro-1,3,5-triazine (RDX) and octahydro-1,3,5,7-tetranitro-1,3,5,7-tetrazocine (HMX). Water Res. 36:1331–1341.

# 6

# Elements of a Decision Protocol for Source Remediation

One perception of source remediation conveyed to the committee by the Army and other responsible parties is that investments in source depletion technologies have often failed to achieve the desired reductions in risk and/or site care requirements. There is no doubt that investing time and money into remedies without apparent progress creates a scenario that is sure to frustrate all parties. Although part of this challenge arises from the fact that historical releases of DNAPLs and explosives are technically difficult to clean up, the problem is also attributable to how source zones are managed.

The design and implementation of a successful source remediation project involves the iterative characterization of the source zone, development of remediation objectives, and evaluation of technologies, each of which is emphasized in previous chapters of this report. The resultant process is sufficiently complex to warrant a formal protocol to ensure that future projects do not skip essential elements—a need that was explicitly recognized by the Army in requesting this study. This chapter describes the elements of a protocol to assist project managers in designing, implementing, and assessing the effects of source remediation.

A protocol is defined as a strategy and methodology to be followed for accomplishing a stated purpose—in this case, the remediation (through removal, transformation, or isolation) of source material from the subsurface. The elements of a protocol presented in this chapter focus on decision making rather than on how to collect information, which is different from other commonly used protocols that provide extensive appendixes on field-sampling techniques, analytical methods, and data interpretation (e.g., the natural attenuation protocol of

306

Wiedemeier et al., 1996). Rather, this chapter is intended to help standardize the conceptual process for evaluating source remediation, including data gathering and analysis, setting objectives, and selecting remedial actions. The guidance is general, in that all possibilities are examined and no technology or endpoint is advocated over others.

Decision tools and protocols for general site cleanup date back to the early days of the Comprehensive Environmental Response, Compensation, and Liability Act (CERCLA or Superfund) and the Resource Conservation and Recovery Act (RCRA). For example, the nine criteria of CERCLA discussed in Chapter 4 are meant to enable the remedial project manager to select among various alternative remedies. Since the early 1980s, many detailed cleanup protocols have been created, most of which focus on specific types of contamination problems (for example, underground storage tanks, fuel hydrocarbon sites, LNAPLs) other than the recalcitrant chlorinated solvent sites that led the Army to request this study. There are also numerous protocols for using individual cleanup technologies, such as pump-and-treat, soil vapor extraction, or monitored natural attenuation, as well as for using engineering controls (such as containment) and institutional controls.

Protocols for remediation of DNAPL sources are still under development and are a focus of a number of ongoing studies. For example, the Texas Risk Reduction Program's NAPL management decision process, on its eighth draft, will be useful for remedy selection. The U.S. Environmental Protection Agency (EPA) recently sponsored a white paper outlining key issues for DNAPL cleanup, authored by an independent panel of scientists (EPA, 2003). The Air Force Center for Environmental Excellence and the Strategic Environmental Research and Development Program (SERDP) of the Department of Defense (DoD) are both supporting research into decision tools for source remediation, notably the SERDP project on Decision Support Systems to Evaluate the Effectiveness and Cost of Source Zone Treatment (Newell, 2003). A review of these ongoing efforts reveals that flexibility to incorporate the diverse conditions that exist at individual sites (natural and anthropogenic) is a common theme in many of the protocols. Furthermore, rigorous predictions of the performance of source remediation technologies are not available, such that the predictive tools under development provide only a general sense of how things work and an ability to make relative comparisons between options (e.g., the LNAPL Dissolution and Transport Screening Tool— Huntley and Beckett, 2002). Finally, none of the current protocols under development outline what to expect in a given setting, from a given technology, and with a specific contaminant. These limitations, as well as the lengthy examination of the necessary elements of a natural attenuation protocol described in NRC (2000), were kept in mind as the committee developed the elements described below.

The decision protocol for source remediation takes the form of a six-step process (Figure 6-1) that includes activities (white boxes), data and information collection (gray boxes), and decision points (gray diamonds). The steps are pre-

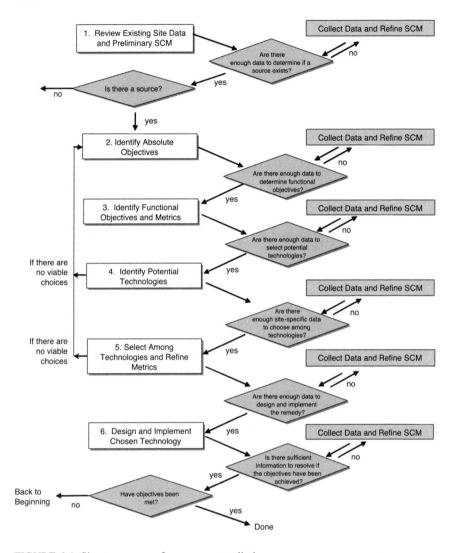

FIGURE 6-1  Six-step process for source remediation.

sented as sequential; in actuality, however, there may be multiple iterations of
each step until a decision can be made to proceed to the next step. As an example,
the limits of what can be achieved with proven technologies may require a revisit-
ing of the objectives. One of the distinguishing features of Figure 6-1 is its focus
on identifying absolute and functional objectives that are clearly articulated and
verifiable (Steps 2 and 3). The attention given to these steps is a reflection of the

observation that the absence of focused objectives can result in solutions that fall short of both expectations and needs.

A second theme identified in Figure 6-1 is managing data gaps and uncertainty (as exemplified by "collect data and refine site conceptual model" shown in gray boxes). As discussed in Chapters 2 and 3, geologic complexities, the subsurface behavior of contaminants, and historical changes in land use contribute to the considerable uncertainty that exists at almost every hazardous waste site with respect to the location and extent of contamination. Thus, the protocol focuses users on recognizing the limitations of their current understanding, on the importance of continually collecting the necessary information to effectively make decisions, and on managing plausible variation from perceived conditions. Although the iteration of data collection activities shown throughout Figure 6-1 may seem administratively burdensome, complex, and time consuming, the committee cannot envision source remediation occurring successfully in its absence. Indeed, the committee's experience at many dozens of sites is that not making this iteration explicit in the remedy selection process merely adds to transaction costs, as remedies are found to be unsatisfactory to major stakeholders, Records of Decision (RODs) are reopened, new contractors are hired, new studies are conducted, and new remedies are chosen. Such a mode of operation must be more transactionally complex than relying on an iterative approach from the start. A primary goal of this report is to portray iterative site characterization as essential to the success of source remediation, and thus increase its acceptability to regulators and the regulated community.

Although one could potentially limit use of Figure 6-1 to only the most complex DNAPL sites, it was designed to encompass all sites with source zones, regardless of their complexity. The potential technologies considered in Step 4 can include all those discussed in Chapter 5, including presumptive remedies like excavation and physical or hydraulic containment. While the protocol was intended to be applicable to all DNAPL and explosives source sites, this does not mean that aggressive remediation has to occur at every site. Indeed, the degree of complication posed by the protocol is dependent on the desired objectives and the feasibility of finding technologies that can meet them. For example, at the thousands of contaminated dry cleaner sites across the United States, going through the elements of Figure 6-1 could be a one-day exercise, and it could be consistent with state programs that list presumptive remedies. Thus, although the questions in Figure 6-1 should be asked at every site, the amount of effort spent on each step will vary by orders of magnitude depending on the site conditions.

It is important to note that the suggested elements for a source protocol shown in Figure 6-1 are consistent with relevant CERCLA regulations and guidance documents. The suggestions presented herein reflect shifts in emphasis from and logical refinement of currently used approaches, as opposed to wholesale modifications of them. As a general matter, the National Contingency Plan (NCP) contemplates that the development of remedial objectives, the collection and analysis of pertinent data, and the selection of an appropriate remedy will be accomplished in the context of a remedial investigation/feasibility study (RI/FS). Thus, most of the steps called for in Figure 6-1 should (and permissibly may) take place in that phase of the CERCLA process. At the same time, however, Step 6 in Figure 6-1 is likely to be carried out during the remedial design/remedial action (RD/RA) stage of a CERCLA-based cleanup. Under the NCP, all RD/RA activities are generally expected to conform with the remedy set forth in the Record of Decision (ROD). Nonetheless, the NCP very clearly allows for flexibility during the RD/RA, with respect to important details of remedial activities at sites. Thus, where a remedial action "differs significantly from the remedy selected in the ROD with respect to scope, performance or cost," the lead agency must publish an explanation of all significant differences that do not "fundamentally alter" the remedy previously described in the ROD. In contrast, where differences in an RD/RA stage remedial action will fundamentally alter the basic features of an earlier-selected remedy, the lead agency is required to follow a prescribed procedure for amending the ROD. These NCP provisions clearly allow for the continuing analysis and supplementary data collection, referred to in Figure 6-1, that will take place during the RD/RA phase of source remediation. The iterative nature of source characterization shown in Figure 6-1 is consistent with current EPA guidance on site characterization (EPA, 2004; http://clu-in.com/download/char/dynwkpln.pdf) and with the Triad approach (EPA, 2001) to environmental data collection.

Not all elements of cleanup (particularly those not specific to source remediation) are discussed in this chapter. For example, how to conduct long-term monitoring or the need for contingency plans are not discussed. This should not be interpreted as suggesting that certain activities be omitted.

Although not explicitly shown in Figure 6-1, early and ongoing involvement of potentially affected parties is a necessary part of source remediation to gain consensus on appropriate actions. Without adequate participation, critical elements of solutions may be missed, a subset of the involved parties may feel that their needs have been ignored, and/or false expectations may develop as to what can be achieved. Furthermore, remedial actions may be viewed as being less aggressive and less costly alternatives that offer advantages to responsible parties without fully protecting human health and the environment. Another concern is that the presence of a source area may affect community property values. For these reasons, a comprehensive source remediation protocol needs to reflect ongoing input from the affected community at all stages. In some cases, no nearby residential or commercial community per se may exist, so flexibility on this issue is warranted.

The following sections describe each of the steps identified in Figure 6-1.

## REVIEW EXISTING SITE DATA AND
## PRELIMINARY SITE CONCEPTUAL MODEL

Whether investigating a new suspected source or a known subsurface source, the first step is always to review existing site data. In the case of a potential new source, historical land use, waste management practices, regional geologic reports, and aerial photos all provide critical initial clues as to the nature and extent of sources. Chapter 3 discusses in greater depth the characteristics of subsurface sites that should be measured at this stage and the necessary tools. The most common scenario will be that the suspected source has been investigated previously and that release-specific documentation including a site conceptual model exists, which should be carefully reviewed.

During the review of existing site data, a key question to ask is: "Are there enough data to determine if a source exists?" If this question cannot be answered in the affirmative, additional data should be collected. If enough data do exist, then it is appropriate to ask *whether* there is a high probability that a source exists. The intent of this question is to ensure that source remediation does not proceed in the absence of a reasonable basis for expecting a source to be present (as has been observed at some sites).

Building on the above, guidances on data mining, on resolving the existence of a source, and on comprehensive source characterization are critical supplements to the source remediation protocol. This could include, for example, guidance on how much information is needed to understand the hydrogeology controlling contaminant transport and fate at the site, criteria for delineating the source zone, tabulations of common data sources, suggestions for the development of preliminary site conceptual models, lists of common exposure scenarios to consider, and descriptions of risk assessment methodologies.

## IDENTIFY ABSOLUTE OBJECTIVES

Identifying absolute objectives is specifically denoted in Figure 6-1 because in many cases examined by the committee, it was not clear that this was done properly or that it was done at all. Although shown in Figure 6-1 as Step 2 because it is critical at that point in the framework, identifying objectives can begin as soon as a site is labeled as a potential concern.

The development of remedial objectives for a site is inherently a social valuation process, to which stakeholders will bring differing (and perhaps irreconcilable) points of view (Presidential/Congressional Commission on Risk Assessment and Management, 1997). A key requirement in this process is to differentiate between absolute objectives (which are not substitutable) and functional objectives (substitutable alternatives to meeting those absolute objectives). Making this distinction requires careful communication during stakeholder discussions. This is especially important where a particular objective, such as attaining maximum contaminant levels (MCLs) at a particular point in time and space, may be a functional objective for one stakeholder (i.e., the stakeholder's real objective can be obtained in another way) but an absolute objective for another stakeholder. During these discussions, it must be kept in mind that some stakeholders may have already begun to translate absolute objectives into particular functional objectives and may not even be aware of having done so. In addition, many stakeholder objectives may be independent of the conclusion that there is a source on the site, while others will correspond directly to the identification of a source contributing to ongoing contamination of the groundwater flowing under the site.

It is not the purpose of stakeholder discussions to necessarily achieve consensus on absolute objectives or even necessarily on a common list of absolute objectives for site cleanup. Rather, the process is intended to ensure that the objectives are explicit and to remove confusion regarding the status of objectives for different stakeholders. For example, it needs to be made clear how certain absolute objectives (e.g., protecting human health) will be defined operationally and how progress will be measured. Other sources of common confusion involve the relevant temporal and spatial scales for certain objectives (e.g., calls to protect the environment must be accompanied by some notion of how much to protect and for how long).

Another issue to be clarified as part of the process of setting absolute objectives is the meaning of partial success. For example, if an EPA objective is to get a ROD signed by the end of the fiscal year, and it turns out that this cannot be done, will it be considered a partial success if the ROD is signed in November or the following August? Does it matter if it is done in the next fiscal year, rather than taking five more years? As another example, is opening a river to catch-and-release fishing (but not to catch-and-eat) valued as a partial success, or does only total removal of restrictions on fishing represent meeting the goal? Finally, to the extent possible, trade-offs among objectives for each stakeholder need to be clarified to the extent possible. Typically, those responsible for cleanup costs will have very different trade-offs of performance vs. cost than will those without such responsibility.

In an attempt to illustrate the principles of developing absolute objectives (including the conflicting objectives that often are voiced at initial stakeholder meetings), a theoretical case study is presented in Box 6-1. The point of the case study is to show that the identification and selection of absolute objectives is a multistakeholder exercise that often reveals conflicting as well as similar desires.

## IDENTIFY FUNCTIONAL OBJECTIVES AND PERFORMANCE METRICS

As discussed in Chapter 4, the relationship between absolute and functional objectives is not always simple. One functional objective may serve multiple absolute objectives. For example, achieving a particular contaminant concentration at

## BOX 6-1
## Development of Absolute Objectives:
## Hypothetical Example of Fort Alpha

Fort Alpha, a hypothetical vehicle maintenance facility, began operation in 1943 and ceased active operations in 2001. The primary cause of concern is the engine maintenance center, which used large amounts of chlorinated solvents as degreasers. An estimated 100 55-gallon drums were disposed of in an open-pit during the 1960s. The contamination is known to be a DNAPL consisting of chlorinated solvents and waste oil (by weight, 70% TCE, 10% 1,1,1-TCA, 5% PCE, 15% waste oil and grease). The Army intends to turn over the site, which is unoccupied, to local authorities as a potential industrial park.

The site has two major hydrogeologic settings—6 m of fluvial alluvium overlying a fractured (weathered) shale. The alluvium (Type III) transitions from fine-grained sand ($k=10^{-12}$ m$^2$, $K=10^{-5}$ m/s) at the water table to coarse channel sands at the base ($k=10^{-10}$ m$^2$, $K=10^{-3}$ m/s) with sparse low permeability silt beds ($k=10^{-15}$ m$^2$, $K=10^{-8}$ m/s). The shale (Type V) has horizontal and vertical fractures associated with release of overburden pressure in marine shale, with a bulk permeability of $10^{-14}$ m$^2$ ($K=10^{-7}$ m/s). The fracture porosity in the shale is 0.001 and the matrix porosity is 0.3. Vertical hydraulic gradients between the alluvium and shale are negligible. The depth to groundwater is 2 m below ground surface (bgs), and the average horizontal hydraulic gradient in the alluvium is 0.001. From the top to the bottom of the alluvium, the apparent range of groundwater flow velocities is 1–100 m/yr.

Limited information is available on the DNAPL source zone, which is estimated to be 0.40 hectare (1 acre). The vertical extent of DNAPL source zone is based on soil samples taken 2 m bgs (at the water table) and 8 m bgs (2 m into the shale). There is a plume of contaminated groundwater extending downgradient from the disposal area that contains not only parent chlorinated hydrocarbons, but also degradation products like DCE and vinyl chloride—all of which exceed standards by orders of magnitude 30 m downgradient of the site. Downgradient from the site is a residential neighborhood, Duke Estates, home primarily to persons formerly employed at Fort Alpha and their families. After passing beneath the residential neighborhood, the plume discharges to Halftrack Memorial State Wetlands Reserve 500 m from the source. At this location, TCE concentrations exceed drinking water standards by less than 1 order of magnitude. Absent the influence of the site, the aquifer would be Class I under state groundwater regulations, suitable as a source of drinking water. There is a well field 5 km upgradient of the site serving the regional water company, but no exceedances of standards for site-related contaminants have been observed in this location.

The various stakeholders at the site have come to the first meeting with vastly different priorities for site cleanup, as listed below. This list is not intended to be complete, and other stakeholders with additional interests could be identified as well.

**Army Remedial Project Manager (RPM)** – Prime responsibility is to terminate Army responsibility for the site. This includes providing for protection of public

*continued*

## BOX 6-1 Continued

health (i.e., no complete exposure pathways to the population at unacceptable concentrations) and protection of the environment (i.e., no demonstrable impact on endangered or threatened species). Cleanup must meet Army standards for transfer to civilian authority. Also responsible for overall costs of cleanup, within context of an annual budgeting process.

**State Environmental Agency Site Manager** – Prime responsibility is to ensure closure/transfer consistent with state law. Groundwater classification as Class I (suitable for drinking water) is key driver. Thus, per existing state regulations, there is no alternative other than to meet MCLs at all aquifer locations, including source areas. There is flexibility only in the time to achieve the objective.

**U.S. EPA Site Manager** – Responsible for consistency with the nine Superfund criteria and Applicable or Relevant and Appropriate Requirements (ARARs). For the protection of public health, there can be no complete exposure pathways to the population at unacceptable concentrations. For protection of the environment, there must be no demonstrable impact on endangered or threatened species; EPA also wants to satisfy natural resources trustee. With respect to ARARs, MCLs are not applicable because contaminant concentrations are not exceeded in drinking water supply. The Superfund Amendments and Reauthorization Act (SARA) dictates a preference for source removal, but this is not a decision driver.

**State Fish and Wildlife District Officer** – Natural Resource Trustee responsible for the wetlands. Protection of the environment is the primary concern, including the reversal of existing adverse effects on wetlands ecosystem and prevention of recurrence. Here, the agency's criterion of protection embodies permanent removal of threat.

**County Commissioners** – Concerned about continuing tax base from active site use, avoidance of incurring liability by acquiring site, and employment opportunities. In particular, if transfer requires more than three years, the economic viability of site will be endangered. The commissioners defer to others on criteria for protection of health and environment, meaning that both the electorate in the county and relevant regulators must accept.

**Jane Doe** (Nearby Resident) – Concerned about protection of family health. She lacks confidence in active remedies; her concern remains as long as contamination in groundwater exceeds MCLs. Not convinced that use of public water supply removes risk. She is also concerned about property values and wants to be able to sell her house for same price as "comps" in nonaffected parts of community.

**T. P. Jones** (Nearby Farmer) – Concerned about use of aquifer water on farm in the future; potential impacts on certification of produce as "organic." As long as contaminants are identified in water, his concern remains. He is unable to run the business he wants, or pass on the farm to his grandchildren as a viable enterprise (in his terms).

*continued*

## BOX 6-1 Continued

At the first stakeholders meeting following the identification of a source area at Fort Alpha, a long list of absolute objectives was put forward in discussions among the stakeholders, touching on each stakeholder's concerns. These included clean-up to background, attainment of MCLs, and prevention of contaminant intrusion into the wetlands, as well as more globally expressed objectives such as protection of human health. Some of the confusion noted in Chapter 4 is experienced during the stakeholder meetings. For example, it is not immediately clear if the call for meeting MCLs is an absolute or functional objective or if it is viewed differently by the federal and state EPA representatives. It is not clear how to measure success of the objective of "giving our grandchildren a clean environment." Nor are the time elements associated with the absolute objectives clarified.

Through further discussion, the following clarifications were made that allowed the various objectives to be redefined in the following way:

**Meeting MCLs.** This is an absolute goal only for the state site manager, as the other stakeholders have used it as a metric for public health protection. The EPA site manager is able to determine that it is not "appropriate" under SARA. The state site manager may have flexibility regarding when it is met in order to approve a remedial decision.

**No contaminant intrusion into wetlands.** This turns out to be both unnecessary in the short term because the Trustee can accept intrusion that does not exceed the "capacity" of a healthy ecosystem. Over the long term, the objective is inadequate because the Trustee cannot consider the wetland protected as long as there is a source. Accordingly, this is not included in the list of absolute objectives for the site.

**Having the Army clean up the mess it made.** This objective is determined to be equivalent to removal of the source by some stakeholders and to protecting human health and the environment by others. In either case, it is redundant and is not included in the final list of absolute objectives.

**Buy-out of homes sitting over the middle of the plume.** This is viewed by some stakeholders as one of many ways to remedy economic damage due to presence of groundwater contamination. Similarly, it is one of many ways to prevent adverse effects on health and is agreed to be deferred to the consideration of functional objectives.

Stakeholder discussions led to the following list of noneconomic absolute objectives, with the understanding that not every objective is absolute for each stakeholder, that there remains considerable overlap, and that not every stakeholder's objectives are viewed as attainable by the others:

• Restoration to background conditions by achieving MCLs throughout the aquifer (at least at some point in the future)—and demonstrable progress toward this

*continued*

**BOX 6-1 Continued**

- No complete exposure pathways to the population at unacceptable concentrations
- No demonstrable impact on endangered or threatened species
- Prevention of additional contaminant intrusion into wetlands, and repair of damage already noted to this ecosystem
- Transfer/redevelopment of site within three years

Functional objectives that address some these absolute objectives can be defined and are the subject of Box 6-2. Any given functional objective may address a larger or smaller set of these absolute objectives.

a point in time and space may serve to protect human and environmental receptors, to meet an absolute statutory requirement, and to meet the programmatic requirements of an agency. Similarly, a given absolute objective may be capable of translation into alternative functional objectives. For example, prevention of human exposure to site contaminants can be achieved by removing the contaminants or by removing the people from a given area.

The difficult part of setting functional objectives, and identifying alternative functional objectives, is generally the integration of multiple absolute objectives. When a given functional objective (e.g., preventing migration of contaminants off-site) meets multiple absolute objectives (e.g., protecting human health and preventing damage to identified environmental receptors, by interrupting the exposure pathway), alternative functional objectives may be fungible with regard to one absolute objective but not with regard to others. For example, a buy-out of neighbors could protect human health, but not environmental receptors.

In specifying functional objectives, it is important to continue elaborating until a metric can be identified by which it will be possible to measure progress toward achieving the functional objective and, hence, the absolute objective. For example, in the case of a property buy-out that was noted above as a way to interrupt an exposure pathway by moving receptors away from the contaminant, one might need to specify the secondary functional objectives of (1) moving current residents out and (2) preventing anyone else from moving in. The first of these could have as metrics such things as signed purchase contracts for each property and documented vacating of the property. The second might be more complex, involving factors such as destruction of residential structures, backfilling of wells, fencing the property, posting warning notices, establishing deed restrictions applicable to any future transfers of the property, and establishing a procedure for ensuring that these land-use controls remain in effect.

This last point illustrates an aspect of functional objectives that is often given insufficient attention—how well the functional objectives are matched to the temporal dimension of the absolute objectives they are intended to serve. If the characterization of a site suggests that conditions will persist for an extended period (predictions for some sites have extended to many centuries), then either the functional objectives must be stated so as to address that timeframe or it must be made clear to all stakeholders that the absolute objective is only being addressed within a limited period.

As shown in Figure 6-1, prior to determining functional objectives, it should be asked whether there are enough data to make such a determination. If not, then additional data collection and analysis are warranted.

The translation of absolute objectives into functional objectives is illustrated for the hypothetical Fort Alpha hazardous waste site in Box 6-2.

---

## BOX 6-2
## Translating Absolute Objectives into
## Functional Objectives at Fort Alpha

*Functional Objectives to Achieve Interruption of the Human and Environmental Exposure Pathways*

Most or all of the noneconomic absolute objectives for many of the stakeholders in Box 6-1 can be obtained by any functional objective that interrupts the pathway between the source and various receptors. For example, the absolute objective specified by some of the stakeholders was to protect the public from excessive concentrations of site contaminants by disrupting human exposure pathways (at present, this is only relevant to residents of Duke Estates). This can clearly be accomplished in several ways: moving the humans, decreasing the concentrations to which they might be exposed, interspersing barriers to contaminant transport, or preventing release of contaminants from a confined area.

Barriers to exposure could be implemented at any of several levels and locations, depending upon site conditions. At one extreme, residents could be supplied with personal protective equipment or (as has occurred elsewhere) have vapor strippers installed on their wells. However, there are no wells at present in Duke Estates, and public water supplies are in place.

For Fort Alpha, the barrier approach could address the limited human health risk, but it does not address other absolute objectives (like protecting the environment, specifically the wetlands). Alternative functional objectives, focused on reducing concentrations or transport of contaminants in groundwater, may be capable of addressing both absolute objectives. Any of a number of objectives can be specified for reducing concentrations in the exposure zone (or, alternatively, instituting a barrier to contaminant transport). These could include defining con-

*continued*

## BOX 6-2 Continued

centration limits in the exposure area, specifying a geographic boundary to the contaminant plume (e.g., the property fence line), or, in a more nuanced fashion, limiting the entrance of contaminants to the capacity of the ecosystem to absorb it. Each alternative definition will impose a different set of associated metrics and inference chains in order to establish that (1) the defined functional objective meets the absolute objective and (2) the measured quantity is in fact reflective of the functional objective.

Interestingly, previous cleanup efforts at Fort Alpha illustrate some of the potential problems in clearly specifying functional objectives and metrics. In an attempt to decrease contaminant concentrations under Duke Estates, a pump-and-treat system was installed, and it was carefully documented that several hundred kilograms of solvents were recovered. However, subsequent investigations showed no reductions in plume concentrations under the houses or in the wetlands, the ostensible goals of the technology. Thus, the chain of inference between metric and objective, however reasonable a priori, apparently involved some inappropriate assumptions for the site. The metric of mass removal was deemed to be inappropriate for documenting concentration reduction in this case.

### *Functional Objectives to Achieve Source Elimination*

As can be seen from Box 6-1, only complete elimination of the source area will meet the absolute objectives of Mr. Jones, of the Natural Resources Trustee, and of the state site manager. There are likely significant differences in the expected time to elimination required to fulfill their absolute objectives (with Mr. Jones having the more immediate need). Clearly, functional objectives that address concentrations of contaminants in Duke Estates or the wetlands are not relevant to this absolute objective.

As in the case of interrupting exposure pathways, there are multiple alternatives for specifying functional objectives and metrics in support of the objective of eliminating site contamination. One alternative is to focus on mass removal from the source. A straightforward metric is mass of contaminant removed. It must be remembered, however, that it may be very difficult to relate this objective/metric to other objectives, such as the time during which source material will remain in the subsurface.

An alternative functional objective could be defined in terms of source strength—that is, the ability of the source to continue to contribute to groundwater contamination. This, in turn, could be further specified in terms of downgradient concentrations, some measure of mass flux, or as performance in a series of extraction tests. As in the case of interrupting exposure pathways, it is important that the connection between the stated absolute objective and the actual measurements is well established.

## IDENTIFY POTENTIAL TECHNOLOGIES

Once functional objectives have been decided on, the next issue is to identify technologies that are viable means of achieving those objectives (Step 4 in Figure 6-1). As described in Chapter 5, the potential efficacy of a given technology is dependent on the contaminant type, the hydrogeologic setting at the site, and the chosen functional objectives. This suggests that a multidimensional screening approach can be used to identify promising technologies. This is represented by the cube shown in Figure 6-2, which has independent variables of hydrogeologic setting, functional objectives, and source remediation technologies. The details on these three factors are presented in Chapters 2, 4, and 5, respectively. Obviously, there must be sufficient information on these three factors to enable use of the cube. Otherwise, further data collection and analysis should be undertaken prior to using the cube, as shown in Figure 6-1. The dependent output of the cube is a "potential for success" rating of high, medium, low, or unknown. These

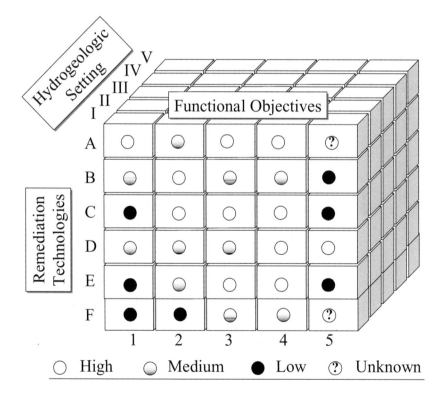

FIGURE 6-2 Multidimensional screening cube for identifying candidate technologies based on their potential for success (high, medium, low, unknown) relative to specific functional objectives and in different hydrogeologic settings.

320                                            *CONTAMINANTS IN THE SUBSURFACE*

ratings could correspond to those given in Table 5-7, such that the entries to that table can be considered as elements of the cube, if the physical objectives used to create Table 5-7—mass removal, concentration reduction, mass flux reduction, reduction of source migration potential, and change in toxicity—are relevant. If not, then other functional objectives can be used. No matter what the chosen objectives on the cube axis are, the cube entries are not necessarily site-specific, but rather are gleaned from the literature, from other case studies, etc.

Use of the cube begins by considering the source contaminant or contaminants. As conceptualized in Figure 6-3 (and as noted in the second column of Table 5-7), contaminant type can constrain the list of applicable technologies. For example, soil vapor extraction is a technology best applied to volatile compounds and not to semivolatile or nonvolatile compounds.

Next, the physical setting is imposed. The efficacy of specific technologies is often highly dependent on the environmental setting. For example, many flushing technologies (e.g., surfactant-enhanced recovery) are not applicable to low-permeability media due to slow rates of fluid delivery and recovery. The output envisioned in Figure 6-4 describes the potential of those technologies appropriate for a given contaminant and in a given setting to achieve different functional objectives. The example in the figure indicates that for both technologies, objective 5 would be unlikely to be achieved, while objective 3 is far more likely to be achieved. What can be achieved provides a rational point for resolving what should be undertaken. In the event that none of the applicable technologies can

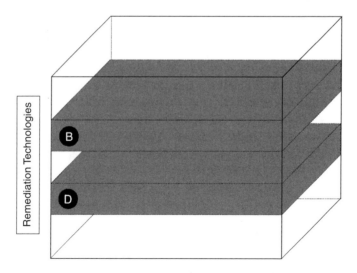

FIGURE 6-3 Constraining applicable technologies by contaminant type. The grey slabs indicate those technologies that are appropriate for the contaminant of concern. The technologies indicated by B and D were chosen for illustrative purposes only.

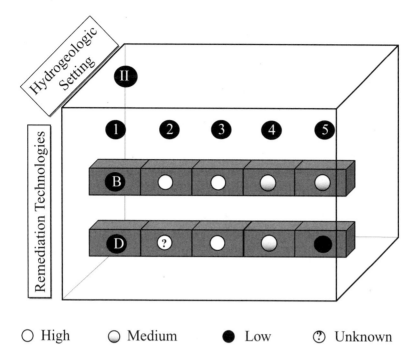

○ High     ○ Medium     ● Low     ⑦ Unknown

FIGURE 6-4 Further narrowing the possible remedial choices by considering the physical setting.

achieve a desired objective, then the user must cycle back to Step 2, as indicated in Figure 6-1.

It is important to understand the scope of this exercise. The cube is intended only to identify those technologies that could be used to achieve certain objectives in certain hydrogeologic settings, and thus it is a screening tool rather than a definitive solution. Furthermore, the limitations and caveats associated with the data in Table 5-7 must be kept in mind when using the cube. Not all of the table entries are based on published performance data. If a technology that is not well documented in the scientific and engineering literature is chosen, then the primary functional objective may need to shift to demonstrating an innovative technology as opposed to achieving specific objectives for the site itself. Ideally, only technologies that are rated "high" should be carried further through the evaluation process. As a fallback position, given an inability to achieve a desired outcome, technologies that are rated "medium" can be considered with the recognition that the likelihood of success is less certain.

Returning to our hypothetical case study, an example of using the cube is presented in Box 6-3.

**BOX 6-3**
**Using the Multidimensional Screening Cube at Fort Alpha**

Box 6-1 describes contamination at Fort Alpha as being a chlorinated solvent DNAPL and the hydrogeologic setting as a combination of Type III and Type V. Taking these factors into consideration, the following table was developed from Table 5-7 and screened for options that include a score of "low-high," "medium-high," or "high" for Type III and V settings. The table is equivalent to the plane of the cube shown in Figure 6-4 (although the exact table entries do not correspond with circles in Figure 6-4).

A critical feature of the table is that there appear to be no applicable technologies for achieving any of the five physical objectives for the fractured shale other than physical containment. Due to the adjacency of the shale to the alluvium, not treating the shale will diminish the likelihood of achieving the noted local aqueous concentration reduction and/or mass flux reduction in overlying adjacent alluvium. One notable exception to this is the soil mixing/chemical reduction technology. In this case, clay is usually used with the iron to aid in delivering the iron and reducing the horsepower required to mix the soil. Use of the soil mixing/chemical reduction

| Technology | Media Settings |
| --- | --- |
| Physical Containment | III |
| | V |
| Hydraulic containment (Including Pump-and-Treat) | III |
| Excavation | III |
| Surfactant/Cosolvent Flushing | III |
| Chemical Oxidation | III |
| Soil Mixing/Chemical Reduction | III |
| Steam Flushing | III |
| Conductive Heating (ISTD) | III |
| Electrical Resistance Heating | III |

NOTE: only those table entries that were low-high, medium-high, or high are included in the table in order to highlight the technologies that might be moved forward into Step 5.

technology might cap the contaminated shale with low-permeability reactive media that could prevent back diffusion of contaminants from the shale into the alluvium.

Next, the functional objectives identified in Box 6-2 are compared to what can be achieved. None of the identified options will "eliminate" the source because at a minimum, some source material will be left in the fractured shale. This objective may need to be eliminated or modified to reflect mass removal to the extent practical using best available technology. The objective of interrupting human exposure pathways was discussed in Box 6-2 as being achievable in several ways: moving the humans, decreasing the concentrations to which they might be exposed, interspersing barriers to contaminant transport, or preventing release of contaminants from a confined area. While none of the objectives used in the cube and Table 5-7 precisely match the two absolute objectives at Fort Alpha, mass removal and mass flux reduction were interpreted to be closest to source elimination and elimination of the human exposure pathway, respectively. At least for the portion of the subsurface that is Type III, this narrows the list of technologies to excavation, surfactant flushing, conductive heating, and chemical reduction.

| | | Likely Effectiveness at Appropriate Sites | | |
| Mass Removal | Local Aqueous Concentration Reduction | Mass Flux Reduction | Reduction of Source Migration Potential | Change in Toxicity |
| --- | --- | --- | --- | --- |
| | | High | High | |
| | | Low-High | Low-High | |
| | | High | | |
| High | High | High | High | |
| Medium-High | | Medium-High | High | |
| | | Medium-High | | |
| High | High | High | High | Medium-High |
| | | Medium-High | | Medium-High |
| Medium-High | Medium-High | High | High | Medium-High |
| | | Medium-High | | Medium-High |

## SELECT AMONG TECHNOLOGIES AND REFINE METRICS

Step 4 of the protocol may have revealed three or four different remedies or combinations of remedies that have the potential to succeed in meeting the functional objectives at a given site. For example, one option might be containment coupled with institutional controls, while another might be enhanced bioremediation followed by monitored natural attenuation. Given the suite of promising technologies revealed by the cube, the next step is to conduct site-specific evaluations of these options so that an informed decision on which technology to choose can be made. This is essentially a data collection, data analysis, and modeling exercise to better determine whether the various technologies will work at a given site. In addition to technical information, site-specific evaluations commonly involve estimates of cost, time to complete, endpoints, vendor performance, and side effects (e.g., noise, air emissions, or vehicle traffic). Depending on the scale of the exercise, this may be akin to CERCLA Feasibility Studies and RCRA Corrective Measure Studies.

A potential pitfall at this point is for the "best available technology" to be chosen without rigorously determining whether it will work at the site or questioning whether the absolute objectives will be addressed. Decisions of this nature typically reflect a desire to do the "best thing." Unfortunately, actions that are not tied closely to functional objectives often fall short of expectation and/or needs. This can lead to the all-too-common scenario in which multiple technologies are sequentially deployed with no sense of meaningful progress toward closure. Thus, the site-specific data collection exercise should be considered an investment of time and resources that can prevent future problems associated with a faulty or inappropriate technology selection.

Once site-specific data on the technologies revealed by the cube have been collected and analyzed, a decision has to be made about which remedy to pursue (assuming that more than one possibility exists after Step 4). One approach to reaching this decision in a systematic way is by constructing a matrix of objectives vs. candidate technologies, as shown in Table 6-1. Technologies that were identified using the cube are listed in the vertical column, and functional objectives are listed across the top of the matrix. Each intersection represents the ability of a particular technology to meet a particular objective. The entries used in the matrix should be informed primarily by the site-specific evaluation exercise discussed above. All of the entries in a given row should be considered to give a total rating for each technology (although how this should be done—that is, qualitatively, via simple addition, or using weighted combination rules—is not specified here). The two or three highest-rated technologies are considered the leading candidates for the remedial action. Creation and use of this matrix should involve all stakeholders and it should be well documented.

The listed objectives in Table 6-1 should be those used in the cube as well as others that could not be meaningfully included in the cube but which are refined

TABLE 6-1  Source Remediation Alternatives Analysis Matrix

| Technical Alternatives | Functional Objectives | | | | | Rating |
|---|---|---|---|---|---|---|
| | *1* | *2* | *3* | *4* | *5* | |
| A | | | | | | |
| B | | | | | | |
| C | | | | | | |
| D | | | | | | |
| E | | | | | | |

to the site-specific situation and/or are more practical. These might include minimizing life cycle costs, complying with regulations, maintaining positive relationships, eliminating liability, and ensuring worker safety. If a functional objective is found to have similar ratings for all remedial actions under consideration, then it is likely to have little impact on the decision to be made and might be removed from the matrix. Box 6-4 illustrates the use of such a site-specific matrix for the Fort Alpha site, using qualitative (low, medium, high), narrative ratings rather than numerical scores.

One of the concerns that could be taken under consideration when using Table 6-1 is how robust technologies are during implementation. Robustness refers to the ability of a technology to operate properly despite unfavorable or unexpected conditions. This requires understanding many complicating factors, such as the potential impacts of site heterogeneity, mixtures of contaminants, the potential formation of transformation byproducts, variability with climate and/or groundwater temperature, and the potential adverse effects of other remedial actions or land-use changes.

As for any complicated technical issue, making and implementing good decisions about source remediation requires a considerable amount of training, experience, and expertise. These are also essential to ensure that the protocol elements recommended in this report are implemented properly. To maximize success, it is important that the chosen vendors have documented experience using the specific technologies being considered, ideally in hydrogeologic settings similar to the site being studied. Qualifications for the major stakeholder groups are described in more detail in Box 6-5.

One possible outcome of Step 6 is that none of the specific technologies will remain viable, given the objectives of Table 6-1 and site-specific conditions. In this case, it is necessary to revisit the absolute and functional objectives (Steps 2 and 3) and subsequently develop new options. As a general matter, EPA regulations allow for changes in remedial objectives, although this requires that site managers engage the relevant stakeholders to ensure that the new objective is

## BOX 6-4
## Site-Specific Analysis of Source Remediation Alternatives at Fort Alpha

Four technologies were brought forward for treating contamination in the overburden at Fort Alpha, while one (containment) was deemed to be potentially applicable for the fractured shale bedrock. In addition to the functional objectives of mass removal and mass flux reduction chosen for the site, the Army also desires to have a safe working environment, to be in compliance with all regulations, to have the community accept the remedial action, and to minimize the life cycle cost of the remedy. All of these objectives appear in the matrix below, which is filled in

| Technical Alternatives | Objectives | | |
|---|---|---|---|
| | *Mass Removal* | *Mass Flux Reduction* | *Life Cycle Cost Minimization* |
| Excavation | *High* | *High* | *Medium* |
| | All mass should be removed if excavation is properly designed | Flux out of source zone should be 0. Flux from downgradient sorbed material will continue | |
| Surfactant/ Cosolvent Flushing | *Medium-High* | *Medium-High* | *Low* |
| | Removal probably 80% to 90% | Flux reduction depends on effectiveness of sweeping low-perm units | Depends on ability to recycle surfactant |
| Soil Mixing/ Chemical Reduction | *High* | *High* | *Medium* |
| | Chlorinated solvents will react quickly and be destroyed | Permeability will be lowered dramatically, flux also | |
| Conductive Heating (ITSD) | *Medium-High* | *High* | *Low* |
| | Depends on water content | Expected to be good, depends on extent of removal in clay zones | |
| Hydraulic Containment | *Low* | *Low* | *Medium-High* |
| | Mass removal only from plume | Flux reduction depends on the degree of plume capture | |

with site-specific data and information. Note that the table entries are strictly hypothetical and were created using the information gleaned from Table 5-7 and the committee's best professional judgment.

A qualitative reading of this matrix leads the Army and its advisors to believe that excavation and soil mixing/chemical reduction appear to be most likely to meet the Army's list of objectives. Once all stakeholders agree to this evaluation, those two technologies should then be carried further through a highly detailed feasibility analysis. A laboratory evaluation and a field pilot of the soil mixing/chemical reduction technology should seriously be considered if that technology is likely to be the Army's final selection. Because excavation is a mature technology, there would be no need for a field pilot project if it were the final selection.

| Worker Safety | Community Acceptance | Regulatory Compliance |
|---|---|---|
| Medium | High | High |
| Potential exposure during excavation and transportation | Highly visible, may be concerns about transportation, perceived as final | Reduces mobility, toxicity, volume |
| High | Medium | High-Medium |
| Little contact with media, separation step might expose | Community may not understand the technology | Reduces mass, volume; low impact on toxicity |
| Medium | High | High |
| Standard risks of large construction equipment | Highly visible, perceived as final treatment | Rapidly reduces mass, toxicity, volume |
| Medium | Medium | Medium |
| High current flows, collection and treatment of vapor pose an exposure hazard | Community may not understand the technology | Reduces volume, possibly mobility; does not change toxicity |
| High | Medium | Low |
| Some exposure during installation and operation | Highly visible, results are long-term | Controls mobility, little impact on volume, none on toxicity |

**BOX 6-5**
**The Need for Knowledge**

There is a substantial body of knowledge needed to conduct successful source remediation. The needed expertise differs depending on an individual's role—be it regulator, responsible party, remediation consultant, or community-based organization.

*Responsible Parties*

Responsible parties need to be able to evaluate the work of remediation consultants and make remediation decisions based on the consultant's work. These individuals should have sufficient technical background and experience to actively manage their consultants and to negotiate responsibly with environmental regulators. Responsible party representatives should ideally have a relevant technical degree and understand the fundamentals of risk evaluation and community involvement.

*Contractors and Consultants*

Remediation companies should have an expert in source remediation with a thorough understanding of the relevant science and engineering principles. Experience in related technologies and similar site conditions is invaluable.

*State and Federal Regulators*

State and federal regulators are charged with evaluating the merits of the various remediation proposals they receive and making judgments on whether source remediation proposals have sufficient technical justification. Because these regulators must be able to understand a wide variety of technical information, desirable skills include knowledge of environmental engineering, geology, chemistry, and biology principles, as well as how to use conceptual models, mathematical tools and models to estimate contaminant movement and degradation, and fundamentals of risk evaluation.

*Community-Based Organizations*

The members of communities affected by contaminated sites where source remediation is being considered should be included in the decision-making process as early as possible and should have the resources necessary to participate in this process. Potential training might include an introduction to contaminant behavior in the environment, basic risk assessment, a review of strengths and weaknesses of source remediation technologies, a review of how site conditions and contaminants affect source remediation, and a discussion of verifying the effectiveness of the various source remediation methods.

SOURCE: Adapted from NRC (2000)

acceptable and compatible with the expected future land use at the site. One option that is sometimes pursued at DNAPL sites in heterogeneous (Type III and V) media is a determination of Technologically Impracticability (TI). TI waivers, which are granted for groundwater contamination only, result in the selection of a less strict remedial objective (such as a higher allowable contaminant concentration). However, in order to be granted a TI waiver, the responsible party must provide their best estimation of a technically practicable alternative remedial strategy that will prevent migration of contamination beyond the source zone (see EPA, 1993, and NRC, 2003 for greater explanation of technical impracticability).

## DESIGN AND IMPLEMENT CHOSEN TECHNOLOGY

In large part, the design and implementation of chosen remedies follow standard engineering practice, accompanied by essential administrative elements such as qualified staff, sufficient funds, and appropriate schedules. In both design and construction, it is advisable to be as adaptive and flexible as possible, which is the premise of the observational approach developed by Karl Terzaghi (e.g., Bjerrum et al., 1960) and Ralph Peck (e.g., Peck, 1962, 1969, 1980). This suggests that remediation systems should be designed and built per known and predicted future conditions, while at the same time anticipating plausible variations in site conditions and having contingency plans in place. This is similar to the adaptive site management concept espoused in NRC (2003), which suggests that higher risk, less certain remedies be explored in parallel with the chosen remedy in the event that the remedy fails to achieve cleanup objectives.

It is important to ask prior to designing a remedy whether there are enough data to design and implement the remedy. Further information is frequently needed to develop specifications, verify the extent of the source zone, and/or optimize treatment processes, similar to the conducting of a treatability study under CERCLA. Under select circumstances, this activity may include extensive field investigations and/or implementation of a pilot system or "first module."

Once a remedy has been implemented, the monitoring data that accumulate will eventually be used to determine whether a remedy has performed as expected and whether objectives have been met. These monitoring data should provide the agreed upon scientific evidence needed to determine whether the remedy is meeting the objectives. Figure 6-1 highlights the question "is there sufficient information to resolve if the objectives have been achieved?" because this is a test of the adequacy of the monitoring system. If there is insufficient information after the remedy is allowed to operate for the prescribed period of time, then the monitoring program must be redesigned to allow the correct type of information to be collected. (This generally should not be a problem if practical performance metrics were identified in the previous step.) A second interpretation of the question involves how much time has passed since implementation of the remedy. Although it is highly specific to the chosen technology (and ranges from weeks to decades), sufficient time must have passed before the remedy's efficacy can be judged. For example, in the case of bioremediation, enough time must pass for the microorganisms to acclimate to changes in redox conditions and seasonal conditions (in temperature, for example). Aggressive source remediation strategies like in situ chemical oxidation produce changes within weeks. However, in most cases additional monitoring (on the order of months) is needed to determine the permanence of the result and the potential for rebound. Years might be required for hydraulic containment and air sparging.

If there is sufficient information, then the final step is to determine whether the absolute and functional objectives have been achieved. If the objectives have not been achieved, then it may be necessary to repeat the steps outlined in the above text and in Figure 6-1. The likelihood of this undesirable setback can be greatly reduced by employing a protocol that addresses the critical elements outlined herein.

* * *

As mentioned previously, the elements of a source remediation protocol described above should be applied to all current hazardous waste sites, regardless of their state of maturity. The amount of time spent at each step will be a reflection of the sufficiency of the information that has been acquired to date, such that those sites where source characterization has occurred commensurate with the chosen objectives and remedy can proceed rapidly through the protocol.

Documentation is not discussed in a separate section in this chapter. However, good documentation is an essential part of the decision-making and remedy selection process. The performance of a remedy can be properly evaluated only in relation to the objectives that were set before it was implemented. Likewise, the quality of decisions can be fairly evaluated in retrospect only if the context in which they were made is preserved. Therefore, each stage of discussions and each selection—whether of objectives or of technologies—should be documented along with the reasons that the selections were made. Establishing a central and complete record of the performance of the Army's source zone remedies vs. their objectives would be very helpful for understanding how source zone technologies perform in a wide range of settings, and it would be a good reference for Army program managers faced with new source zone remediation decisions. When a remedial action is evaluated, it would be desirable for its degree of success to be evaluated by a party with no stake in the outcome.

## CONCLUSIONS AND RECOMMENDATIONS

The six critical elements of a source remediation protocol are (1) review of existing site data, (2) identification of absolute objectives, (3) identification of functional objectives and metrics, (4) elucidation of potential technologies given site hydrogeologic and contaminant characteristics, (5) selection of an appropriate technology, and (6) design and implementation of the chosen technology. Site-specific data collection informs each step of the process and is used to refine the site conceptual model. If all of these steps are not included, source remediation at an individual site will have a low probability of success.

**The Army should develop and use a detailed protocol consistent with the elements prescribed in this chapter.** A protocol specific to source zones is needed to aid stakeholders in optimizing the benefits derived from investments in remediating source zones. The key attributes that need to be addressed are pursuing actions that effect intended changes, understanding the extent to which objectives are attainable, and being able to measure progress toward desired objectives. The protocol will need to be integrated into the existing remedy selection frameworks used by the Army at individual sites, including Superfund, RCRA, relevant state laws, or the Base Realignment and Closure program.

**Improved technology transfer and guidance on DNAPL source remediation technologies are desirable.** Guidance is needed to help responsible parties determine whether source remediation is appropriate for the sites under consideration and to aid in the selection of technical approaches that are most appropriate for the site-specific conditions and remedial action objectives. Army personnel should be thoroughly trained in the use of this guidance. Particular attention should be paid to justifying and documenting remedial decisions as they are made, establishing success metrics at the time of remedy selection, and explicitly documenting the degree of success or failure of remedies by using those preestablished metrics.

**Involvement of potentially affected parties is essential to the success of source remediation.** Stakeholder participation is needed to better understand the range of absolute objectives at a given site, to develop functional objectives, and to gain consensus on appropriate actions. Without adequate public participation, critical elements of solutions may be missed, a subset of the involved parties may feel that their needs have been ignored, and/or false expectations may develop as to what can be achieved. As for all relevant stakeholders, knowledge acquisition by the public is essential to making decisions about source remediation.

One of the goals of this study was to be able to make definitive statements about the future use of source removal as a cleanup strategy. An important conclusion that can be made from reviewing source zone remediation attempts to date is that the data are inadequate to determine how effective most technologies will be in anything except the simpler hydrogeologic settings. (Indeed, many entries in Table 5-7 were based on the committee's best professional judgment in lieu of having relevant data from field studies.) Furthermore, it is unlikely that available source remediation technologies will work in the most hydrogeologically complex settings such as karst.

**The committee believes that by following the elements of a source remediation protocol illustrated in Figure 6-1, project managers will be able to make critical decisions regarding whether and how to attempt source remediation and thereby accomplish a more beneficial distribution of resources.** It is evident from the review of source zone remediation projects at Army facilities and elsewhere that the steps presented in Figure 6-1—determining whether a source exists; developing clear absolute and functional objectives and their metrics; selecting, designing, and implementing a technology; and collecting data to support all these decisions—have seldom been conducted in the manner described in this report. The efforts of potentially responsible parties to date suggest that in some cases, source remediation technologies are being prematurely scaled up at poorly characterized sites, at sites where there is known complex hydrogeology, and at sites where there is no clear reason for proceeding with the project.

Finally, Chapter 5 suggests that several technologies show enough promise, in terms of demonstrated mass removal and concentration reduction in monitoring wells, to warrant further investigation to determine their long-term effects on water quality—especially if objectives other than MCLs, such as mass flux reduction, become more prevalent. For almost all of the technologies discussed, their effectiveness is more uncertain in the more complex hydrogeologic settings. Thus, future work should attempt to determine the full range of conditions under which these technologies can be successfully applied, and to better understand how mass removal via these technologies affects water quality.

## REFERENCES

Bjerrum, L., A. Casagrande, R. B. Peck, and A. W. Skempton (eds.). 1960. From theory to practice in soil mechanics: selections from the writings of Karel Terzaghi. New York: Wiley.
Environmental Protection Agency (EPA). 1993. Guidance for evaluating the technical impracticability of ground-water restoration. OSWER Dir. No. 9234.2-25. Washington, DC: EPA.
EPA. 2001. Using the Triad approach to improve the cost-effectiveness of hazardous waste site cleanups. EPA-542-R-01-016. Washington, DC: EPA.
EPA. 2003. The DNAPL Remediation Challenge: Is there a Case for Source Depletion? EPA 600/R-03/143. Washington, DC: EPA Office of Research and Development.
EPA. 2004. Improving Sampling, Analysis, and Data Management for Site Investigation and Cleanup. EPA 542-F-04-001a. Washington, DC: EPA Office of Solid Waste and Emergency Response.
Huntley, D., and G. D. Beckett. 2002. Evaluation of Hydrocarbon Removal from Source Zones and its Effect on Dissolved Plume Longevity and Concentration. Publication 4715. Washington, DC: American Petroleum Institute.
NRC. 1994. Alternatives for Ground Water Cleanup. Washington, DC: National Academy Press.
NRC. 1997. Innovations in Ground Water and Soil Cleanup: From Concept to Commercialization. Washington, DC: National Academy Press.
NRC. 1999. Groundwater Soil Cleanup. Washington, DC: National Academy Press.

NRC. 2000. Natural Attenuation for Groundwater Remediation. Washington, DC: National Academy Press.

NRC. 2003. Environmental Cleanup at Navy Facilities: Adaptive Site Management. Washington, DC. National Academies Press.

Newell, C. 2003. Presentation to the NRC Committee on Source Removal of Contaminants in the Subsurface. April 14, 2003. Washington, DC.

Peck, R. B. 1962. Art and Science in Subsurface Engineering. Geotechnique 12(1):60–66.

Peck, R. B. 1969. Advantages and limitations of the observational method in applied soil mechanics, Ninth Rankin Lecture. Geotechnique 19(2):171–187.

Peck, R. B. 1980. Where has all the judgment gone? The Fifth Laurits Bjerrum Memorial Lecture, Norwegian Geotechnical, Oslo, Norway.

Presidential/Congressional Commission on Risk Assessment and Risk Management. 1997. Framework for environmental health risk management. Volumes 1, 2. Washington, DC: U.S. Government Printing Office.

Wiedemeier, T. H., M. A. Swanson, D. E. Moutoux, E. Kinzie Gordon, J. T. Wilson, B. H. Wilson, D. H. Kampbell, J. E. Hansen, P. Haas, and F. H. Chapelle. 1996. Technical protocol for evaluating natural attenuation of chlorinated solvents in groundwater. Report prepared for Air Force Center for Environmental Excellence.

# Appendix A

# Tables on Contaminants at Army and Other Facilities

TABLE A-1 Contaminants Reported at the Most Severely Contaminated Army Sites

| Installation | Number of Sites | Contaminants | Media Affected | Source Remediation |
|---|---|---|---|---|
| Aberdeen Proving Ground, MD | 254 | VOCs, SVOCs, metals (Hg), PCBs, explosives, pesticides, UXO, radiation, chemical weapons/munitions, biological warfare material | Groundwater (GW), surface water (SW), Sediment (Sed), Soil | Removal of soils, underground storage tanks (USTs), dump debris, and UXO |
| Alabama Army Ammunition Plant, AL | many | Nitroaromatic compounds, heavy metals (Pb), munitions-related waste (explosives), PCBs | GW, SW, Sed, Soil | Incineration and cell storage of contaminated soils |
| Anniston Army Depot, AL | 47 | VOCs, heavy metals, phenols, petroleum HCs, acids, caustics | GW, Soil | Removal of soils; hydrogen peroxide injection treatment; SIA GW treatment system |
| Camp Bonneville, WA | 14 AOCs | POLs, solvents, UXO | Soil | Removal of USTs, debris (drums), and soil |
| Fort Chaffee, AR | many | POLs, DDT, chlordane, TCE | GW, Soil | Removal of USTs, oil–water separators, wash racks, fuel fill stands, and soil; open burning and detonation |
| Fort Dix, NJ | many | Heavy metals, POLs, chlorinated solvents, PCBs | GW, SW, Sed, Soil | Removal of USTs and soil |
| Fort Eustis, VA | 27 | Petroleum HCs, PCBs, VOCs, pesticides, heavy metals | GW, SW, Sed, Soil | Removal of USTs and soil; use of pneumatic pumps and passive skimmers for petroleum removal |

| Installation | | Contaminants | Media | Actions |
|---|---|---|---|---|
| Fort George G. Meade, MD | many | Heavy metals, petroleum HCs, VOCs, UXO, asbestos | GW, Soil | Removal of aboveground storage tanks (ASTs), buildings, pits, petroleum drums, and soil |
| Fort Lewis, WA | many | VOCs (TCE), PCBs, heavy metals, waste oils and fuels, coal liquefaction wastes, PAHs, solvents, battery electrolytes | GW, Soil | GW extraction and treatment systems; P&T systems; removal of TCE drums (air sparging and soil vapor extraction systems shut down in 1999) |
| Fort McClellan, AL | 67 | VOCs [(TCE and pentachloroethane (PCA)], SVOCs, pesticides, explosives, metals (Pb), UXO, radioactive sources, nonstockpile chemical material | GW, Soil | Removal of USTs; removal actions at pistol ranges |
| Fort Ord, CA | 61 | VOCs (carbon tetrachloride); petroleum HCs, heavy metals (Pb), pesticides | GW, Soil | Removal of "waste"; recycling of Pb; GW P&T systems |
| Fort Richardson, AK | 38 | White phosphorus, PCBs, heavy metals, POLs, solvents (TCE), dioxins, chemical agents, UXO, explosives, pesticides | GW, Soil | Removal of USTs, buried drums, and soil; thermal desorption; heat-enhanced soil vapor extraction; pond draining (to reduce white P) |
| Fort Riley, KS | many | Solvents, pesticides, Pb | GW, SW, Sed, Soil | Removal of soil and incinerators; performed soil vapor extraction pilot test and free-product recovery pilot test |
| Fort Ritchie, MD | many | UXO, heavy metals, (Pb), asbestos | GW, Soil | Removal of USTs, lead paint, and asbestos |
| Fort Sheridan, IL | many | Fuel HCs, PAHs, metals, UXO | GW, Soil | Removal of USTs and soil; performed UXO "clearances" |
| Fort Wainwright, AK | many | POLs, heavy metals, solvents, pesticides, paints, UXO, ordnance compounds, chemical agents | GW, Soil | Removal of soil, drums, fire training pits, and structures; use of air-sparging curtain; removed and recycled old air sparging/soil vapor extraction systems |

continued

TABLE A-1 Continued

| Installation | Number of Sites | Contaminants | Media Affected | Source Remediation |
|---|---|---|---|---|
| Fort Wingate, NM | many | Explosive compounds, UXO, PCBs, pesticides, heavy metals, asbestos, lead-based paint, nitrates (from TNT) | GW, Soil | Removal of soil; disposal of PCB-contaminated building materials; clearance of UXO from Indian tribal lands |
| Hamilton Army Airfield, CA | many | Metals, VOCs, SVOCs, fuel HCs, POLs, PCBs, PAHs, pesticides | GW, SW, Sed, Soil | Removal of USTs and soil; removal of onshore fuel lines; flushing, sealing, and abandoning offshore fuel lines |
| Hingham Annex, MA | many | POLs, heavy metals, VOCs, PCBs, asbestos | GW, SW, Sed, Soil | Removal of USTs, ASTs, oil–water separator, and soil; use of asphalt batching technology |
| Iowa Army Ammunition Plant, IA | 42 | Explosives, heavy metals (Pb), VOCs, pesticides | GW, SW, Sed, Soil | Removal of ASTs and soil; excavation and off-site incineration of soils; excavation of explosives-contaminated sumps; creation of wetlands and use of phytoremediation; use of low-temp. thermal desorption; "treatment" of explosives- and metal-contaminated soils |
| Jefferson Proving Ground, IN | many | Solvents, petroleum HCs, VOCs, PCBs, heavy metals, depleted uranium, UXO | GW, Soil | Removal of USTs; "treatment" of soil; UXO removal operations |
| Joliet Army Ammunition Plant, IL | 53 | Explosives, heavy metals, VOCs, PCBs, TNT | GW, Soil | Removal of sludge and soil; removal of UXO debris |

| Lake City Army Ammunition Plant, MO | 73 | Explosives, heavy metals, solvents, POLs | GW, Soil | Removal action for sumps; P&T system; extraction wells |
| Letterkenny, PA | many | VOCs, POLs, PCBs, metals (Hg), explosives, asbestos | GW, SW, Sed, Soil | Removal of fire-training pits, oil burn pits, and other soils; low-temp. thermal treatment |
| Lexington-Blue Grass Army Depot, KY | 67 | VOCs, SVOCs, heavy metals (Pb), PCBs, pesticides, herbicides, asbestos | GW, SW, Sed, Soil | Removal of USTs, transformers, asbestos, soil, and lagoon sludge |
| Lone Star Army Ammunition Plant, TX | many | VOCs, petroleum, heavy metals, explosives | GW, Soil | Removal of soil, cisterns, and fuel storage areas; soil "decontamination" |
| Louisiana Army Ammunition Plant, LA | 7 | Oils, grease, degreasers (TCE), phosphates, solvents, metal plating sludges, acids, fly ash, TNT, RDX, HMX | GW, SW, Sed, Soil | Incineration of explosives-contaminated soil |
| Milan Army Ammunition Plant, TN | 38 | Munitions-related wastes | GW, Soil | Soil excavation; bioremediation of soil; constructed landfill for bioremediated soil; GW treatment plant (granular activated carbon) |
| Military Ocean Terminal, Bayonne, NJ | 67 | Petroleum HCs, BTEX, VOCs, SVOCs, dieldrin, heavy metals, PCBs, asbestos | GW, Soil | Removal of USTs, ASTs, soil–water separators, asbestos from buildings, and soil |
| Picatinny Arsenal, NJ | 156 | VOCs, explosives, PCBs, heavy metals, arsenic | GW, SW, Sed, Soil | Removal of USTs, buried drums, contaminated pipe, and soil; use of GW extraction and treatment system; use of GW P&T system |

*continued*

TABLE A-1 Continued

| Installation | Number of Sites | Contaminants | Media Affected | Source Remediation |
|---|---|---|---|---|
| Pueblo Chemical Depot, CO | many (29 potential UXO sites) | Heavy metals, POLs, VOCs, SVOCs, pesticides, explosives, PCBs, and UXO | GW, Soil | Removal of soil; decontamination, demolition, and removal of buildings; GW extraction and treatment systems; sheet-pile barriers; soil bioremediation |
| Red River Army Depot, TX | many | TCE, pesticides, metals | GW, SW, Sed | Removal of soil and lagoon sludge |
| Redstone Arsenal, AL | 298 (216 Army; 82 Marshall Space Flt.) | Heavy metals, solvents, chemical weapons/munitions, pesticides | GW, SW, Sed, Soil | GW extraction and treatment system; air strippers; soil vapor extraction systems; construction of industrial septic tank system |
| Riverbank Army Ammunition Plant, CA | 6 | Chromium, cyanide, zinc | GW, Soil | Removal action at evaporation and percolation ponds; ion exchange system in GW treatment system; GW extraction system |
| Rocky Mountain Arsenal, CO | 209 | Pesticides, chemical agents, VOCs, chlorinated organics, PCBs, UXO, heavy metals, solvents, asbestos | GW, Soil | Removal of drums, soil, and asbestos materials; GW excavation and treatment systems; chemical and sanitary sewer plugging; off-post soil tillage |
| Sacramento Army Depot, CA | 16 | Waste oil and grease, solvents, metal plating wastes, caustics, cyanide, metals | GW, Soil | Removal of soil; soil vapor extraction systems; air sparging; GW extraction and treatment systems |
| Savanna Army Depot, IL | many | Explosives, metals (Pb), solvents POLs, VOCs, UXO, pesticides, TNT | GW, SW, Sed, Soil | Removal of soil; high-temp. thermal treatment; incineration of TNT-contaminated sediment |

| Site | No. | Contaminants | Media | Action |
|---|---|---|---|---|
| Seneca Army Depot, NY | 36 | Chlorinated solvents (TCE), radioactive isotopes, heavy metals, petroleum HCs | GW, SW, Sed, Soil | Removal of USTs and soil |
| Stratford Army Engine Plant, CT | many | PCBs, asbestos, fuel-related VOCs, solvents, metals (hexavalent chromium), PAHs | GW, SW, Sed, Soil | Removal of USTs |
| Sudbury Training Annex, MA | 74 | VOCs, PCBs, pesticides, heavy metals, arsenic, asbestos, UXO | GW, Soil | Removal of drums, USTs, debris, and soil |
| Sunflower Army Ammunition Plant, KS | many | Nitrates, sulfates, lead, chromium, propellants | GW, SW, Sed, Soil | Removal of USTs and soil; cleanup of asbestos dump |
| Tobyhanna Army Depot, PA | many | Heavy metals, VOCs, PCBs, POLs, UXO | GW, SW, Sed, Soil | Removal of soil and sewage drying beds |
| Twin Cities Army Ammunition Plant, MN | 25 | VOCs, PCBs, heavy metals | GW, SW, Sed, Soil | Removal of soil; soil vapor extraction system; SVE air sparging system |
| U.S. Army Soldiers System Center, MA | many | Pesticides, herbicides, pentachlorophenol, solvents, VOCs (TCE) | GW, SW, Sed, Soil | Removal of waste oil storage tanks, pavement, and soil |
| Umatilla Chemical Depot, OR | 80 | Explosives, UXO, heavy metals (Pb), pesticides, nitrates | GW, Soil | Removal of USTs; bioremediation (by windrow composting) of contaminated soils |

SOURCE: Defense Environmental Restoration Program: Annual Report to Congress FY2001

TABLE A-2 Contaminants Reported at Other Military Installations

| Site | Number of Sites | Contaminants | Media Affected | Source Remediation |
|---|---|---|---|---|
| Andrews AFB, MD | 22 | Heavy metals (Ni), SVOCs, VOCs PAHs, PCBs, pesticides | SW | Soil removal |
| Bangor Naval Submarine Base, ME | 22 | Residual TNT, RDX, Otto fuel, VOCs | GW, Sed, Soil | Removal of soils and USTs; P&T systems; closed loop passive soil washing system with granular activated carbon |
| Barbers Point Naval Air Station, HI | 17 | PCBs, heavy metals (Pb), petroleum HCs, pesticides, solvents, asbestos | GW, Soil | Removal of USTs |
| Barstow Marine Corps Logistics Base, CA | 38 | Heavy metals, PCBs, petroleum HCs, pesticides, herbicides, chlorinated VOCs (TCE) | GW, Soil | Removal of USTs and industrial waste sludge |
| Bergstrom AFB, TX | 30 + 454 AOCs | Chlorinated VOCs (TCE), pesticides, petroleum HCs, metals, low-level radioactive wastes | GW, Soil | Removal of USTs, ASTs, soil, radioactive wastes; soil vapor extraction and air sparging systems |
| Camp Lejeune MCB, NC | 176 | Battery acid, fuels, used oils, paints, thinners, PCBs, pesticides, solvents, metals | GW, SW, Sed, Soil | Remediation systems at 23 sites |
| Castle AFB, CA | many | TCE, PCBs, POLs, pesticides, cyanide, cadmium | GW, Soil | Removal of soil and USTs; soil vapor extraction systems; oil–water separators; P&T systems; bioventing systems |

| Installation | Number | Contaminants | Media | Remediation Actions |
|---|---|---|---|---|
| Hanscom AFB, MA | 22 | VOCs, chlorinated solvents, gasoline, jet fuel, tetraethyl PB, PCBs, Hg | GW, SW, Sed, Soil | Removal of soil and USTs; GW/product recovery and soil vapor extraction |
| Hill AFB, UT | 106 | Solvents (TCE), sulfuric and chromic acids, metals, petroleum HCs | GW, SW, Sed, Soil | P&T systems |
| Homestead, AFB, FL | 26 before/ 540 after hurricane | Metals (Pb), VOCs, cyanide, pesticides, solvents, PCBs | GW, Soil | Removal of USTs and soil; GW extraction and treatment; removal of oil–water separators; remedial bioventing systems |
| Lakehurst Naval Air Engineering Station, NJ | 45 | Fuels, PCBs, solvents (TCE), waste oils | GW, Soil | Removal of soil, drums, tanks, and debris; soil washing, asphalt batching, solar-powered soil irrigation and sparge treatment systems; P&T; vapor extraction; soil bioventing |
| Langley AFB, VA | 45 | Petroleum HCs, chlordane, PCBs, heavy metals, solvents | GW, SW, Sed, Soil | Removal of USTs and soil; soil vapor extraction system |

SOURCE: Defense Environmental Restoration Program: Annual Report to Congress FY2001

TABLE A-3 Prevalence of Organic Contaminants of Concern at Army Installations

| Installation | Chlorinated Solvents | | | | | Explosives | | | |
|---|---|---|---|---|---|---|---|---|---|
| | PCE | TCE | cis-1,2-DCE[a] | 1,2-DCA | TCA[b] | DNT | TNT | HMX | RDX |
| Aberdeen Proving Ground | X | X | X | X | X | X | X | X | X |
| Alabama AAP | | | | | | X | X | X | |
| Anniston AD | | X | | | | | | | |
| Badger AAP | | | | | X | X | | | |
| Blue Grass Army Depot | X | | | | | | X | X | X |
| Blue Grass Army Depot-Lexington | | | | | | | | | |
| Cameron Station | | X | | | | | | | |
| Camp Bonneville | | | | | | | | | X |
| Camp Bullis | | X | | | | | | | |
| Camp Crowder | | X | X | | X | | | | |
| Camp Kilmer | X | X | X | | | | | | |
| Camp Navajo | | | | | | | X | | |
| Camp Roberts | X | X | | | | | | | |
| Cold Regions Research Lab | | X | | | | | | | |
| Cornhusker AAP | | | | | X | | X | | X |
| Devens Reserve Training Facility | | | | | X | | | | |
| Dugway Proving Ground | X | | | | X | X | X | X | X |
| Floyd Wets Site | | | | | X | | | | |
| Fort Belvoir | | | | | | | | | |
| Fort Benning | | X | | | | | | | |
| Fort Bliss | | | | | X | | | | |
| Fort Bragg | X | X | | | | | X | | |
| Fort Campbell | | X | | | | | | | |
| Fort Carson | X | X | | X | | | X | | |
| Fort Des Moines | X | | | | X | | | | |

continued

| | | | | | | | | | |
|---|---|---|---|---|---|---|---|---|---|
| Fort Detrick | X | | | X | | | X | X | X |
| Fort Dix | | | | | | | X | X | X |
| Fort Drum | | | | | | X | | X | |
| Fort Eustis | | | | | X | | | X | X |
| Fort Gillem | | | | | X | X | X | X | X |
| Fort Gordon | | | | X | X | X | X | X | X |
| Fort Jackson | | | | | | | X | X | X |
| Fort Knox | | | | | | | | X | X |
| Fort Leavenworth | | | | | X | | X | X | X |
| Fort Lee | | | | | | | X | X | X |
| Fort Leonard Wood | | | | | | | X | X | X |
| Fort Lewis | | | | | | X | | X | |
| Fort McClellan | X | | X | X | X | | X | X | X |
| Fort McCoy | | | | | | | | | |
| Fort McNair | | | | | | | | | |
| Fort Meade | | | | | | | X | X | X |
| Fort Monmouth | | | | | | | X | X | X |
| Fort Richardson | | | | X | | X | X | X | X |
| Fort Riley | | | | | | | X | X | X |
| Fort Ritchie | | | | | | | X | X | |
| Fort Rucker | | | | | | | X | X | |
| Fort Sam Houston | | | | | | | X | | |
| Fort Shafter | | | | | | | | X | X |
| Fort Stewart | | X | | | | | | X | |
| Fort Story | | | | | | | | X | X |
| Fort Wainwright | | | | | X | X | | X | X |
| Fort Wingate Depot Activity | X | X | X | X | | X | | | |
| Haines Pipeline | | | | | | X | | | X |
| Hamilton Army Airfield | | | X | | X | | | | |
| Hawthorne Army Depot | X | X | X | X | | | | X | X |

TABLE A-3 Continued

| Installation | Chlorinated Solvents | | | | | Explosives | | | |
|---|---|---|---|---|---|---|---|---|---|
| | PCE | TCE | cis-1,2-DCE[a] | 1,2-DCA | TCA[b] | DNT | TNT | HMX | RDX |
| Holston AAP | X | X | | | X | | | X | |
| Hunter Army Airfield | X | X | X | | X | | | | |
| Indiana AAP | X | X | | | X | X | | | |
| Iowa AAP | | X | X | | X | X | X | X | X |
| Jefferson Proving Ground | | | | | X | | | | |
| Joliet AAP | | | | X | X | X | X | | |
| Kansas AAP | | | | X | X | | X | X | X |
| Kelly Support Facility | | | | | X | | | | |
| Kunia Field Station | | | | | | | | | |
| Lake City AAP | X | X | X | X | | | | | |
| Letterkenny Army Depot | X | X | X | X | X | | X | X | |
| Lincoln AMSA | X | X | | | | | | | |
| Lone Star AAP | | X | | X | | | X | | |
| Longhorn AAP | | X | | X | X | X | X | | X |
| Los Alamitos Armed Forces | X | X | X | X | | | | | |
| LTA, Marion ENGR Depot | | X | X | | | | | | |
| McAlester AAP | | X | | | | | | | |
| Middletown USARC | | X | | | | | | | |
| Milan AAP | | | | | | | X | X | |
| Military Ocean TML Sunny Point | | | X | | | | | | |
| Newport Chem Depot | X | X | X | | | X | X | X | X |
| NTC and Fort Irwin | X | X | | X | X | X | X | X | X |
| Oakland Army Base | X | X | X | | X | | | | |
| Papago Military Reservation | X | | | | | | | | |
| Phoenix Military Reservation | | X | | | | | | | |

| Installation | | | | | | | | |
|---|---|---|---|---|---|---|---|---|
| Picatinny Arsenal | X | | | X | X | | X | X |
| Presidio of Monterey | X | | X | X | X | | X | X |
| Pueblo Chemical Depot | X | X | X | X | X | X | X | X |
| Radford AAP | X | X | X | X | X | X | X | X |
| Ravenna AAP | | | X | | X | | X | X |
| Red River Army Depot | X | | X | X | X | X | X | |
| Redstone Arsenal | X | | X | | X | | X | X |
| Rock Island Arsenal | | | X | X | X | | | |
| Rocky Mountain Arsenal | X | | | | | | | |
| Sacramento Army Depot | X | | | | | | | |
| Savannah Depot Activity | X | X | X | X | X | X | X | X |
| Seneca Army Depot Activity | X | X | X | X | | | | |
| Sierra Army Depot | X | | X | X | | | | |
| Soldier Systems Center | X | | | X | X | | | |
| Stratford Army Engine Plant | X | X | | X | X | | | |
| Sunflower AAP | X | | | | | | | |
| Tacony Warehouse | X | | | | | | | |
| Tarheel Army Missile Plant | X | X | X | | | | | |
| Tobyhanna Army Depot | X | | | | | | | |
| Tooele Army Depot | X | X | | X | X | | | |
| Twin Cities AAP | X | X | X | X | X | | | |
| Umatilla Chem Depot | | | | X | | X | | |
| USARC Fort Nathaniel Greene | | | | | | | | |
| USARC Kings Mills | X | | | | | | | |
| Vint Hill Farms Station | X | | | | | | | |
| Volunteer AAP | | | | X | | | X | |
| Walter Reed Army Medical | | | | X | X | | | |
| Watervliet Arsenal | X | X | X | X | X | | | |
| White Sands Missile Range | X | | X | X | X | | | |
| Yakima Training Center | X | | | | | | | |
| Yuma Proving Ground | | | X | | | | | |

continued

TABLE A-3 Continued

| Installation | Chlorinated Solvents | | | | | | Explosives | | | | |
| | PCE | TCE | cis-1,2-DCE[a] | 1,2-DCA | TCA[b] | DNT | TNT | HMX | RDX |
|---|---|---|---|---|---|---|---|---|---|
| Totals | 51 | 74 | 32 | 24 | 35 | 26 | 30 | 19 | 14 |
| Percentage of all installations with this contaminant | 37 | 54 | 23 | 1 | 25 | 19 | 22 | 14 | 10 |

Number of BRAC installations: 23
Number of active installations: 115
Total number of installations: 138

[a]Does not include other DCE isomers.
[b]Includes 1,1,1-TCA and 1,1,2-TCA

SOURCE: Compiled by Laurie Haines, Army Environmental Center.

# Appendix B

# Abbreviations and Acronyms

| | |
|---|---|
| AAP | Army ammunition plant |
| AD | Army depot |
| AFB | Air Force Base |
| ARAR | Applicable or relevant and appropriate requirements |
| AS/SVE | Air sparging with soil vapor extraction |
| AST | Aboveground storage tank |
| | |
| BRAC | Base realignment and closure |
| BTEX | Benzene, toluene, ethylbenzene, and xylene |
| | |
| CERCLA | Comprehensive Environmental Response, Compensation, and Liability Act |
| CMC | Critical micelle concentration |
| CPT | Cone penetrometer |
| CT | Carbon tetrachloride |
| CTE | Central tendency exposure |
| CVOC | Chlorinated volatile organic compound |
| | |
| DCA | Dichloroethane |
| DCE | Dichloroethene |
| DERP | Defense Environmental Restoration Program |
| DMPL | Dense miscible phase liquids |
| DNAPL | Dense nonaqueous phase liquids |
| DNT | 2,4-dinitrotoluene |

| DoD   | U.S. Department of Defense |
| DOE   | U.S. Department of Energy |
| DPT   | Drive point screening |
| DWEL  | Drinking water equivalent level |

| ERH   | Electrical resistance heating |
| EPA   | U.S. Environmental Protection Agency |

| FLUTe | Flexible Liner Underground Technologies Everting |

| GPR   | Ground-penetrating radar |
| GW    | Groundwater |

| HC    | Hydrocarbon |
| HMX   | High Melt Explosive or Her Majesty's Explosive; octahydro-1,3,5,7-tetranitro-1,3,5,7-tetrazocine |
| HRS   | Hazard ranking system |
| HVOC  | Halogenated volatile organic compound |

| ISCO  | In situ chemical oxidation |
| ISTD  | In situ thermal desorption |

| LNAPL | Light nonaqueous phase liquids |

| MCB   | Marine Corps Base |
| MCL   | Maximum contaminant level |
| MCLG  | Maximum contaminant level goal |
| MIP   | Membrane interface probe |
| MNA   | Monitored natural attenuation |
| MNT   | Mononitrotoluene |
| MTBE  | Methyltertbutylether |

| NAPL  | Nonaqueous phase liquid |
| NAS   | Naval Air Station |
| NCP   | National Contingency Plan |
| NFESC | Naval Facilities Engineering Service Center |
| NPL   | National Priorities List |
| NRC   | National Research Council |
| NRD   | Natural resource damage |

| OU    | Operational unit |
| O&M   | Operation and maintenance |
| OMM   | Operations, maintenance, and monitoring |

| | |
|---|---|
| PAH | Polycyclic aromatic hydrocarbon |
| PCB | Polychlorinated biphenyl |
| PCE | Perchloroethylene |
| PITT | Partitioning interwell tracer test |
| PRP | Potentially responsible parties |
| P&T | Pump-and-treat |
| | |
| RAGS | Risk Assessment Guidance for Superfund |
| RCRA | Resource Conservation and Recovery Act |
| R&D | Research and development |
| RDX | Royal Demolition Explosive/Research Demolition Explosive or hexahydro-1,3,5-trinitro-1,3,5-triazine |
| RD/RA | Remedial design/remedial action |
| RIP | Remedy in place |
| RI/FS | Remedial investigation/feasibility study |
| ROD | Record of decision |
| RPM | Remedial project manager |
| RME | Reasonable maximum exposure |
| | |
| SARA | Superfund Amendments and Reauthorization Act |
| SCM | Site conceptual model |
| SEAR | Surfactant-enhanced aquifer remediation |
| SERDP | Strategic Environmental Research and Development Program |
| SVE | Soil vapor extraction |
| SVOC | Semivolatile organic compound |
| | |
| TCA | Trichloroethane |
| TCE | Trichloroethylene |
| TI | Technical impracticability |
| TNB | 1,3,5-trinitrobenzene |
| TNT | 2,4,6-trinitrotoluene |
| TSDF | Treatment, storage, and disposal facilities |
| TSCA | Toxic Substances Control Act |
| | |
| UCL | Upper confidence limit |
| USACE | U.S. Army Corps of Engineers |
| UST | Underground storage tank |
| UXO | Unexploded ordnance |
| | |
| VC | Vinyl chloride |
| VOC | Volatile organic compound |
| | |
| ZVI | Zero valent iron |

# Appendix C

# Biographical Sketches of Committee Members and NRC Staff

**JOHN C. FOUNTAIN (CHAIR)** is professor and head of the Department of Marine, Earth, and Atmospheric Sciences at North Carolina State University. He was formerly a professor of geochemistry at the State University of New York at Buffalo. He has worked continuously on DNAPL remediation since 1987, including both lab work and field studies on using surfactants as a remediation tool at a number of different sites including military facilities (Hill Air Force Base and the Canadian Forces Base Borden), industrial sites (DuPont Plant in Corpus Christi, Texas, and the PPG plant in Lake Charles, Louisiana), and Department of Energy sites (in Paducah, Kentucky, and Portsmouth, Ohio). In recent years he has concentrated on remediation of DNAPL contamination in fractured bedrock, including studies of fracture location and flow in fractures. He has served on four previous NRC committees, including the Committee on Technologies for Cleanup of Subsurface Contaminants in the DOE Weapons Complex. He received his B.S. from California Polytechnic State University San Luis Obispo, and his M.A. and Ph.D. in geology from the University of California at Santa Barbara.

**LINDA M. ABRIOLA** is dean of engineering and professor of civil and environmental engineering at Tufts University. Prior to this appointment, she was the Horace Williams King Collegiate Professor of Environmental Engineering at the University of Michigan. Her primary research focus is the integration of mathematical modeling and laboratory experiments to investigate and elucidate processes governing the transport, fate, and remediation of nonaqueous phase liquid organic contaminants in the subsurface. Dr. Abriola's numerous professional activities have included service on the U.S. Environmental Protection Agency Science

Advisory Board, the NRC's Water Science and Technology Board, and the U.S. Department of Energy's NABIR (Natural and Accelerated Bioremediation Research) Advisory Committee. Dr. Abriola served on the NRC's Committee on Ground Water Cleanup Alternatives, which was the first NRC committee to investigate the efficacy of pump-and-treat technologies. An author of more than 100 refereed publications, Dr. Abriola has been the recipient of a number of awards, including the Association for Women Geoscientists' Outstanding Educator Award (1996) and the National Ground Water Association's Distinguished Darcy Lectureship (1996). She is a fellow of the American Geophysical Union and a member of the National Academy of Engineering. Dr. Abriola received her B.S. in civil engineering from Drexel University and her M.A., M.S., and Ph.D. in civil engineering from Princeton University.

LISA M. ALVAREZ-COHEN is the Fred and Claire Sauer Professor of Environmental Engineering in the Department of Civil and Environmental Engineering at the University of California, Berkeley. She received her B.S. in engineering and applied science from Harvard University and her M.S. and Ph.D. in environmental engineering and science from Stanford University. Her current research interests are the biotransformation of contaminants in the subsurface, including chlorinated solvents, MTBE, and NDMA, and innovative methods for evaluating in situ bioremediation, including molecular biology, isotope use, and direct microscopy. Part of her research on natural attenuation took place at Alameda Point Naval Air Station. Dr. Alvarez-Cohen is an associate editor of *Environmental Engineering Science*. Her previous NRC service includes the Committee on USGS Water Resources Research and the Committee on In Situ Bioremediation.

MARY JO BAEDECKER is a scientist emeritus at the U.S. Geological Survey. Previously she was the chief scientist for hydrology at the USGS where she oversaw the National Research Program in the hydrologic sciences. Her research interests are the degradation and attenuation of organic contaminants in hydrologic environments. She was the Darcy Lecturer for the Association of Groundwater Scientists and Engineers in 1993 and served on the board and as chair of the Hydrogeology Division of the Geological Society of America in 1999. She has been a professorial lecturer at the George Washington University. She received an M.S. from the University of Kentucky in organic chemistry and a Ph.D. from the George Washington University in geochemistry. She served on the NRC Committee on Ground Water Cleanup Alternatives.

DAVID E. ELLIS is an environmental scientist at DuPont with expertise in a wide variety of remediation technologies and field testing of these technologies. As the remediation technology leader for DuPont's Corporate Remediation Group, he currently focuses on bioremediation, in situ treatment, sediments, explosives and unexploded ordnance, hydrogeology, and modeling. He is also on the board of directors of the Interstate Technology and Regulatory Council, the goal of which is to advance innovative remediation technologies by identifying and overcoming regulatory barriers. Finally, he is the founder and chair of the RTDF Bioremediation Consortium, which is an industry/government consortium developing safe and effective bioremediation techniques for treating chlorinated solvent contamination. Dr. Ellis received his B.S. in geology from Allegheny College, and his M. Phil. and Ph.D. in geology and geochemistry, respectively, from Yale University. He served on the NRC Committee on Intrinsic Bioremediation.

**THOMAS C. HARMON** is an associate professor and founding faculty member in the School of Engineering at the University of California, Merced. Prior to this, he was in the Department of Civil and Environmental Engineering at UCLA. He received his B.S. in civil engineering from the Johns Hopkins University, and his M.S. and Ph.D. in civil engineering from Stanford University. He currently directs contaminant transport monitoring research efforts in the Center for Embedded Networked Sensing (CENS), a National Science Foundation science and technology center housed at UCLA. Dr. Harmon's research focuses on chemical fate and transport and biogeochemical cycling in the subsurface environment. He has published recently on measuring and modeling DNAPL dissolution rates, as well as on locating DNAPL as dissolution sources using inverse modeling techniques.

**NANCY J. HAYDEN** is an associate professor in the Department of Civil and Environmental Engineering at the University of Vermont. She received her B.S. in forest biology from the SUNY College of Environmental Science and Forestry, and her M.S. and Ph.D. in environmental engineering from Michigan State University. Her basic and applied research is in the area of remediation strategies for soils and groundwater contaminated with water-immiscible solvents (particularly alcohol flushing of DNAPLs in clay-containing porous media). She has used both bench-scale and field-scale studies to investigate the feasibility of new techniques and the optimization of technologies already in use. More recently, Dr. Hayden has become involved in research on phytoremediation and plant-based wastewater treatment systems.

**PETER KITANIDIS** is a professor of environmental fluid mechanics and hydrology in the Department of Civil and Environmental Engineering and (by courtesy) professor in the Department of Geological and Environmental Sciences, both at Stanford University. He specializes in the analysis of data, uncertainty, scale issues, and the development and implementation of mathematical models that describe flow and transport rates in the environment. He has devised methods for the analysis of spatially distributed hydrologic and water-quality data, the calibration of groundwater models, the optimization of sampling and control strategies when the available information is incomplete, and methods for the study of dilution and mixing in geologic media. He has authored or coauthored about 130 papers on these topics and is the author of the book *Introduction to Geostatistics*. He received his Ph.D. from the Massachusetts Institute of Technology.

JOEL A. MINTZ is a professor of law at the Nova Southeastern University Shepard Broad Law Center. Prior to joining the faculty, he represented the federal government in litigation involving hazardous waste disposal sites that pose an imminent hazard to health or the environment. Well published in the area (he has authored four books including *Environmental Law: Cases and Problems* and numerous journals articles), he is considered one of the nation's leading authorities on environmental enforcement in general. While at the U.S. Environmental Protection Agency (EPA), he received the bronze medal for commendable service and a special service award, and he has been listed in *Who's Who in America, Who's Who in the World, Who's Who in American Law, Who's Who in the South and Southwest, Contemporary Authors,* and the *Directory of American Scholars*. He received his B.A., LL.M., and J.S.D from Columbia University.

JAMES M. PHELAN is a distinguished member of the technical staff at Sandia National Laboratories. Since joining Sandia in 1983, he has been involved with laboratory and field-testing of environmental technology for the characterization and restoration of polluted soils and groundwater. Most recently, he has focused on the in situ treatment of explosives in soils and groundwater, including bio-treatment, chemical oxidation, and chemical reduction technologies. He has past experience with treatment of chlorinated organics by air sparging and vacuum extraction; treatment of an unlined landfill organic waste disposal cell by thermal enhanced vapor extraction; and technology program management for the containment and/or treatment of radioactive/mixed-waste landfills. His explosives experience extends to environmental factors that affect chemical sensing of buried landmines and mass transfer of solid phase energetics to soil pore water. Mr. Phelan received his B.S. in environmental toxicology from the University of California Davis and an M.S. in environmental health from Colorado State University.

GARY A. POPE is the director of the Center for Petroleum and Geosystems Engineering at the University of Texas at Austin, where he has taught since 1977. He holds the Texaco Centennial Chair in petroleum engineering. Previously, he worked in production research at Shell Development Company for five years. Dr. Pope earned a Ph.D. from Rice University and a B.S. from Oklahoma State University, both in chemical engineering. His teaching and research are in the areas of chemical and thermal methods of source zone removal, groundwater modeling, groundwater tracers, enhanced oil recovery, chemical thermodynamics, reservoir engineering, and reservoir simulation. He has authored or coauthored more than 200 technical papers on these research subjects and has supervised more than 100 graduate students at the University of Texas. Dr. Pope was elected to the National Academy of Engineering in 1999 for his contributions to understanding multiphase flow and transport in porous media and applications of these principles to improved oil recovery and aquifer remediation.

**DAVID A. SABATINI** is a professor and is the Sun Company Chair of Civil Engineering and Environmental Science; the director of the Environmental and Ground Water Institute; and the associate director of the Institute for Applied Surfactant Research—all at the University of Oklahoma. His areas of research include subsurface transport and fate processes, subsurface remediation technologies, development of environmentally friendly products and processes, and innovative educational methods. He is also a partner at Surfactant Associates, Inc., and the cofounder and coprincipal at Surbec Environmental, which designs and implements innovative subsurface remediation technologies. He has served as an associate editor for *Ground Water, Journal of Contaminant Hydrology*, and *Journal of Surfactants and Detergents*. Dr. Sabatini received his B.S. from the University of Illinois, his M.S. from Memphis State University, and his Ph.D. from Iowa State University, all in civil engineering.

**THOMAS C. SALE** is an assistant professor in civil engineering at Colorado State University and an independent consulting hydrogeologist. He has been actively involved in the characterization and remediation of subsurface releases of NAPLs since 1981. In the early 1980s Dr. Sale worked on the design, construction, and operation of free product recovery systems at petroleum refineries in the central United States using water flooding and horizontal drains. During his ten years with CH2M HILL, his primary focus was on characterization and mitigation of risks posed by subsurface contaminants and development of groundwater resources. He is currently working as a consultant at DOE's Rocky Flats Facility near Golden, Colorado (solvents contamination) and at six major U.S. refineries where the technical practicability of mobile NAPL recovery and appropriate endpoints for mobile NAPL recovery systems are issues. Dr. Sale received his Ph.D. from Colorado State University, his M.S. in watershed hydrology from the University of Arizona, and B.A. degrees in geology and chemistry from Miami University, Ohio.

**BRENT E. SLEEP** is a professor of civil engineering at the University of Toronto. He received his B.A.Sc., M.A.Sc., and Ph.D. from the University of Waterloo. Dr. Sleep's expertise is in the thermal treatment of organics in the subsurface, particularly via steam and hot air injection and thermal extraction. He also has experience with the sequential in situ bioremediation of solvents in soil and groundwater and with in situ chemical oxidation followed by bioaugmentation. His research includes lab-scale and pilot-scale experiments in both saturated and unsaturated soil systems as well as fractured rock. Finally, he has experience with contaminant source delineation in the subsurface using isotopes.

JULIE L. WILSON is a senior associate at EnviroIssues, Inc., in Portland, Oregon. She has over 20 years of experience in environmental investigation, risk assessment, and industrial hygiene/health and safety. She has managed environmental projects, including site investigations and cleanups, environmental audits for industrial and clinical facilities, and human health and ecological risk assessments for numerous sites across the country including Superfund sites, military facilities, and private industrial sites. She has worked at sites ranging from landfills to sites with organics-contaminated soil and groundwater to sediment sites. Dr. Wilson has held appointed positions in Oregon and Washington to help develop site investigation and cleanup policies and regulations. Dr. Wilson, a certified industrial hygienist, received her B.S. in biology from Michigan Technological University, her M.S. in health physics from Purdue University, and her Ph.D. in environmental medicine (toxicology) from New York University.

JOHN S. YOUNG serves as a risk assessor and toxicologist with the Food and Nutrition Service of the Israeli Ministry of Health. He also holds a faculty appointment at the Institute for Earth Sciences of the Hebrew University of Jerusalem. He was formerly the president of Hampshire Research, where he directed the development of government-sponsored risk assessment software systems for evaluation of contaminated sites and of multipathway exposure to pesticides, and served as a technical advisor for remediation at diverse contaminated sites. Prior to joining Hampshire, he served as a faculty member at the Johns Hopkins University School of Hygiene and Public Health in the Department of Environmental Health Sciences. Dr. Young received his B.S. in psychology from Georgetown University, his Ph.D. in psychology from Brown University, and his postdoctoral training at Johns Hopkins and the Naval Medical Research Institute. He recently served on the NRC Subcommittee to Review the Updated Radioepidemiology Tables from the Centers for Disease Control and Prevention.

KATHERINE L. YURACKO is the founder and CEO of YAHSGS, a company that provides technical, analytical, and management support services and conducts innovative research in the physical, engineering, and life sciences. YAHSGS performs research in toxicology and information technology through research grants from the National Institutes of Health, National Science Foundation, U.S. Air Force, and Department of Energy. YAHSGS also provides services in life cycle analysis, value engineering, and strategic planning; risk analysis and risk mitigation; innovative technology application analysis; regulatory analysis and planning; and environmental cleanup. Prior to founding YAHSGS, Dr. Yuracko was director of the Center for Life Cycle Analysis at Oak Ridge National Laboratory, held positions with the U.S. Office of Management and Budget and the White House Office of Science and Technology Policy, and was a Fulbright Scholar. Dr. Yuracko received her S.M. in health physics and her Nucl.E. and Ph.D. in nuclear engineering from the Massachusetts Institute of Technology, her

M.P.P. in energy and environmental policy from Harvard University, and her A.B. in physics from Harvard University.

LAURA J. EHLERS is a senior staff officer for the Water Science and Technology Board of the National Research Council. Since joining the NRC in 1997, she has served as study director for 11 committees, including the Committee to Review the New York City Watershed Management Strategy, the Committee on Riparian Zone Functioning and Strategies for Management, and the Committee on Bioavailability of Contaminants in Soils and Sediment. She received her B.S. from the California Institute of Technology, majoring in biology and engineering and applied science. She earned both an M.S.E. and a Ph.D. in environmental engineering at the Johns Hopkins University. Her dissertation, entitled RP4 Plasmid Transfer among Strains of *Pseudomonas* in a Biofilm, was awarded the 1998 Parsons Engineering/Association of Environmental Engineering Professors award for best doctoral thesis.